Marina Technology

Acknowledgement is made to E. Ozhan *et al.* for the use of
Figure 9 on p.280, which appears on the front cover of this book

Marina Technology

Proceedings of the Second International Conference, held in
Southampton, UK, 31 March - 2 April 1992

Editor: W.R. Blain

Computational Mechanics Publications
Southampton Boston

Co-published with

Thomas Telford
London

Dr. W.R. Blain
Wessex Institute of Technology
Ashurst Lodge
Ashurst
Southampton SO4 2AA UK

British Library Cataloguing in Publication Data

A Catalogue record for this book is available
from the British Library

ISBN 1-85312-161-4 Computational Mechanics Publications, Southampton
ISBN 1-56252-089-X Computational Mechanics Publications, Boston, USA
ISBN 0-7277-1689-1 Thomas Telford London

Library of Congress Catalog Card Number 91-77630

Co-published and distributed by Thomas Telford Services Ltd, Thomas
Telford House, 1 Heron Quay, London E14 4JD, U.K.

First published in 1992

Printed in the United Kingdom by Bell and Bain Ltd., Glasgow

PREFACE

Following the success of the first international conference on the planning, design and operation of marinas, a second conference was convened to look at further developments and trends within the marina field.

Again, the conference has attracted contributions from many different countries, covering a multitude of topics from the feasibility stage to final construction and operation. Reference is made to a number of existing marinas throughout the world which illustrate particular techniques and new ideas.

This volume contains edited papers related to the planning, design and operation of marina developments and has been subdivided into the following sections:

Section 1: Overview of Yacht Marinas
Section 2: Planning and Feasibility
Section 3: Environmental and Water Quality Aspects
Section 4: Site Investigation and Water Level Control
Section 5: Wave Modelling and Analysis
Section 6: Flushing Characteristics and Water Circulation
Section 7: General Modelling of Marinas and Marine Structures
Section 8: Marina Developments
Section 9: Design and Construction
Section 10: Breakwaters
Section 11: Pontoon Services and Marina Maintenance
Section 12: Management and Marina Income

The conference was organised by the Wessex Institute of Technology and held in Southampton from 31st March to the 2nd April, 1992.

The editor would like to thank the following:

- Dr C A Brebbia, Director of the Wessex Institute of Technology and all staff of the Institute who helped make this conference a success.

- Jenny Mackenzie and staff at Computational Mechanics Publications.

- Members of the Organising Committee.

- The co-sponsors, namely:
  The Institution of Civil Engineers, UK; The American Society of Civil Engineers, USA; Hydraulics Research, UK; Permanent International Association of Navigation Congress, UK; Yacht Harbours Association, UK and Marina Management International, UK.

W.R. Blain
Editor
March 1992

CONTENTS

SECTION 5: WAVE MODELLING AND ANALYSIS

SECTION 6: FLUSHING CHARACTERISTICS AND WATER CIRCULATION

SECTION 7: GENERAL MODELLING OF MARINAS AND MARINE STRUCTURES

SECTION 11: PONTOON SERVICES AND MARINA MAINTENANCE

SECTION 12: MANAGEMENT AND MARINE INCOME

SECOND INTERNATIONAL CONFERENCE MARINA II PLANNING,
DESIGN AND OPERATION
MARINA 92

INTERNATIONAL ADVISORY COMMITTEE

SECTION 1: OVERVIEW OF YACHT MARINAS

Marinas: An Overview

P. Lacey

Maritime Engineering Department, Ove Arup & Partners, 13 Fitzroy Street, London, W1P 6BQ U.K.

INTRODUCTION

Marinas have changed extensively over the last thirty years. However, this change has not been uniform throughout the marina industry. Open moorings in small harbours, in rivers and estuaries have been supplanted slowly but surely by hard berthing with easy access direct onto land.

This has been caused by an enormous upsurge in yacht ownership. The new owners, however, have required a more comfortable type of marina with plenty of amenities. This requirement in turn has been fed by developers springing out of a very healthy residential boom accentuated by waterfront sites.

Unfortunately, the present downturn of the economic climate has stopped almost in its tracks the developer lead boom for large, high quality, waterfront complexes in many countries.

The last International Marina Conference held in 1989 at Southampton also contained an overview paper and it is sad to see after the acceleration of marina leisure activity in the late eighties that the subjects touched on then are in the main still with us.

The table "Guideline to Boating Numbers" sets out some statistics. The number of marinas, properly defined, would be of interest as they can vary from 14 No. in Mexico, 370 No. in Japan up to 10,000 No. in the USA.

GUIDELINE TO BOATING NUMBERS

COUNTRY	NO OF PERSONS IN BOATING INDUSTRY	PER CAPITA BOAT OWNERSHIP	TOTAL SAILBOATS	TOTAL MOTOR BOATS
Belgium	2,000	450	-	-
Canada	6,000	10	215,000	200,000
Denmark	4,000	120	29,500	13,000
Finland	2,000	7	17,500	90,000
France	13,000	76	229,422	503,670
Germany	25,000	120	146,602	83,171
Greece	-	198*	7,500	4,040
Ireland	4,000	200	-	-
Italy	20,000	90	-	-
Japan	270,000	430	-	-
Netherlands	5,000	25	107,835	100,522
Norway	2,500	7	48,000	388,000
Portugal	-	-	-	-
Sweden	3,000	6	133,000	1,015,000
Switzerland	1,250	63	41,225	59,025
United Kingdom	22,000	7	118,000	51,200
USA	550,000	16	1,282,000	2,009,200

*   Taken from a paper produced by Mr Toshiro Takai, Japan in 1991.

SOURCE: BOATING INDUSTRY STATISTICS, 1990, ICOMIA, UK

SITES

The choice of sites is still difficult especially in the United Kingdom, United States of America and Spain especially where environmental rules and unenlightened controls by authorities have all but stopped medium to large marina developments already subject as they are by the roller coaster economics of funding most usually provided by housing and offices.

In other parts of the world such as Japan, Turkey, Eastern Europe and the Far East, there are many sites still available, as described earlier and including derelict docks and marginal berthing strips, which can be developed with proper master planning, concept and engineering evaluation.

ENGINEERING

Quite correctly there is now a trend to investigate and study a proposed site to assess what particular problems, and there are particular problems, there are to each site. Much money can be saved by bringing in Engineers at concept stage. Developers and Architects have begun to realise this. Problems must be assessed in three dimensions and not by drawing outline layouts on paper.

The retaining of water levels which is now an aesthetic and safety consideration as well as navigational means that practical engineering is required especially in the case of old docks which have old walls of uncertain safety, dredging problems and contamination.

An important link is locks and lock gates. I suspect that more research must be carried out on lock gate performance, lock throughput and the necessary modelling of these functions.

In the past many marinas have suffered from the lack of a professional team at concept stage. This has led to schemes which look promising at concept stage being delayed and even stopped by costs and lack of support by planning authorities due to poor appraisal.

We should be looking for innovation in concept and design and as an example I refer to the Baia dei Gabbiani Marina, Marano Lagoon where dry boat storage is provided not for longterm, but for normal usage. The berths are not floating but are raised precast bridge type construction.

FUNDING AND ECONOMIC CLIMATE

Large village harbour marinas and even large boat park marinas have felt the economic recession in that the upfront costs of marinas i.e. the essential civil engineering to establish the marina is too great without the revenue of housing to support them.

In addition, the initial costs of feasibility studies, investigations, demand surveys, hydraulic modelling and numerous costing exercises are extremely large but important if a bankable project is to be produced.

It is difficult to see that the present downturn in available funding, which is pretty much international, will change within a 2 to 5 year period.

BENEFITS

There is more often than not much bad press attached to marina development, mostly on traffic or environmental grounds.

Clearly there are benefits accruing from marinas, either in employment or spend power by local or visiting yachtsmen.

More research and published data should be produced to show what benefits there are not only to the marina operator but to the environs and the populace.

ENVIRONMENT

Much has been heard of the environmental dangers of marina development ranging from erosion of beaches, to noise and pollution.

However, there would appear to be little statistical record of pollution at or near marinas. Naturally enclosed water spaces if not managed properly can attract litter, oil and organic enrichment. The land side of large developments offers some environmental concern due to housing, traffic and infrastructure requirements.

STANDARDS AND QUALITY

I believe the time has come for national if not international standards to be prepared and recognised. There are some standards available now, Reference 3, 4 and 9 and I understand that others are under preparation in the United States.

I also believe that it is time to improve the quality of the fixtures and fittings in marinas. By this I mean not provide the bare essentials and the cheapest that is on the market but look towards durability and reduced maintenance costs.

Boaters will be encouraged to find an acceptable quality of amenity and berth which should then encourage longer stays and perhaps a bigger spend within the marina and environs.

FUTURE

Some of the following is reproduced from my paper given at the World Marina Conference, Longbeach, California in September 1991 and maybe of interest to new readers.

Subject to the fluctuation of interest rates, and therefore the role of residential building, the construction of village style marinas will continue either on greenfield sites or in existing disused docks or harbours but at a slower rate which will be dictated by funds and planning approvals.

Planning and feasibility studies will still be carried out pending the time when large residential schemes become viable again. Internationally there is a great interest in leisure engineering which encompasses village harbour marinas, marinas and waterside properties with limited mooring facilities.

Boating and interest in boating is increasing at the same time boat sizes are increasing. This means that mooring layouts must be flexible to accommodate what appears to be a changing population.

The length and possibly the width of finger piers might well increase to cope with the larger boats. Fairway and entrance dimensions will also need to be adjusted to suit.

The interest in boating has brought its own problems in that whereas before the yachting community was small and well experienced in sailing the community now consists of less experienced sailors and motor cruiser enthusiasts. The ideal sailing distance for amateur sailors is probably now approximately a day without slipping into a well set up marina with good landside facilities.

Available sites even some that are clearly not the most attractive will now receive attention. Sites requiring reduced upfront civil engineering costs with access circulation mostly in place and with infrastructure networks established such as docks, canal or river sites will be eagerly contested. The large village harbour developments which have received great publicity will run into a series of problems.

Land acquisition will become more difficult as Acts of Parliament, inquiries and protest groups will slow the advance of projects. Naturally developers will need to assess their financial risks more carefully. There is already a case where, after spending considerable time and cash effort, a marina scheme has been stopped by an efficiently mounted protest group.

The attention, quite rightly, paid to the architecture and the aesthetics of large schemes often is a surprise to a developer and there will be much more care taken with approvals for developments in the next few years on these grounds. Above all the ambience of large residential marinas must be established at a high level of quality.

Marina security will take a much higher profile. Access will need to be controlled as the boat owner wants to feel safe while still being able to move around.

The coastal village harbour marinas and their development will also receive much more scrutiny to cover the needs of the statutory instruments covering environmental issues.

This is seen as becoming a major hurdle for developers as the authorities struggle with implementing a reasonably new set of rules. The implications on sea defences, littoral drift, erosion, pollution and aesthetics will inevitably mean a long hard look by the money men before developments commence.

Developments not on the coast but those that retain a body of water by means of locks or flap gates will have to pay particular attention to water strategy. Pollution by the boats or the development itself which would effect water quality cannot be ignored. Many large schemes use as a selling point the water leisure activities available to house or flat dwellers.

Land, near water, which will become part of the development may well contain contaminated fill. Removal or containment of such material is subject to strict legislation and this will have an increasing effect on the prefeasibility study of marinas especially those in the urban context.

The use of a professional team which can foresee and deal with these potential problems will become even more essential. I do not mean a quantity surveyor and an estate agent who will at best only speak of possible costs and profits but cannot forecast difficulties or alternative master planning changes particular to that site, for this you need an engineer and architect preferably one who has worked on large marina schemes before.

The future could be concerned with the following:

(a)   Modernisation and reassessment of existing marinas wherever possible.

(b)   Conversion of swinging and piled moorings into marinas. This will almost certainly require subsidisation but the increase in people safety will more than compensate.

(c)   Continued regeneration of old docks and harbours using marinas/leisure use as a catalyst to attract public grants and redevelopment.

(d)   A few, carefully planned, large village harbour marina projects comprising housing, commercial uses and leisure pursuits.

(e)   Larger involvement of the boating industry to encourage provision of affordable berths.

(f)   Much more emphasis by coastal towns encouraging and promoting marinas to attract tourists and income.

However there are other aspects to consider. At one stage I thought that dry stack storage was the coming panacea for land storage of boats. Apart from the United States and a few in the rest of the world it does not seem to have caught on.

Maybe the problem is that by and large the buildings are large and unattractive at the waters edge and could dwarf other quality buildings. The initial costs and lack of boat standardisation make them, at present, unattractive.

Berth leasing which at the inset looked attractive and a good buy does not appear to have caught on as expected and I believe that the time period will eventually fit the market and this form of capital expansion will become more acceptable.

Maintenance of marinas, those of the large development type, will become extremely important. Berth holders and the general public will expect to see a clean, well maintained marina both berthside and landside. The "live aboard" type resident will probably increase and cause maintenance to increase.

The importance of choosing materials of good and durable quality, especially the berth pontoons, will become even more paramount. This aspect of marina design is often glossed over in some development schemes where initial economy is often chosen in a short sighted policy.

The boom in marinas and leisure boating has increased in such a short period that I believe that the designers and planners are short of good researched statistics.

Research on worldwide trends in boat sizes, which maybe regional, must be encouraged and disseminated, preferably at conferences such as this.

A subject of growing importance is the spending power caused, enjoyed or encouraged by marinas of all sizes. We need realistic surveys to be undertaken and published to establish marina utilisation and spending power. Do "boaters" spend the amount of money we believe? Does employment increase in the region and is it annual? The village harbour marinas, of course, must be separated in such a survey as of course 1000 - 2000 houses with families will produce an enormous increase in spend as well as a larger drain on surrounding services.

In the United Kingdom we are seeing that supermarkets are either planned or spring up at large marina sites. As far as I can see this is not a worldwide trait and I would be pleased to receive data from other countries as to what urban infrastructure follows marinas of the larger type.

SUMMARY

Marinas are serious business and it is good to see that in general demand exceeds supply.

As the economic climate varies so will the development of large residential marinas. However, small marinas will carry on to serve the demands created by the growth in leisure, especially boating.

To establish many of the parameters made necessary by environmental impact studies, greater attention will be given to pollution and water quality.

I believe that there will be greater resistance to large marina schemes and that now is the time to form an overall plan for the United Kingdom looking at a considered siting of new marinas to enhance the complete amenity picture around the coastline.

Similar initiatives are happening in other countries such as large strategic marinas with satellite marinas close by in easy sailing distance to encourage a greater use of the facility.

The village harbour marina will gradually expand to become a place where the full range of leisure activities are available plus hotels-cum-conference centres, restaurants and shopping outlets. As a general statement these larger schemes with more attention paid to the amenities are likely to prevail.

It is important that national associations produce statistics and research material to enable better planning to be carried out and allow funders of developments to feel comfortable concerning financial outlay.

BIBLIOGRAPHY

1.   PIANC.   Standards for the Construction Equipment and Operation of Yacht Harbours and Marinas with Special Reference to the Environment. 1979

2.   PIANC. Dry Berthing of Pleasure Boats. 1980.

3.   National Yacht Harbour Association Ltd. Code of Practice for the Construction and Operation of Marinas and Yacht Harbours. 1980.

4.   BS6349:Parts 1 to 7:Code of Practice for Maritime Structures, 1984.

5.   LACEY P.   Changing Face of Marina Development in the United Kingdom. PIANC Bulletin No. 59. 1987.

6.   BLAIN W C and WEBBER W B. Marinas 1989 Conference. Southampton. 1989.

7.   HIRST M and LACEY P. Village Marina Developments. PIANC Conference, Osaka, Japan. 1990.

8.   World Marina Conference. Longbeach, California. 1991.

9.   Australian Standard. Marinas - Design Practice. 1991.

10. PIANC. Guidance on Facility and Management Specification for Marine Yacht Harbours and Inland Waterway Marinas with Respect to User Requirement. Draft Final Report of Working Group No. 5. 1991.

SECTION 2: PLANNING AND FEASIBILITY

The Influence of Private Legislation on a Development Brief, Port Edgar, South Queensferry, Edinburgh

A.S. Couper

Landscape Development Unit, Department of Planning, Lothian Regional Council, Castlebrae Business Centre, Peffer Place, Edinburgh, EH16 4BB U.K.

ABSTRACT

Preparing a Provisional Parliamentary Order is complex. Preparing one without knowing precisely how a developer will approach the redevelopment of Port Edgar is even more difficult. Consulting the public on the draft development brief prepared to guide the developers before commencing on the Order helped to produce a clearer solution but the process of translating that into the detail demanded by the legislation equally influenced the development brief.

INTRODUCTION

Following the preparation of an extensive feasibility study, eg. Landscape Development Unit [1], Halcrow [2], Wimpey [3], L & R Leisure Group [4], for a waterside village redevelopment package for Port Edgar, the Regional Council accepted that the best option for the future of Port Edgar was for the Council to approach the market directly and to invite developers to submit tenders for the development of a marina village in line with a development brief for the site, based upon the various studies done to date.

On 13th March 1990 the Council approved the draft development brief, eg. Department of Planning [5], for the Port Edgar Marina Village for consultation with affected parties and requested, following conclusion of the consultation period, that it be finalised and application made thereafter for outline planning consent for a marina village. The Council also approved the preparation of a draft

Provisional Order for submission to Parliament in November 1990 and to invite developers to tender for the development of Port Edgar as a marina village according to the development brief, with the exception that the Council retain the sailing school, once outline planning consent had been obtained and the Parliamentary procedure has commenced. It also approved capital allocations over a two year period, estimated at £100,000, to meet the costs of obtaining planning consent, the Parliamentary Order, and the selection of a developer.

BACKGROUND

The Regional Council acquired Port Edgar from the Ministry of Defence in 1978 for development as a marina and water sports centre.
In doing so it had incurred a major capital debt which, together with running costs, constituted a significant annual deficit to the Planning Account.

In the purchase of the site and subsequent development of the marina facilities, there was a total capital debt of some £649,000 which was due to be fully repaid by March 2014.   Including capital repayments and interest, this loan was charged as a cost to Port Edgar of around £80,000 pa.   The outstanding debt at the end of 1988/89 had been £486,000. In addition to these financing costs, Port Edgar had an annual trading deficit which in recent years had been in the range of £100,000-£156,000.

The purpose of the feasibility study was to fully evaluate the possibility of reversing this situation through a substantial redevelopment and reclamation project.

OUTCOME OF THE FEASIBILITY STUDY

Various stages had been gone through in progressing the proposal for a waterside village redevelopment package for Port Edgar [1].   In the present circumstances Port Edgar, although a highly successful watersports centre, had little opportunity to increase income.   In addition substantial maintenance was required which would add to the annual loan charge.   The situation was further complicated by recent legislation on compulsory competitive tendering (CCT) for sports and leisure activities.

In the recommended option, approved by the Regional Council on 13th March 1990, to invite developers to submit tenders for the development of a marina village, the Council would sell the land for housing, including reclaimed land, to the chosen developer and would

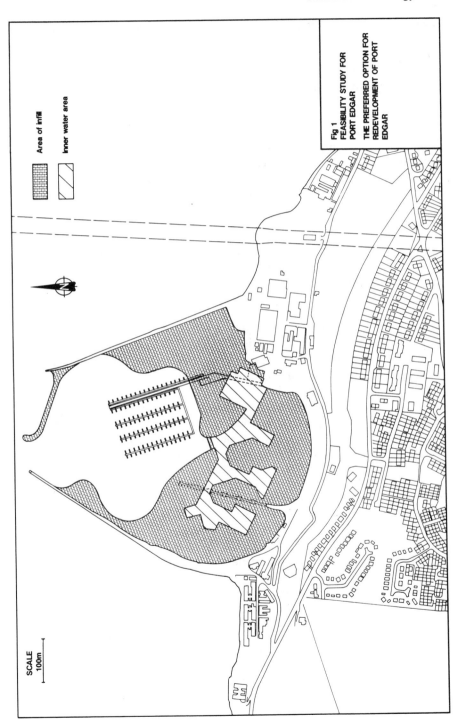

SCALE
100m

Area of infill

Inner water area

Fig 1
FEASIBILITY STUDY FOR
PORT EDGAR

THE PREFERRED OPTION FOR
REDEVELOPMENT OF PORT
EDGAR

have the option of retaining ownership of the marina and harbour area and the land to be developed for commercial uses. The developer would be required to undertake the dredging and other major capital works. The Regional Council would also retain the option of whether or not to continue to run the marina and/or sailing school subject to CCT legislation.

The advantages would be; firstly, that the financial risk to the Council would be minimised as this would be carried largely by the developer. Secondly, capital expenditure on Regional Council services would be paid out gradually as the development progressed. Thirdly, a capital receipt would be obtained by the Council, part of which could be used to discharge the capital debt. The disadvantages would be; the lack of certainty that the reclamation would be completed to an adequate standard, and the limited scope for the Council to take advantage of increases in land values during the development period.

NEED FOR A DEVELOPMENT BRIEF

A development brief is fundamental to a development by tender approach. A draft development brief was prepared based upon the feasibility studies carried out to date and following discussion with officials of Edinburgh District Council Planning Department the local planning authority, who informally indicated support in principle. It covered site description, planning context, proposed development, constraints, engineering works, design and layout, access and parking, services, terms of disposal and tendering procedure. It particularly described what the developer would be required to meet by way of capital costs and what area of land would be retained by the Council for the development of marina related activities. The Council, for its part, would undertake to meet the costs of obtaining the Parliamentary Provisional Order. The brief also underlined the desirability that the marina should continue to offer berthing facilities throughout the redevelopment period, in order not to lose custom.

The Council gave a commitment that all comments from local community organisations and other affected parties would be fully taken into consideration in the preparation of the final development brief.

NEED FOR PARLIAMENTARY PROVISIONAL ORDER

The feasibility study proposed that the redevelopment include substantial land reclamation and the extension of the eastern breakwater. In order to undertake these works within tidal waters, the

Council first had to obtain special powers from Parliament by petitioning for a Provisional Order under the Private Legislation Procedure (Scotland) Act 1936. This could only be done in either March or November.

In addition to powers to undertake the engineering works, the Order will also confer on the Council general powers of harbourmaster at Port Edgar. It will not grant planning permission for the proposed development which will be sought from the local planning authority, City of Edinburgh District Council, through the normal channels.

A firm of Parliamentary Agents was engaged to draft the text of the Order which will be known as the Lothian Regional Council (Port Edgar) Order 1990, eg. Welsh and Dyson Bell Martin [10], and to advise on the drawings illustrating the proposed engineering works that form part of the Order. The target date for submission of the draft Order and accompanying drawings to the Secretary of State for Scotland, in accordance with the Private Legislation Procedure, was 27th November 1990. Before submission could be made, prior ratification authorising the Council to construct works in connection with the improvement of the harbour at Port Edgar, as described in the Provisional Order, was required. Only six months were available to achieve this target. If there were no substantial objections to the Order, confirmation might be expected by the end of 1991 and marketing and selection of developer should be undertaken during the following Spring/Summer, with development commencing early the next again year.

CONSULTATIONS ON THE DRAFT DEVELOPMENT BRIEF

Following the decision of the Regional Council on 13 March 1990, the draft development brief was circulated to 32 local community organisations and other affected parties. Two months were given for the consultation period which in most cases was adequate. Written responses were received from 22 of the bodies consulted. In addition, meetings to explain the proposals were held with the Queensferry and District Community Council, the Port Edgar Yacht Club, the commercial tenants at Port Edgar and Regional Council employees. The consultation process elicited a number of useful observations, comments and criticisms which will be taken into account in re-drafting the development brief and at the detailed design stage.

Two major concerns were raised by several bodies which required to be addressed before submission of the planning application. These related to (a) the impact of traffic generated by the new development on the local area, and (b) the impact of the development on the physical environment. In order to address these concerns, two studies were commissioned: a Traffic Impact Analysis and an Environmental Statement. Both were in any event required to support the planning application. Again, because there was no actual development proposal, both were based on the design concept described in the development brief.

The Traffic Impact Analysis, eg. Kirpartick [6, 7] was carried out in two stages. Stage 1 was concerned with evaluating the impact of the development on the strategic and local road networks, taking account of three alternatives for the main access road [2]. Stage 2 assumes construction of the Council's preferred access and recommended a series of traffic management measures to mitigate the impact of traffic on the local road network.

The Environmental Statement, eg. EAG Montague Evans [8], evaluated a wide range of environmental impacts, including water quality, noise and landscape issues. Although the planning application was for outline consent, the design concept of the development brief had to be viewed on the basis of how would it actually be constructed which presumed a certain sequence of operations. Fortunately the Parliamentary Order process helped evolve the concept into a defined entity that could be confidently tested for such impacts. The Environmental Statement concluded that with appropriate safeguards and good working practices the environmental impact of the development would be acceptable. These safeguards and practices have been incorporated in the final development brief.

EVOLUTION OF THE DESIGN THROUGH THE CONSULTATION PROCESS

Evolving the development brief in parallel with the preparation fo the Parliamentary Order was of benefit to the development of the project. The iteration between the consultation process on the brief and the development of the Order drawings helped refined the design concept into an entity that could stand the scrutiny of the parliamentary process, yet would be sufficiently flexible and appealing to a developer. Initially there was concern that a developer would not wish to be tied to too defined a concept but in reality the process produced a solution that was pragmatic and helped in turn overcome most of the local concerns, isolating those remaining to land based issues, which were addressed in detail by the two additional studies, the Traffic Impact Analysis and the Environmental Statement. Having this more defined proposal also

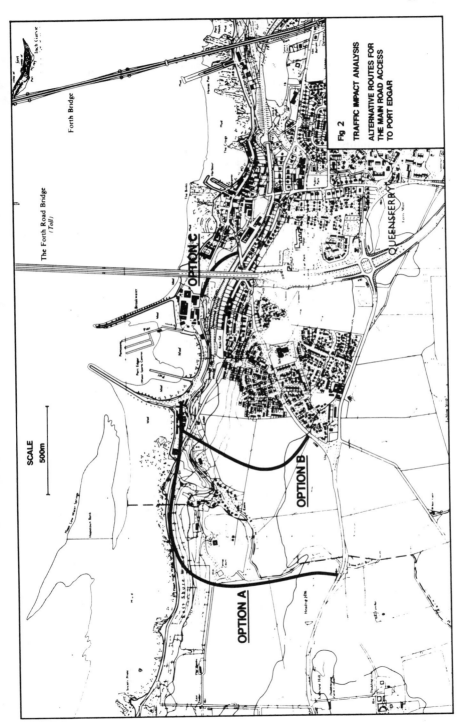

Fig 2

TRAFFIC IMPACT ANALYSIS
ALTERNATIVE ROUTES FOR
THE MAIN ROAD ACCESS
TO PORT EDGAR

Forth Bridge

The Forth Road Bridge
(Toll)

OPTION C

OPTION B

OPTION A

QUEENSFERRY

SCALE
500m

B

helped the preparation of the outline planning application. How did this come about?

Initial work started on the Order after the conclusion of the two month consultation period on the development brief. Concerns relating to the engineering aspects of the redevelopment seemed to fall into four areas:-

1.   The continuity of local businesses at the marina, the chandler and yacht sales, the boat and engine repairers and the electronics sales and servicing businesses.

2.   The expansion of the existing marina berths and location of the commercial berths.

3.   Access to the water for the sailing school and for boat launching and repair.

4.   The need for the inner basin, the feasibility of its shape, boat access to it and quality of water and berthing within it.

It was also clear from consultation with the Parliamentary Agent that the Order drawings had to be very accurate which posed the immediate problem of how did the Council know what a developer would actually construct. If the developer constructed the works as shown in the feasibility study but due to unforeseen difficulties of either a physical or market nature, had to make changes, how could such be accommodated? The proposal contained in the feasibility study had to be able to stand the test of scrutiny if there was a Parliamentary Inquiry into the Order. On the advice of the Parliamentary Agent, and his broad understanding of the Council's proposal and chosen development route, the works described in the Order were evolved into a series of six related works [3] but not necessarily obligatory or dependant in each other. This could give the developer scope for flexibility but it did mean that each work became very precisely described in detail. By close examination of the four areas of concern mentioned above, the detailed product became a practical and realisable proposal and one that was costed more accurately than in the feasibility study. The Estimate of Expense required for the Order estimated the cost of the works to be £17.6 M.

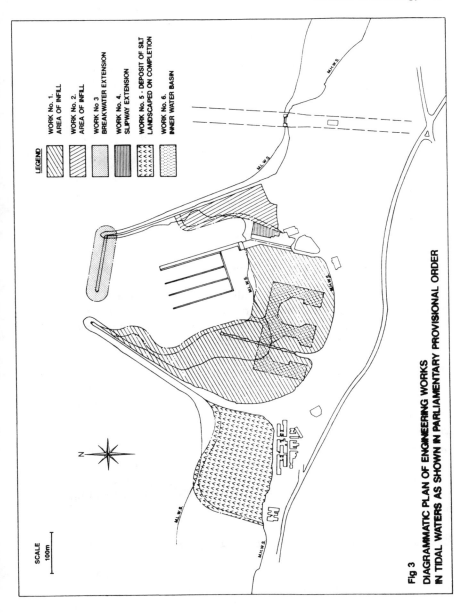

SCALE
100m

LEGEND

WORK No. 1.
AREA OF INFILL

WORK No. 2.
AREA OF INFILL

WORK No 3
BREAKWATER EXTENSION

WORK No. 4.
SLIPWAY EXTENSION

WORK No. 5 - DEPOSIT OF SILT
LANDSCAPED ON COMPLETION

WORK No. 6.
INNER WATER BASIN

Fig 3
DIAGRAMMATIC PLAN OF ENGINEERING WORKS
IN TIDAL WATERS AS SHOWN IN PARLIAMENTARY PROVISIONAL ORDER

Continuity of Marina Services

The development brief envisaged the developer would provide an area of land for the Council for the development of marina related activities to include:

- administration building/office and ancillary facilities;
- chandlery;
- boat repair facilities including yard and slipway;
- electronics sales and servicing;
- sailing school;
- yacht clubroom.

The developer would be required to construct a dock for boat hoisting and to provide a dinghy park and slipway.

As these activities take place beside or within the existing naval buildings that are grouped on the south east side of the harbour, the question was could any of these buildings remain and could they be refurbished to fit in with the kind of architectural character that was being suggested in the development brief. Fuller examination of their potential showed that two, the sailing/training school shed, and the No. 4 shed could be retained and refurbished/developed, allowing both the sailing school and Port Edgar Marina Services Limited, who operate many of the activities above, to continue without major expenditure on new premises. What this did was to open up the area and allow consideration of it for, not only marina related activities like a boat yard or boat parking, but for redevelopment of a related commercial use, like small business units/offices suites and/or craft workshops. It also raised the question whether it was absolutely essential to infill the harbour to the extent proposed in the feasibility study. The outcome was finally influenced by the decisions taken on the second and third areas of concern.

Expansion of marina berths and provision of commercial berths

If the developer decided not to construct an inner basin which could achieve an increase in berths of 100, the extent of infill proposed in the feasibility study would preclude any organic expansion of the existing berths. As the developer will be required to remove the steel structure of the west pier, which presenting provides commercial berthing, a new berth would be required so that "Maid of the Forth" pleasure cruiser and the Forth River Purification Board's survey ship could continue to operate at Port Edgar.

When the feasibility study was being prepared, demand for berths in the marina increased to a point where a waiting list began to exist. Since the development timescale is some way into the future and the solution to extra berthing was linked to the provision of an inner basin, the question of whether the existing berth layout should be expanded by another 'leg' of pontoons on the west side was addressed. Given an element of uncertainty over whether a developer would actually make a commitment to an inner basin, however desirable, on the grounds of cost, it was decided to alter the extent of the infill to allow for an increase in marina berths within the 'outer' harbour to a total of 500. The knock-on effect of catering for this expansion was to push the proposed entry point into an inner basin, south westwards towards the shore, but in doing this it also created space for a commercial berth at the very heart of the marina, just where the public would be. Pulling the infill back to the Loop Shed and incorporating a steel sheet piled edge from there to the inner basin entry point, would enable boats like the "Maid of the Forth", to pick up passengers from the very heart of the new village. More vessels could be accommodated than the two existing, perhaps a feature ship of historical interest off the quayside.

The consequence of making these changes was that the new berths and inner basin entry point were placed in more direct tidal influence by being in line with the open harbour mouth. The testing of this decision was the subject of a further hydraulic study, eg. Halcrow [9], that examined the design of the eastern breakwater extension and the inner basin in more detail.

Access to the water for boats

The feasibility study proposed the replacement of the existing slipway. The development brief requires the developer to construct a new slipway and dock for a boat hoist, within the eastern part of the 'outer' harbour. The slipway is for the sailing/training school and has to be accommodated within the land the developer is required to provide for the Council. Existing means of recovering a yacht from the water comprise a fixed crane on the north end of the east pier which requires tractor and trailer operation to subsequently move to it the boat yard or park. If the commercial berth had to be moved into the same area of the 'outer' harbour, there would be conflict and congestion between dinghies, yachts and commercial vessels.

By retaining the sailing/training school shed it seemed more sensible to keep the existing slipway and widen it so that the full width is available at all states of the tide. This was after all the marina's best asset. This could enable a boat hoist to operate without a dock by running up and down this slipway. Commercial traffic would not interfere as it would be berthing on the other side of the east pier. There would be no need therefore to create an expensive concrete structure for a hoist or to consider an equally expensive slipway, perhaps on the east side of the east breakwater. The requirements of the Council are less, but the decision to retain the existing slipway enhances the practicality of the development proposal.

Feasibility of the inner basin

Most concerns about the inner basin centred on the quality of water impounded within it. Would it be less than desirable, how would it be flushed out, how would boats gain access, was it's shape suitable for mooring boats, what level would the water be at, would house owners have private berthing and who would manage it were the kind of questions raised during the consultation period?

Many of these points had been answered in the feasibility study but it was clear that further study by a specialist knowledgeable in dealing with water quality problems would be required. As there is an inter-relationship between the 'inner' and 'outer' harbours a further hydraulic study combining the detail design of the breakwater and the location of the inner basin entry was commissioned, eg. Halcrow [9].

The second hydraulic study confirmed the design parameters for extending the east breakwater which is a requirement a developer must meet to reduce siltation and to give protection to the marina pontoons. The alignment, construction and length to ensure acceptable wave conditions within the 'outer' harbour were defined. The study also defined the most appropriate point of entry to the inner basin in terms of navigation, in and out off the entrance, and possible build-up of silt at it or within it. Having set out boat usage requirements of the inner basin and possible water depth scenarios in the hydraulic study brief, the specialist consultant tested various options of water control versus retained water level versus tidal influence versus induced water flow and concluded that a flap gate arrangement operating on a tidal cycle with the water retained at 0.5M O.D. would be a satisfactory solution in terms of ensuring a high quality of water persisted at all times [4].

One of the decisions taken following this hydraulic study was a reappraisal of the inner basin's shape and its fitness of purpose for berthing yachts. With a doubt about a developer's intentions, it was decided to simplify the inner basin in a way that construction could be phased and phased to suit the construction timetable for infill

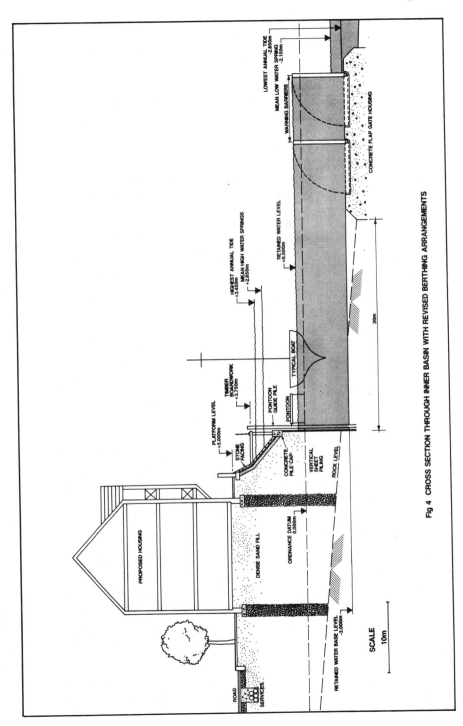

Fig 4 CROSS SECTION THROUGH INNER BASIN WITH REVISED BERTHING ARRANGEMENTS

operations. The resultant 'E' shape achieves an element of flexibility and by rotation of the shape around the entry point [3] minimises expensive excavation of shallow outcropping rock at the foreshore and extensive demolition of the concrete dolphins of the west pier. It is conceivable that a developer could construct half of the inner basin with or without the flap control, thus deferring infilling of the remainder of the harbour until it is considered viable.

FINALISATION OF THE DRAFT ORDER

While the result of examining these four areas undoubtedly helped allay local concerns, the process did not stop there. It crystalised how the works could be sequenced and as a result six separate works were delineated and described with a great deal of accuracy, using the special convention of the legislation, continuous cross sections through individual works. The most appropriate alignment for these sections through the complex form of the infilling works, was decided on the advice of the Parliamentary Agent, because the written description in the Order had to explain very precisely where they started and finished.

The process of "pulling" back the extent of infill originally proposed in the feasibility study, resulted in the idea of depositing dredged silt on an area of foreshore to the west of the harbour, to form a landscaped amenity area.   This should benefit the reclamation making it more efficient and economic. Since the land is currently in Crown ownership and cannot be acquired compulsorily through the Order, it will have to be acquired by agreement.

The volume of silt required to be dredged by a developer from the harbour and the excavation of soft materials below the silt is based on the extensive site investigation carried out as part of the feasibility study, eg. Wimpey [3]. A 30 m horizontal limit of deviation was adopted to cater for and to overcome any local variations in these results that may arise.

During the preparation of the drawings describing Works No. 6, the Inner Basin, the retained water level was lowered by 0.5 m by incorporating a separate floating pontoon berth from the original board-walk idea, which segregates public and boat users without imposing restrictions on public access. It will also help management of the inner basin berths and enables the marina to be managed as one unit.

The consequences of rationalising all boat related uses, has benefited the layout of land uses within the new marina village. There

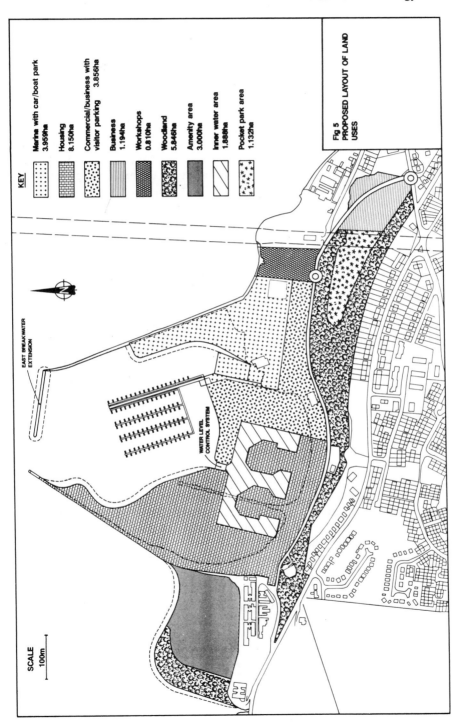

KEY

Marina with car/boat park
3.959ha

Housing
8.150ha

Commercial/business with
visitor parking 3.856ha

Business
1.194ha

Workshops
0.810ha

Woodland
5.846ha

Amenity area
3.000ha

Inner water area
1.888ha

Pocket park area
1.132ha

Fig 5
PROPOSED LAYOUT OF LAND
USES

EAST BREAKWATER
EXTENSION

WATER LEVEL
CONTROL SYSTEM

SCALE
100m

is simple logic in having marina uses grouped by the slipway, commercial uses centred on the commercial quay and housing set around the inner basin. It has also made the housing land use less dominant, but still gives more than sufficient space to accommodate the 500 house types described in the development brief [5].

SUBMISSION AND FUTURE PROGRAMME

The Draft Order and Petition were deposited at the Scottish Office in London on 27 November 1990. From then until 23rd January 1991 there was a period for lodging petitions against the Order. During that period public notices were published and the drawings displayed.

One petition against the Order has been received and is being decided upon by the Secretary of State. Unless it is withdrawn a Parliamentary Inquiry is likely to be held.

On 28th March 1991, the Council reaffirmed its original instruction to submit an application to the City of Edinburgh District Council for outline planning consent for the development of a marina village at Port Edgar and agreed to submit copies of the Traffic Impact Analysis and the Environmental Statement in support of the planning application.

Consultations on the application have been completed by the City of Edinburgh District Council. A decision on the application is expected soon.

CONCLUSION

Preparing a Draft Parliamentary Order is undoubtedly complex. The legislation demands that the works shown are very precisely described both on the drawings and in the text. Undertaking public consultation on the draft development brief greatly assisted the process because it highlighted areas of concern and only by detail examination of them was it then possible to prepare solutions of sufficient detail for the Draft Order.

The needs and conventions of the Draft Order strongly influenced how far the detail design solutions had to be taken and that in turn influenced the development brief. It took commitment but the advantages of this degree of consideration and testing before any developer has been invited to express interest, are that the redevelopment has been refined and made more viable.

ACKNOWLEDGEMENTS

This paper represents the views of the author which are not necessarily those of Lothian Regional Council. Gratitude is expressed to the Director of Planning for allowing publication of this paper and to Chris Bushe, David Sillence and John Inman for their helpful assistance and comments.

REFERENCES

Reports

1   Landscape Development Unit, Department of Planning, Lothian Regional Council, Port Edgar Development Phase II Study, Volume 11, February 1989.

2   Sir William Halcrow and Partners Scotland Limited, Port Edgar Development Project, Hydraulic Study, Glasgow, August 1988.

3   Wimpey Laboratories Limited, Proposed Redevelopment of Port Edgar, Report on Site Investigation, Lab. Ref. No S/25863, Broxburn, May 1988.

4   L & R Leisure Group, in association with CASCO, GRM Kennedy and Partners, W J Cairns and Partners, Tozer Gallacher, LRC Landscape Development Unit, Price Waterhouse, Port Edgar Development Phase II Volume 1, Edinburgh, November 1988.

5   Department of Planning, Lothian Regional Council, Development Brief, Port Edgar Marina Village, Queensferry, Edinburgh, Draft for Consultation, December 1989.

6   Kirkpatrick and Partners, Port Edgar Redevelopment Traffic Impact Analysis, Edinburgh, December 1990.

7   Kirkpatrick and Partners, Port Edgar Redevelopment Traffic Analysis Stage 2 Report, Edinburgh, May 1991.

8   EAG Montague Evans, Environmental Implications of Proposals to Redevelop Port Edgar Marina, Edinburgh, June 1991.

9   Sir William Halcrow and Partners Scotland Limited, Port Edgar Marina Village Additional Marine Studies, Glasgow, November 1990.

Draft Provisional Order

10   Welsh, G.F.G. and Dyson Bell Martin & Co., Lothian Regional Council (Port Edgar) Draft Provisional Order, Edinburgh and London, November 1990.

Paper in Conference Proceedings

11   Couper, A.S., A Feasibility Study of Port Edgar, South Queensferry, Edinburgh, in Marinas: Planning and Feasibility (Ed Blain, W.R., Webber, N.B.), pp 189-211, Proceedings of the International Conference on Marinas, Southampton, UK, 1989. Computation Mechanics Publications, 1989.

12   Couper, A.S. and Montgomery, H., Minestone Fill for a Maritime Village Development at Port Edgar, South Queensferry, Scotland, in Reclamation, Treatment and Utilisation of Coal Mining Wastes (Ed Rainbow, A.K.M.), pp 377-390, Proceedings of the Third International Symposium on the Reclamation, Treatment and Utilisation of Coal Mining Wastes, Glasgow, UK 1990. Balkema, Rotterdam, 1990.

Leisure Harbours as an Economic Resource: Planning Problems in Sardinia

G.P. Ritossa, N. Migliavacca, P. Sanna

Transport Institute, University of Cagliari, via Palestrina 72, 09100 Cagliari, Italy

INTRODUCTION

The aim of this paper is to determine what constraints influence the size and choice of sites of a network of tourist ports in Sardinia.

Owing to its geographical location and the admirable environmental characteristics of its coastline, the island of Sardinia offers excellent and desirable sites for this kind of "marine" tourism. However, given the lack of services and infrastructure, it cannot attract many holidaymakers in this sector of the tourist trade.

The objective proposed should be reached by means of planning supported by quantitative and qualitative analyses which, given the high investment involved, guarantee reasonable economies of scale and an adequate return for community.

First and foremost, it is essential to know the effective demand for boat moorings in terms of both number and boat length, as well as availability in the various ports.

In order to determine demand, market research was carried out on the boat market, on the population which uses boats for holiday purposes, on future trends and also on the existing "supply" of marinas and ports, including projects not as yet finalised.

The final objective was to determine what we may call "fleet-type" by means of statistical diagrams and calculated estimates, this being the basic guideline in planning a port's size.

GENERAL CRITERIA

The model for economic development followed so far in Sardinia has undermined its socio-economic structure to a point of crisis.

The situation has been exacerbated by the general
crisis affecting Italy as well as many other
countries. Faced with this scenario, it is clear
that, with such a chronic lack of financial resources
and at the same time numerous demands to be satisfied,
priority should be given to those investments which
guarantee the highest return. Additionally, existing
infrastructures should be improved and local resources
exploited to the best advantage.
This could be done not with sweeping developmental
changes, but rather with a series of sensible and
controlled projects of infrastructure improvement,
where such resources already exist "ready made".
The coastline and the sea can be considered
immeasurable resources of production and will be the
cornerstore of any plan to stimulate development and
provide new jobs in the tourist and fishing
industries.
Existing infrastructures in Sardinia are heavily
concentrated in the North-Eastern area where a
thriving tourist industry has made necessary the
creation of a number of mooring points in order to
satisfy the high demand for boat moorings.
It is little wonder that this privileged area has
remained almost immuned to the general economic crisis
and its consequent problems.
The fact that almost 90% of tourist port business is
concentrated in this part of Sardinia must lead us to
conclude that a wider-spread distribution of supply
(port facilities) would produce greater overall
demand. This, in turn would generate further wealth in
as much as the creation of tourist landing places at
the present moment in time would give high investment
returns both in direct terms and in terms of the
number of jobs created in relation to the number of
boat moorings.
Therefore, we envisage a port system with a network of
tourist ports in each coastal village, all meeting the
usual standard nautical regulations, but with
consideration given to local socio-economic
conditions.
All this should be carried out on the basis of a
feasability study which will take account of socio-
economic conditions characterised by the frustrating
inability to satisfy basic demands, due to a chronic
lack of financial resources.
The construction of harbours should be undertaken
within the context of a general project that allows
for the immediate initial exploitation of local
infrastructures, with an approach which will encourage
sound planning and altogether dynamic operations.
It seems clear that proceeding along these lines will
help to avoid the kind of risks which often produce

inaccurate forecasts for long-term demand. Thus, not
only should the first working harbour be built with
the results of profitability calculations in mind
[these can be compromised by innacurate evaluations of
cost-benefit] but construction costs ought to be kept
to a minimum.
Further steps will be carried out once there are
reliable indications as to the increase in demand for
mooring points as well as reliable evidence that a
project will be worthwhile in terms of the number of
jobs created and the amount of income generated. In
this way, those infrastructures which turn out to be
the most efficient economically will be rewarded with
further project-works while it is likely that others
will be exploited only to the extent required by local
needs.
From a technical viewpoint, the project engineers will
have to be aware of the need to co-ordinate
infrastructure development in such a way as to
facilitate any further phases of development.
It is also of prime importance that costs in operation
techinques and of materials are kept to a minimum. In
any case the viability of any project will be judged
in terms of suitability to a particular area of
coastline.

EVALUATION OF SUPPLY AND DEMAND

Information and statistics gathered from the local
maritime Authorities, various Port Authorities and
the Local Sardinian Government provide a global
picture of port facilities (including those under
construction) and of the presence of Italian and
foreign craft.
After sub-dividing the whole perimeter of Sardinian
coastline into five tracts, data relating to
availability (and non-availability) of moorings in
each port was tabled according to the size of port and
the level of services offerred. Subdividing further
into residents and visitors the demand for harbour
facilities and accessory services in Sardinian ports
was calculated by quantifying the fleet of craft
currently existing in Italy and Sardinia.
Sources for the collection of data were the Ministry
of Shipping and National Association of shipyards and
Nautical Industries.
The Ministry of Shipping did not provide any
information on small unregistered craft which, while
constituting a large part of the overall fleet, do not
usually use the ports.
A graph was then constructed using linear regression,
between the number of pleasure craft and G.D.P.
relative to the holiday population, and based on

statistics for the years 1980/88, so as to evaluate stability of future trends in pleasure craft tourism and thus potential demand for port services up to the year 2000.

POTENTIAL DEMAND IN SARDINIA
(units in thousands)

	1990	1995	2000
PEAK SEASONAL DEMAND	22.1	25.8	29.6
YEAR-ROUND DEMAND	8.9	10.3	11.7
TOTAL	31.0	36.1	41.3

CRITERIA FOR PROJECTING PORT SIZE

The basic tenets for project development are:
- knowing the number of potential future users of a port
- defining the fleet-type.
With this information, the number of mooring points needed in the port can be determined.
A study carried out set a minimum limit of 400-500 units for the number of moorings (C) which an economically viable tourist port would require. It's also suggested calculating "C" in terms of the number of inhabitants (n) that utilise the port, thus:
$C = 0.04 n$
C is the sum of craft permanently stationed (N_p), those temporarily stationed (N_t), and those under repair. Once the number of boats is known, the berth surface area can be calculated approximately in this way: Sb (berth surface area) $= 110 N_p = 11 C$
It is essential to determine percentages for the different sizes of craft expected to make up the fleet of a port, so that the required surface area of a harbour can also be determined.
Our research was carried out in regional ports with the aim of ascertaining the changing nature of users'needs and requirements. Results show that there has been a tendency for average boat length to increase.
By using the percentages method, graphic diagrams relating to Sardinian, French and American ports can be traced and an estimated curve for projected fleet-type may also be traced (Figure 1).

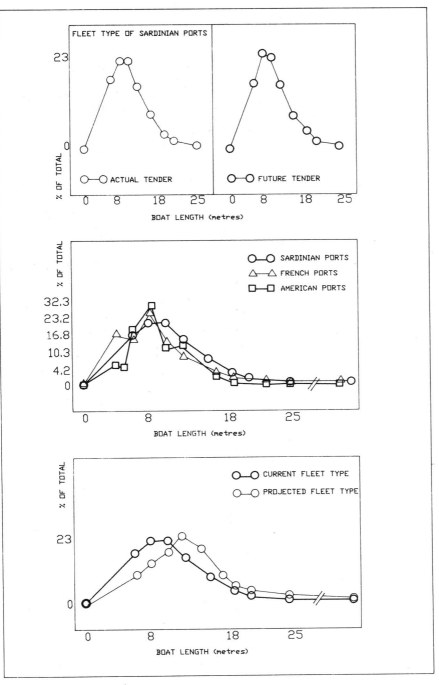

FIGURE 1.

BUDGETING

An optimal planning strategy can only be found by
considering factors linked purely to Marine (coastal)
engineering, as well as considering the effects of new
infrastructures on the environment.
However, when conceiving the project for a port there
should be an assurance that the investment return will
justify such an undertaking.
One way to ensure a rapid return of capital is to sell
rights of access of some of the mooring points.
The number to be sold to ensure maximum profit, will
vary depending on whether a short or long term
analysis is made. In the former case, it would be more
profitable to sell the majority of moorings, in the
latter, temporary moorings would be more lucrative.
In a cost-benefit analysis, it is important to
evaluate accurately any derived advantages, even if
they are not easy to quantify.
Tourist ports generate business activities such as
trade in nautical equipment and related services,
restaurants, boat hire and attractions like Nautical
clubs and sailing schools which generate interest from
non craft-owning tourists as well.
Obviously, a general project would take into account
all the business that a project such as outlined above
would generate; but it would be preferable to divide
expansion and development into two or more working
phases.

COMPARISON WITH OTHER MEDITERRANEAN PORTS

The graphs represent in the best way possible, the
results obtained from a comparison on pricing policies
in Sardinian, French, Jugolsav and Adriatic ports.
Sardinian ports are the most sought after and so have
the highest tariffs in peak season, while the prices
are comparatively low during low season because the
"year round" permanent fleet is not very large
(Figure 2).

CONCLUSION

The study made so far of port facilities currently
operative and of those under construction or in
project phase, has shown how the criteria for planning
and management have been derived from analyses of
specific local conditions.
Our study has however revealed that there are criteria
for planning and management which will ensure the
correct utilisation of a port facility.
Taking into consideration the changing nature of
demand and the minimum size that a port facility

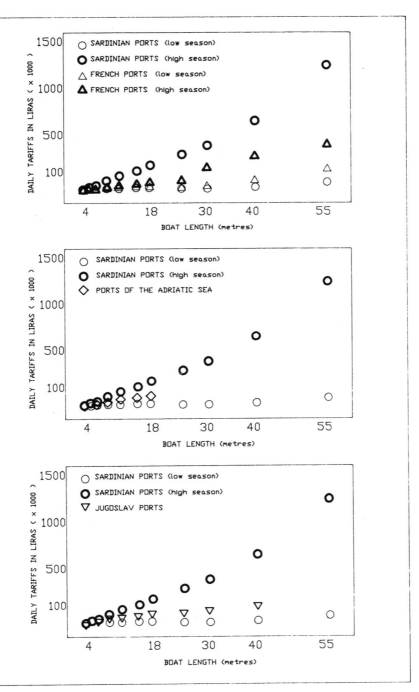

FIGURE 2

should be in order to make it a profitable on-going concern, we were able then to define particulars of size and management criteria such as fleet-type, utilisation of the harbour area and management structure of a typical model infrastructure in Sardinia.

One of the most important factors to be underlined is the reduction of the optimal number of boat-moorings from 400-500 (as indicated above) to 250-300, and the fact that the choice of sites for the port network can be made irrespective of the distance between one port and another.

In other words, every coastal town might have its own landing place, regardless of size, this giving greater importance to the economic benefits for the hinterland rather than to other factors usually taken into consideration.

We must aim, however, to have a global development plan, in which all development areas directed at the various markets are in someway integrated. We have already seen that it takes time to attract a good-sized fleet to a new pleasure craft docking area. It seems true to say that the establishment of a network of Sardinian ports will require high level sales and publicity campaigns for each single port.

For potential users, the type of services offered by a port will be an important factor in choosing where to go. A central organising body will be able to help provide technical assistance in each individual port, as an inherent part of each port's organization.

Aside from any decision on the suitability of creating an unified structure, it will certainly be opportune to adopt a well organized trade policy, particular for dealing with non-local users.

The project's management and organization will geared towards obtaining total economic self sufficiency. The greatest economic benefits will initially be reaped by activities auxiliary to the port (see above), outside the system itself. It is therefore in the interest of the local economy to keep tariffs to a minimum, in order to encourage to formation of significant sized fleet.

REFERENCES

1. Pennisi, Tecniche di valutazione degli investimenti pubblici, Istituto Poligrafico e Zecca dello Stato, Roma, 1985.

2. Berriolo, Sirito, Spiagge e porti turistici, Hoepli.

3. Association international permanente des Congres de

navigation, Commission International pour la
Navigation de sport et de plaisance, Rapport final
de la Commission Internationale pour la Navigation
de sport et de plaisance, Bulletin n. 25,
Bruxelles, 1976.

4. Association international permanente des Congres de
navigation, III Commission International pour la
Navigation de sport et de plaisance, Normes pour la
construction, l'equipement et exploitation des
ports de plaisance et des marinas, Bulletin n. 33,
Bruxelles, 1979.

5. Association international permanente des Congres de
navigation, Commission International pour la
Navigation de sport et de plaisance, Rapport sur:
La tracé et la structure des avourage estérieurs
des ports de plaisance, Bullettin n. 38, Bruxelles,
1981.

6. Società Bonifica, Sistema di approdi nel
Mezzogiorno, Ministero della Marina Mercantile,
1986.

7. Cambra Oficial de Comerc, Industria i navigacio de
Barcelona, I jornadas internationales de turismo
nautico en el Mediterranean, Barcelona, 1988.

Evolution of a Harbour: Planning, Construction, and the Resulting Impacts of a Recreational Boating Facility Upon a Small Community in Canada

G.T. Beaulieu (*), J.C. Stansbury (**)
(*) Regional Planning and Engineering Advisor, Small Craft Harbours, Fisheries and Oceans Canada, Burlington, Ontario, L7R 4A6 Canada (**) Hough Stansbury Woodland Ltd, Etobicoke, Ontario, Canada

ABSTRACT

Canadians enjoy a freshwater resource that is one of the most extensive in the world. In the Province of Ontario, recreational boating is now an industry in itself, reaching over $1 billion per year in direct expenditures. The economic impact of recreational boating goes well beyond such direct spending as communities are turning their attention to waterfront renewal and development. Successful recreational harbour development requires a clear understanding of demand, planning, design and operation.

Even the smallest community, if it acts with sensitivity and foresight, can reap very significant benefits from the increasing demand of recreational boaters. One such community is the Village of Hilton Beach situated at the western end of the North Channel, within Lake Huron, one of Canada's most popular boating areas and considered to be world class cruising waters.

The Village of Hilton Beach, with a population of 205, debt load capacity of about $10,000, and virtually no water or sewer services, recognized that recreational boating development was its primary opportunity for economic and social improvement. The old federally-owned government wharf and harbour were in a state of disrepair and unsafe for public use. It had been designed in the late 1890's for use by small coastal transport vessels and did not meet the present day demands of recreational boating.

This paper outlines the problems and challenges which arose in planning the conversion of the existing harbour to a modern recreational boating facility and describes the sequential steps of the project including the planning, construction and the initial years of operation. The importance of designing for everyone in the community, and not just for boaters, is discussed.

The initial impacts which the redevelopment of the harbour has had on the Village, in terms of economic, social and tourism benefits, will be outlined along with comments on direct and indirect revenue generation and job creation.

1: INTRODUCTION

Recreational boating in Canada, and particularly in the Province of Ontario, is one of the main sports activities of the population. These activities generate many economic spin-offs in the communities surrounding the harbour locations. The recreational boating potential in the Province of Ontario is unlimited because of the abundance of fresh water and the numerous opportunities to take full advantage of natural resources.

The Province of Ontario has one of the greatest natural inland waterways in the world. There are approximately 18 million hectares of fresh water comprised of four of the five Great Lakes, 500,000 smaller lakes, thousands of rivers and several man-made canal systems. It is estimated that there are currently in excess of 1.3 million recreational craft - over one-half of Canada's total. There is an overall average of one recreational vessel (kayak to full size cruiser) for every eight people in the Province of Ontario - one of the highest boat ownership ratio in North America.

The location of the Province of Ontario within Canada is shown on **Figure 1.**

The five Great Lakes constitute one of the largest freshwater systems in the world. Lakes Ontario, Erie, Huron, and Superior form the International boundary between the United States of America and Canada (Lake Michigan is completely within the U.S.). As a result, there are many International agreements controlling the use of these waters. Because of the proximity to the U.S., there is considerable potential for international water-related tourism.

The North Channel is a world class cruising area that lies between a chain of islands and the mainland in Lake Huron as shown on **Figure 2.** The Village of Hilton Beach is positioned at the western end of the North Channel and is therefore ideally situated to take full advantage of both Canadian and American tourist traffic.

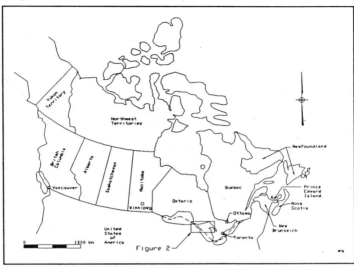

Figure 1- *Map of Canada*

Recreational boating is one of Ontario's fastest growing leisure pastimes and the demand for development of marinas and related facilities is extremely high. Direct economic activity of recreational boaters exceeds $1.0 billion per year in the Province. Allied

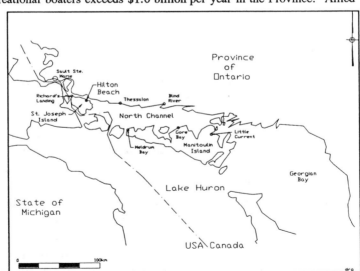

Figure 2 - *Location of the North Channel*

Boating Association of Canada indicated, in 1989, that its members could sell one new boat for every berth created. Many of the older harbours, traditionally commercially-oriented that have fallen into disuse, have been converted in whole or in part to recreational facilities in an attempt to meet this demand. The demand for berths has been reduced in the past several years due to the world wide recession; however, the long term outlook is for further growth.

2: OVERVIEW OF CONDITIONS IN VILLAGE BEFORE REDEVELOPMENT

The Village of Hilton Beach is located on St.Joseph Island, approximately 60 km east of Sault Ste. Marie, Ontario. Historically, St.Joseph Island played an important role in the development of the region. Missionaries and Huron Indians seeking refuge from the Iroquois Indians settled briefly on the Island in 1649, but an outbreak of smallpox in 1658 forced abandonment of the settlement. However, continued development in succeeding years included the construction of a fort in 1796 to protect the lucrative fur trade from the Americans. Three main communities were settled during this period. They were: Sailors Encampment, Richard's Landing, and Marksville which was later called Hilton Beach.

In 1897, the federal government purchased a small wharf in Hilton Beach from a private individual and constructed an addition to permit small coastal transport vessels to visit on a regular basis. Hilton Beach became one of the chief shipping points for St. Joseph Island. Important quantities of lumber and agricultural products passed over the wharf, as well as large numbers of people using marine transportation. Over the years, minor modifications and improvements were carried out for commercial coastal transport vessels.

The use of Great Lakes small coastal vessels declined in the 1940's and 1950's and the last commercial vessel visited Hilton Beach in 1963.   For several decades the Government wharf was not used extensively until recreational boating became popular. Although the commercially oriented wharf was not ideal for use by recreational craft, the growth in recreational boating ensured that any existing facility was used as much as possible. Some recreational berthing and protection was added to the government wharf and shoreline but the wave climate was not acceptable for the new use. The layout of the former government wharf is shown in **Figure 3**. The harbour facilities were leased to the Village of Hilton Beach for management purposes in 1972 to ensure that the Municipality had control over its operation in providing maximum benefits for its citizens. The Village engaged a Wharf Operator to carry out the day to day operations during the summer.   The structures originally constructed for coastal transport vessels continued to deteriorate and became unsafe for use. From a practical point of view, there was no point in reconstructing the existing structures as they had been designed for small coastal transport vessels and the use of the harbour had changed to recreational boating.

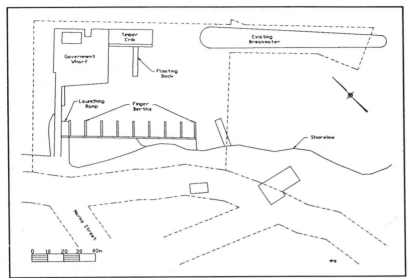

Figure 3 *Government Wharf Before Redevelopment*

In early March of 1983, the Village of Hilton Beach sent a letter to Small Craft Harbours requesting attendance at a local council meeting to discuss possible harbour repair and expansion.   Evidence of the inadequacy and disrepair of the wharf was continuing to mount. In June of the same year, the local Building Inspector threatened to condemn the wharf as unsafe unless repairs were completed to insure that the condition of the facility did not deteriorate further.   Large pot holes were appearing on the wharf and approach road.   Minor repairs were carried out on an interim basis. The Wharf Operator at Hilton Beach wrote a letter to the Minister of Fisheries and Oceans detailing the need to expand the capacity of the marina and improve breakwater protection.   In the words of the Wharf Operator:

> "The economy of Hilton Beach depends almost entirely upon tourism, and the improvement of the Government Wharf in this municipality is essential to that industry."

Federal Government budgetary restraints at the time did not allow the Hilton Beach project to proceed due to the high construction costs involved.

In February of 1986, the Village Clerk, in a letter to the Prime Minister of Canada, indicated the state of disrepair of the wharf and the importance that the facility held for the community. Site inspections by Small Craft Harbours in June of 1986 further reaffirmed the ever increasing dangerous condition of the wharf and the need for repairs and improvements. Two initial conceptual plans were developed internally at Small Craft Harbours to aid in the preliminary discussions on possible redevelopment. After reviewing these conceptual plans, the Village submitted their ideas for consideration. In the interim, the local economy was deteriorating. The general malaise of the main regional industry, the Algoma Steel Corporation in Sault Ste. Marie, was affecting every nearby community, as the world wide demand for steel declined.

The facility was continuing to deteriorate to the point where the main wharf and approach were unsafe and there were only 27 marginally usable slips. Due to record-setting high water levels of the Great Lakes in 1985 and 1986, the existing breakwater (which was marginally acceptable in terms of wave protection under normal water levels) became partially submerged and inadequate in preventing heavy seas from entering the harbour and damaging facilities and boats. Only a few amenities were offered at the wharf such as gasoline and minor retail items. There were no toilets, power hookups, water supply or showers. Excluding fuel sales, revenues in the two years before 1988 averaged about $7,500, or less than $280 per dock.

Of particular concern was the condition of the approach to the wharf which contained a number of large holes posing a threat to both vehicles and pedestrians. Village Council was concerned with this issue on several levels: the negative effect on the tourist business, the liability of the Municipality in the event that someone was injured, and the requirement to get the Municipal fire truck out on the wharf for water during an emergency.

In 1985 the Council of the Village of Hilton Beach acknowledged the waning ability of the community to grow. Symptoms of stagnant assessment, declining population and a shift of the population toward the over 65 age bracket were signs of a depressed community, with little or no growth potential.

The realty assessment level had been static from 1980 to 1985 - the population rose 3.7% from 218 people to 226 between 1977 and 1980 only to drop to 223 people by 1983 with a further decrease to 205 people in 1988. 36% of the population was over 55 years of age, compared to the Provincial average of 19%, and 26% were over the age of 65. These three variables working together could lead to the extinction of the community. Fewer and fewer persons, many on fixed incomes, would have to share the increasing costs of running the municipality. Council realized that action must be taken immediately.

Upon examination of the strengths and weaknesses of the Village, Council determined that tourism would have to play the most important role in revitalizing the economy. However, it was also apparent that the federally owned infrastructure was in serious disrepair and other facilities such as sewage treatment systems and a reliable and safe source of communal water were non existent.

The Council knew that it would be necessary to develop a detailed plan of action. In early 1986, with the assistance of the Provincial Ministry of Municipal Affairs, the Village developed a comprehensive document. This economic revitalization strategy identified the priorities in every area of community life - economic, social, recreational and infrastructure. The priorities were established after an extensive consultative process with the permanent and seasonal residents. This document was officially adopted as the Community Improvement Policies and became part of The Official Plan for the St. Joseph Island Planning Area.

The Council felt that in order to revitalize the community, the causes of the three major symptoms - declining population, stagnant assessment and aging population - would have to be effectively addressed. This could only be done by improving the infrastructure - primarily the wharf but also the water and sewage disposal systems. Other important components were a handicapped-accessible community centre, a public beach and park, and improvements to the ball field and rink. These amenities would be necessary to attract new residents to the Village. The increased residential population along with the marked growth in marine and vehicular tourist traffic would enhance the business climate in the Village significantly.

As the improvements to the waterfront were to be the foundation on which the remainder of the revitalization strategy could be built, these were addressed immediately. Submissions were made to both senior levels of government and meetings were held with personnel from the various Departments and Ministries. Every effort was made to keep all parties informed and to coordinate the development of the project. Indeed, the Village had few financial resources of its own (debt load capacity was $10,000) and saw its role as the facilitator and coordinator of those with financial and advisory resources. The plan identified the projects and phased their implementation over a five year period with waterfront development identified as the number one project. The redevelopment of the Government wharf was seen by the Village as the *catalyst* for the revitalization of the community.

At all times, Council tried to create an atmosphere that would encourage investment by the private sector. This essential component of the revitalization strategy anticipated private sector investment in new or expanded businesses and the creation of better services and jobs.

3: PRELIMINARY FEASIBILITY STUDIES

The Village of Hilton Beach continued to lobby for funds to initiate the preliminary engineering and planning for the harbour repair and re-development and, in January of 1987, funding was provided by Small Craft Harbours for the study. Terms of reference were prepared in January 1987 to determine what repairs and alterations would be necessary to improve the facilities for recreational boating. These required the: 1) examination of the present facilities at Hilton Beach to determine the repairs necessary to restore the functional use of the harbour, 2) preparation of an overall concept plan for harbour restoration and improvements, integrating existing, repaired and new facilities into a workable plan for future development, 3) presentation of the concept and related cost estimates to the Village of Hilton Beach Council, and to the general public. The project was scheduled to commence by February 6, 1987, with a completion date by March 27, 1987. The firm of *Hough Stansbury Woodland Limited* was engaged to carry out the study.

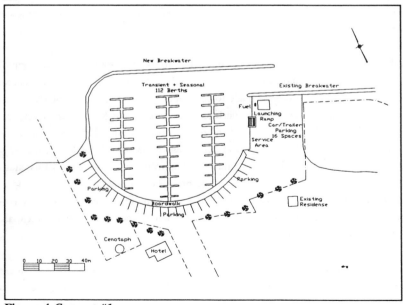

Figure 4 *Concept #1*

The preliminary physical feasibility study suggested two design alternatives, both offering a safe and functional basin for recreational boaters, a capacity of approximately 112 slips, and a launching ramp. The report recommended *Concept #1* as the basis for repair and re-development of the facility. This concept was more compact and centrally located, did not depend upon land acquisition in the initial stages, was less expensive, and could be initiated immediately without having a commitment of provincial funding. The alternate concept required the negotiation and purchase of the land adjacent to Crown property. *Concept #1*, at an estimated cost of $1.4M is illustrated in **Figure 4**. The report also suggested that the repair project be given a high priority in order to minimize the potential consequences of continued rapid deterioration of existing facilities, and the resultant loss of recreational boating opportunities in this strategic location. A number of Village Council and public meetings were held to discuss the project and overcome any concerns of the local citizens. On April 1, 1987, The Council of the Village of Hilton Beach passed a Resolution supporting *Concept #1* for the proposed redevelopment.

As the project awaited funding approval, the Village Council proceeded to meet with representatives of the various Provincial Ministries involved (Municipal Affairs, Northern Development and Mines, and Tourism and Recreation) to determine how each Ministry could assist with technical advice and/or financial support. The various Provincial agencies supported the project and advised that they would be programming their participation as soon as possible. The Village Council also coordinated a second feasibility study funded by the Province and undertaken by Hough Stansbury Woodland Limited with Jack B. Ellis & Associates. This study examined the operational requirements of a new marina and its potential economic, social and financial impacts on Hilton Beach. The study was also designed to assist other North Channel communities in assessing their waterfront potential. The successful replacement of a new swimming beach was also examined: a necessity arising from the expansion of the old harbour facilities. In August of 1987, the Federal Government approved the necessary funding to repair the harbour facility. The Solicitor General of Canada announced on behalf of the Minister of Fisheries and Oceans that the federal government would spend $1 million over the next two years on dock and breakwater construction, dredging and other harbour rehabilitation measures at Hilton Beach. The project called for the removal of the existing pier, dredging and filling, construction of 230 metres of breakwater and 50 floating docks in order to create a protected harbour with an ultimate capacity of 112 berths for recreational vessels. The work was scheduled to start during the fall of 1987 with completion targeted for the 1988-89 fiscal year.

4: DESIGN & CONSTRUCTION OF BASIC HARBOUR INFRASTRUCTURE

In Canada, all major federal government construction projects, including the basic harbour infrastructure at Hilton Beach, are administered by the Federal Department of Public Works. During periods when large numbers of projects are being implemented, the Department of Public Works engage private consultants to handle the overflow. The consulting firm of *Mar-Land Engineering* was hired along with a number of sub-consultants to carry out the engineering design, preparation of plans and specifications, and construction supervision of the Hilton Beach Project.

The basic harbour infrastructure project at Hilton Beach included soundings, wave analysis, soil investigations, breakwater design, final layout, and construction supervision. Surveying and soundings were completed by the Federal Department of Public Works.

The wind-wave hindcast, and preliminary breakwater orientation and design was completed by *W. F. Baird & Associates*. The 20 year inshore design event was estimated as:

Hs (m)	Tp (s)	Direction
1.03	4 - 5	NE - E

As a result, a design water level of $+1.8$ m above chart datum was selected that allowed for future increases in water levels of 0.3 m over the maximum recorded monthly mean. This resulted in a design top elevation of the breakwater of 3.2 m. Two preliminary breakwater designs were identified: a conventional design and a berm design. A cross-section of the final breakwater design is illustrated in **Figure 5**. The design was based upon finding an acceptable source of locally quarried rock. The consultant found 2 sites: a limestone quarry located approximately 5 km from the harbour, for the core material, and a granite quarry located approximately

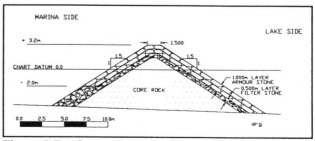

Figure 5 *Breakwater Cross-Section*

14 km from the harbour, for the armour stone. 250 to 500 kg primary armour stone and 25 to 50 kg secondary armour stone were selected based upon the various design

parameters and the requirement to use local quarried stone for cost reasons.   The core had well graded stone with no more than 15 % by weight smaller than 1 kg.

The geotechnical investigation carried out by *Trow Ontario Ltd.* indicated that the subsoil beneath the lake bottom and along the proposed alignment of the new rubble mound breakwater consisted of predominantly dense sand, weakly cemented throughout and "till-like" in structure. "Bouncing" refusal to the casing, suggesting bedrock was reached, was encountered at the base of all four boreholes drilled, ranging in depth from 11.6 m to 4.5 m. The sand was declared competent and capable of supporting the anticipated surcharge load exerted by the proposed breakwater, without any anticipated instability problems nor significant settlements.

During the detailed design process, it became evident that a more efficient marina would result if the entrance and service area was relocated to the western edge of the development. The primary reasons for relocating the entrance included minimizing the transient vessel traffic in the berthing area, acceptability of the service area to tolerate a slightly higher wave climate, utilizing the shallow depths on the west side to minimize the fill required for the service area. This redesign also placed the service area at the foot of Marks Street, the Village's main access road to the waterfront. The east breakwater was also designed for economical expansion of the marina should future demand warrant.

The entrance layout was designed to minimize wave reflections within the harbour. The entrance was modified from the original concept layout to yield a thirty metre entrance width at zero datum.   Due to representations of the Federal Ministry of Transport, the throat of the entrance was widened to 38 metres to facilitate the entry of the 22.5 m Coast Guard vessel "*Caribou Isle*" that is the buoy tender for the entire North Channel. The design of the final layout of the harbour is illustrated in **Figure 6**.

In November of 1987, the detailed specifications and plans, prepared by *Mar-Land Engineering,* were approved by Public Works Canada and Small Craft Harbours. Tenders were announced for the demolition of the old government wharf, dredging and filling, and construction of breakwaters to protect the harbour. Bids were opened and five tenders were received ranging from $897,918 to $2,540,218.18. The low tender from *Howard Avery Contracting Ltd.* was approximately $90,000 over the available budget. After a review of the design, a number of modifications were made including the lowering of the crest of the breakwater by 0.3 metres to reduce the volume, revising the entrance to utilize the existing topography more efficiently and substituting an opening in the east breakwater instead of an expensive culvert. This design element

c

was intended to improve circulation and ensure good water quality in the basin. These modifications reduced the price of the contract by approximately $79,000. The contract was awarded to the low tender in the amount of $818,498. The

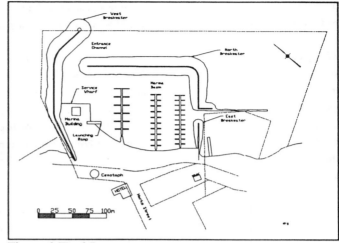

Figure 6 *Final Layout*

reduction in the main contract ensured that the budget of $1.003 million would still accommodate the purchase of the initial 50 floating berths and payment of the consultant's fees. The cost to the Federal Government for the basic infrastructure as outlined totalled $1.0 million. Construction of the project commenced in February of 1988 and was completed by mid-summer in order to be usable for part of the 1988 boating season. Additional floating berths were added in March 1989 and March 1990 to bring the total number of berths to 108. With the present breakwater configuration, there is space for another 12 berths to be installed in the future, for a total of 120 berths.

5: ADDITIONAL SERVICES AND OPERATIONAL REQUIREMENTS

While the construction of the breakwaters and other related work completed by the Federal Government was essential, considerable onshore work was necessary to create a full-service marina. As outlined earlier, the Council submitted proposals to the Provincial Ministries of Municipal Affairs and of Northern Development and Mines for the necessary funding.

In 1988 a total of $342,600 was received from the Provincial Government for the first stage of the onshore work. The Hough Stansbury Woodland team provided site planning and architectural design for improved site services, parking (car and trailer), a community-oriented marina centre, landscaping and a major boardwalk linking the harbour to both the new swimming beach and local shops. The first phase included the construction of the service area, installation of fuel services, construction of the launch ramp and the installation of power and water services to the individual slips. The

Provincial Government also contributed $135,000 toward the purchase of the abutting 3.68 acre parcel of waterfront property that included an old marina building. This building was capable of being converted into industrial space for lease. The Village allocated $41,100 toward the cost of the waterfront development.

In 1989, a further $420,000 was received from the Provincial Government and the Village contributed $10,000. A second phase included the construction of the marina service building that included an office, showers, washrooms, a laundromat and a lobby where boaters could visit with each other. It also allowed for completion of the services to the slips, the boardwalk and landscaping. The marina service building has a treated lake water supply and holding tank for waste water.

Hough Stansbury Woodland Limited has modified the site plan to reflect the purchase of additional property, and desire for other community improvements. Community residents will benefit from a Village "Common", and ice skating facilities. In 1991, the Village used a portion of a $100,000 Provincial grant from 1990 with a further Provincial contribution of $50,000 to renovate the old marina building into commercial space for lease. Renovations were completed in October 1991 and the landscaping will be finished in the spring of 1992. The building will provide prime lease space for eight retail and other users.

Additional services are needed at the marina. A system for putting the boats into, and taking them out of the water, is essential. This will involve some dredging, the installation of a launch ramp and the purchase of a hydraulic trailer. The launch ramp will be on the property purchased in 1988 and in front of the newly renovated building. A boat storage area will be established on the property as well.

The initial response to the marina has been very positive. The allotted number of slips designated for seasonal use were immediately filled, and the transient traffic is pushing the resources to the maximum. It is evident that the marina will have to be enlarged in the short term to effectively handle the demand. Future expansion can be accommodated to the east of the existing marina as the east side of the breakwater system was designed with an extension in mind.

The full benefits of the marina will not be enjoyed until other infrastructure development in the Village is completed. The installation of communal water and waste disposal systems is necessary before several of the proposed expansions of existing businesses and establishment of new ones can take place. These projects are under way and completion is expected by summer 1993.

6: REVIEW OF IMPACTS - ECONOMIC AND SOCIAL

After two seasons of operation, the initial impacts can be identified. The sense of the community believing in itself again became evident with the initial announcement by the Department of Fisheries and Oceans of $1,000,000 for the re-development of the old wharf. It was not long before the private sector recognized that the Village of Hilton Beach had a great deal of potential.

Because of the planned approach to the revitalization of the community, the Village was able to access more financial assistance much more quickly than is usually possible. The senior levels of government realized that the Village had an optimistic but achievable plan that was developed in consultation with the residents and ratepayers. From 1987 to 1991 approximately $2.8 million has been invested in the Village by the Federal, Provincial and Municipal governments. This does not include any private investment. The Village portion has been very small as a percentage of the total, but significant when one considers the resources available. The Village demonstrated that it was able to analyze its problems, develop practical solutions and create a plan for implementation. The senior levels of government responded with the necessary financial support.

The gross revenue from operations in 1990 was $64,546, a significant increase over the $7,800 of 1989 and $6,200 of 1988. In 1991, the revenue was up slightly which is significant because the rates were not raised from 1990, and 1991 has been a poor year in the area for the tourist industry. The main reasons being given are: the recession in both the United States and Canada, the price of fuel in Canada and the implementation of the Canadian Federal Goods and Services Tax.

As was expected, the positive effects of the waterfront development are felt both directly and indirectly. The direct benefit to the municipality is increased operational revenue, and although the importance of this cannot be overstated, the most significant effects are the indirect ones. The business sector is more vibrant, two new tourist-oriented businesses opened in the Village in 1990 and expansions are being planned. The newly renovated building already has two units leased, and three others may be leased by the end of 1991. The rest of the space will be leased in the spring of 1992, for a total of eight new businesses.

The effect is also evident on the three symptoms that were used to diagnose the Village's previous ill health. In 1989, for the first time in several years, there was a real increase in assessment. This was followed with further increases in 1990 and 1991. Actual figures are not very meaningful because the assessment base changed for 1991; however, for the first time in recent history, building permits for four new homes were issued in 1991. Younger families are moving into the Village. With such a small population, any increase is significant. For example, three families of four will increase the Village's population by 6% if there is no outward migration. The 1991 enumeration showed that for the first time in many years the population has not dropped, indeed it rose by 5% to 216. This change is very encouraging. The vitality of the Village and area will increase with the rise in population and change in the demographics.

Thus there is a positive effect on all of the usual economic yardsticks - assessment, number of building permits issued, number of businesses and population.

Although the indirect economic effects are difficult to measure, the social impact is even more elusive. There is a sense of vitality and well being in the Village that was not there four years ago, and one that is not being felt in neighbouring communities.

With the changes that are under way in the Village, there is a potential for negative social impacts. There could be a disruption of the "community" by the increased activity, and a change to the "social fabric". The backlash of the influx of people into the Village - permanent residents, seasonal marine or land residents, and transient tourists, has been minimized by the Council's actions. Council has involved the residents through public meetings, newsletters, questionnaires, and open discussion of the reasons for the economic revitalization strategy. The role of the marina has been emphasized as a major part of the strategy. Therefore, there has not been the outcry from the local people that is often found when they go to the local restaurants for coffee and find that tourists are sitting in their usual seats. The circulation of the Council minutes immediately after the meetings explains the reasons for decisions. There is always the small percentage of people who resist change, but the negative feedback has been minimal to-date.

The efforts of the Council to improve all aspects of community life - social as well as economic are being well received. The new community centre, the public beach and park, the streetscaping, the restoration of a local war memorial, all complement the waterfront development.

One major development that is currently under way is the building of a large recreational complex on the south side of the Village by a private developer. The two hundred acres straddles two municipalities and will include a golf course (designed for eighteen holes, but only nine holes initially developed), all season rental accommodation, cross country ski trails and a club house. This facility will have increased viability because of the marina and vice-versa, as they will attract the same clientele.

The actual number of jobs created because of the waterfront development and the resulting increased economic activity has been estimated to be twelve full time and thirty-four permanent part time. This will be after the infrastructure is in place to allow for such growth. The importance of the permanent part time can not be over emphasized. In this area, any job is important.

The direct net financial return to the Village from the marina and the rental of the renovated waterfront building is conservatively estimated to be $30,000 by 1992. This is in addition to a reserve for major repairs to both facilities. The Village currently raises approximately $65,000 through taxation for municipal purposes, so the impact of this revenue is considerable. The Council has earmarked this revenue for the continued implementation of the revitalization strategy.

It is evident that the benefits of the economic revitalization of the Village of Hilton Beach will be felt on adjacent municipalities. Many of the services on the Island such as welfare, homes for the aged and planning are cost-shared by the area municipalities on the basis of assessment. As the Village's residential and commercial assessment increases, the municipality will bear a greater proportion of these costs.

On another level many of the jobs that are created will be filled by persons outside of the municipal boundaries. These jobs range from unskilled through semi-skilled to highly skilled. Even as the population rises above its present 216, there will not be the necessary pool of people to fill these jobs. Thus, the benefits will accrue to residents of the neighbouring municipalities. They already are enjoying the increased recreational and social amenities - the harbour, community centre and park.

On yet another level, the communities in the North Channel will benefit from the increased tourism traffic generated by the marina. Boaters are, by definition, mobile and the persons with seasonal slips will take trips down the Channel, stopping at the various communities. The Village has initiated a promotion program for the marina and this will attract many new visitors from other parts of the Great Lakes. Boaters from the more congested areas of Lakes Michigan and Huron in particular are very interested in coming north. The Village is an active member in the various economic and tourist development committees that have been established to enhance the image of the North Channel area as a destination point. It is believed that if each community presents itself as a part of the total package, both the community and the area will benefit.

7: CONCLUSIONS

The effects of the initial million dollar investment by the Federal Government are far reaching. It is unusual indeed to see the beneficial effects in such a short period of time. The waterfront development has been, and will continue to be, the foundation on which this community will build its future. The direct and indirect economic benefits are easier to measure than the social spin-offs, but the latter are no less important.

Benefits will continue to accrue to the Village of Hilton Beach, to the adjoining municipalities on St.Joseph Island and to the neighbouring communities in the North Channel in both the short and long term.

Acknowledgement

The authors would like to express their appreciation for the help and assistance provided by Mrs. Gloria Fischer, Clerk and Economic Development Officer, Village of Hilton Beach.

SECTION 3: ENVIRONMENTAL AND WATER QUALITY ASPECTS

Green can mean Gain for Marina Developers

R. Williams

Campaign for the Protection of Rural Wales, Ty Gwyn, 31 High Street, Welshpool, Powys, SY21 7JP U.K.

ABSTRACT

Comfortable, easy-access marina facilities have proved a winning formula to the extent that they are not only promoted by developers but are sometimes also sought by public agencies wanting their area to have a piece of the marina 'action'. However, finding a suitable marina location may prove difficult for various physical, biological, navigational and socio-economic reasons. It will save the developer time and money if the 'green' constraints of a site are given early consideration.

Every potential marina location will have an environmental impact, for example major breakwaters may need to be built on stretches of scenic coastline or estuaries renowned for their wildfowl and migratory fish may need to be barraged. Obviously developers want sites with maximum potential for returns and the minimum of awkward construction costs. However, the environmental value of a site is a very important economic consideration. If either it has a formal wildlife or landscape designation, or if local 'green' groups are keen to protect some aspect of its ecology or appearance, then the developer will face a protracted and costly battle to gain permission for the project. Early rejection of environmentally sensitive sites from a list of possibles makes good economic sense.

If a site appears not to have a high green value, an environmental assessment can be a good investment even when not formally required by the authorities. It can be mutually beneficial if the 'greens' are consulted by the developers and can even be involved in the preparation of the environmental statement, e.g. helping with species surveys. Case studies demonstrate the potential financial benefits to marina developers of working with 'greens' from the outset (1) to avoid expensive, protracted land-use planning and/or legal battles, (2) to adopt mitigating measures at the planning stage rather than be forced to pay for expensive additions post-construction, and (3) to produce the most sensitive and attractive development which will consequently attract the most visitors by sea and by land.

1.   INTRODUCTION

The aim of this paper is to try to persuade marina promoters that a great deal of time and money can often be saved at the planning stage and projects can be made more successful in the long term if environmental issues are confronted at the outset.

Many of the obvious marina sites have already been developed, but when new territory is explored, environmental issues are only some of a range of factors which must be taken into account in the initial feasibility study. However, it is the environmental aspects which often cause a marina proposal to founder, thereby incurring even higher costs for the proponents. Why does this happen so frequently? Is it because conservationists are anti-everything and use every tactic to thwart progress? Or is it because developers avoid adequate environmental assessment hoping to bulldoze their way through the protests of a few "green cranks"? There are examples of both behaviour patterns, but it is important to realize from a practical standpoint that environmental considerations are fundamental to the success of a marina project.

Sadly, this reality is still taking a long time to dawn on many people, as evidenced by papers on the planning and feasibility of marinas presented at the 1989 conference.[1] Other than those speakers solely concerned with environmental aspects of marina development, most either did not refer to green issues or marginalised their importance.

This paper will try to point out where time and money can be saved if environmental aspects are considered at every stage of marina planning, design and operation.

2.   SITE IDENTIFICATION

In England and Wales every stretch of coastline is covered by a County Structure Plan policy and, increasingly, by Local Plan policies. These plans should be the starting point when assessing the "hope value" of a site. Much of the undeveloped coast is protected by conservation designations such as Areas of Outstanding Natural Beauty and Sites of Special Scientific Interest. Proposing a marina in or adjacent to an area of acknowledged conservation value is guaranteed to elicit opposition from both official and voluntary environmentalists. Therefore it is prudent to delete such sites from a preliminary list. Certainly, the Wales Tourist Board advocates such an approach and even since commissioning a survey of potential marina sites in Wales in 1982[2] they have discouraged development in environmentally sensitive areas while supporting marinas elsewhere. This was the approach taken by the Campaign for the Protection of Rural Wales in 1989 when assessing the development potential of thirty-nine sites around the Welsh coastline which had been suggested for marina projects.[3]

3. ENVIRONMENTAL CONSTRAINTS

Marinas impinge on the marine, intertidal and coastal land environments. Historically, man's impacts on the sea (e.g. effluent disposal) or on the seabed (e.g. dredging) have been out of sight and therefore out of mind. However, public interest in the marine ecosystem has been kindled in recent decades by breathtaking underwater TV documentaries – from Jacques Cousteau to David Attenborough. Heightened public awareness coupled with the refinement of marine environmental assessment make it more difficult and foolhardy to disregard impacts below the sea surface. Examples include the reduction of species diversity by the polluting effects of antifouling paints and untreated sewage in confined bodies of water.

Intertidally marinas may either reclaim or flood mudflats. While these areas are often considered unsightly swamps, they have a high nature conservation value. In particular they are important, often internationally important, as vital refuelling stations and wintering grounds for wading birds and wildfowl which feed on the profusion of invertebrates which live in the mud.

Considerable development has been focussed on coastal land throughout man's history which continues today to be a much sought-after place to work, live and play. The decline of many British port and resort economies has left a legacy of run-down coastal areas. Marinas have already been used as catalysts in the regeneration of many of these locations. However, to ensure long-term success of these projects, attention needs to be paid to their design criteria. Planning authorities have little control over design detail unless the land is either in their ownership or is a designated architectural Conservation Area. Therefore it rests with developers to adopt high standards of construction using materials capable of withstanding the rigours of maritime conditions and in sympathy with vernacular styles. The well-designed schemes will undoubtedly be the most successful in attracting and retaining both water-based and land-based custom in the long-term.

Much of the undeveloped coastline is protected from development in various ways such as through planning policies, Heritage Coast definition and ownership, e.g. the National Trust which raises funds through Enterprise Neptune to buy the most attractive parts of the coastline and manage them in perpetuity for the nation. Therefore marina proposals are unlikely to succeed on the open coast and even in small harbours many local authorities are striving to upgrade facilities without any new residential development, e.g. at Pwllheli in northwest Wales.

4.   ENVIRONMENTAL ASSESSMENT

The 1985 European Community Directive: "The assessment of the effects of certain public and private projects on the environment" was implemented by Government Circulars in England and Wales in 1988.[4] These require full Environmental Assessment (EA) of large projects, but on smaller schemes such assessment is at the discretion of the local planning authority and the developer. However, as all parties become more familiar with EA procedures and the benefits which accrue from thorough research, they are increasingly being applied to small schemes.

It is the planning authority which conducts the EA, requiring a detailed Environmental Statement from the applicant. This should include an examination of both the positive and negative impacts of the project on the natural and man-made environment, on amenities, facilities and on the socio-cultural environment. An important part of the EA process is the period of public consultation when the Environmental Statement can be commented upon by all interested parties. Finally the planning authority compiles its EA report, taking into account all the information received, which should result in a structured and integrated evaluation of the potential environmental and social effects of the proposed development. Detailed description of the application of EA to marina projects was ably described to the 1989 Marina Conference by Malcolm McKemey.[5]

Early co-operation and consultation with "green" groups is strongly recommended during the preparation and research for the Environmental Statement. Such organisations will usually be pleased to give freely of their expert local knowledge and may be able to suggest inexpensive mitigating measures. McKemey's observation is worthy of reiteration: "Despite the fact that the consultative approach has proved successful, marina developers still fear that such openness at an early stage will only provide time for the 'stop the marina' group to get organised. There is still a conviction that planning consent by 'ambush' offers the best chance of success. As the UK planning system is specifically designed

to prevent planning decisions being made without consultation
it is more likely that the 'ambush' approach will provide the
animosity it seeks to avoid."[5]

Increasing regulation of discharges to coastal waters is
rendering EA of marine impacts all the more significant.   It
is interesting that the Royal Yachting Association is
advocating a voluntary code of "green" practise for marina
operators.   This implies a realisation that unless marinas act
now to ensure good water quality, legislation is likely to be
forthcoming which will impose even more stringent controls
over discharge of boat effluents, antifouling levels and
diesel and oil spills in marinas and harbours.

5.   CASE STUDIES

Avoidance of protracted battles.

Porthcawl, S.Wales: A marina was mooted in Porthcawl
harbour for over three and a half years.   In that time two
protracted planning appeals have been fought over the marina
and its associated onshore elements.   The principle of marina
development was widely accepted, but there was well-founded
public concern about the impact on the built environment and
about public access issues which were expertly voiced by the
local Civic Society.   Planning permission was finally granted
by the Secretary of State for Wales in November, 1991, but the
excessive delay could undoubtedly have been avoided if the
proposers, objectors and local authorities had sat around a
table at an early stage.

New Quay, Dyfed: A marina has been suggested in the
southern part of Cardigan Bay, West Wales.   The site is
currently an open, undeveloped beach with a caravan site
nearby.   The small, picturesque fishing harbour resort of New
Quay lies just to the south, and to the north are stretches of
Heritage Coast plus National Trust properties.   Cardigan Bay
is a special area because it is one of the few parts of the
world to be a permanent home to a pod of dolphins.   In 1989 a
local businessman and the Crown Estate Commissioners suggested
that new breakwaters be built out into the Bay to shelter a
five hundred and fifty berth marina plus a large onshore
residential development and other facilities.   As soon as
local people heard a whisper about this possible development
there was an immediate outcry and national conservation
organisations expressed their fears that such a development
would have many serious adverse environmental impacts.   No
formal planning application has been submitted for this
location to date.   This provides an example of an open-coast
location where environmental constraints are high and where a
large-scale scheme in an undeveloped area is regarded by many
as totally unacceptable.   If this plan was pursued a long and

costly battle would be guaranteed, with only a slim chance of gaining approval.

Newport, Gwent: As part of their urban regeneration programme, Newport Borough Council propose to manicure the tidal mudflats of the River Usk by constructing a barrage. Under current legislation this has to be pursued as a Private Bill in Parliament because it affects navigation rights. The Bill was presented to the 1990/1 session of Parliament but was not accompanied by an Environmental Statement. The Usk is important for migratory fish and also provides an important scenic focus in the countryside upstream from the industrialised coastal belt. Consequently, the National Rivers Authority and a consortium of "green" groups (including CPRW) petitioned against the Barrage Bill. Belatedly an Environmental Statement did materialize, but was sadly lacking in essential information, therefore the petitions continue to stand against the resubmitted Bill in the 1991/2 Parliamentary session. Although not strictly a marina proposal, this case illustrates the delay and expense that can be incurred by a developer if the requisite environmental homework has not been prepared and if the wider impacts of a project are not considered at an early stage.

Scope for Mitigating Measures:

Hayle, Cornwall: A great deal can be learnt from the experiences of the Hayle Harbour Company Ltd. during its promotion of a Parliamentary Bill to redevelop the largely derelict harbour plus its estuarine mudflats (which included designated Sites of Special Scientific Interest) and also some attractive beaches. The Company pursued a policy of maximum consultation from the outset, (quoting the Port Director): "...from discussion with individual groups to full scale public meetings, in order to produce plans which as many interests as possible could support."[6]

A fifty-four clause Bill was submitted to Parliament in November 1988 and finally received Royal Assent in July 1990. While the principle of the redevelopment was broadly accepted, certain user and conservation groups had specific concerns. Intense negotiation during the twenty months of the Bill's passage through Parliament resulted in a number of key undertakings by the Harbour Company to satisfy all objectors. A fundamentally important concession was the insertion of a clause in the Bill promising the establishment of an Advisory Committee of twenty representatives of relevant parties which would be kept informed of the Company's intentions and would be able to advise the Company. The constitution of the Advisory Committee was confirmed between the Hayle Harbour Company, the then Nature Conservancy Council and the County and District Councils.

Other undertakings included the donation of the most environmentally sensitive areas to the Royal Society for the protection of Birds, a promise to protect the Sites of Special Scientific Interest and an agreement to manage Copperhouse Pool to permit seawater to be impounded for recreational purposes in summer months while allowing the mudflats to remain tidal for birds to feed and roost during the rest of the year. Thus the Hayle experience demonstrates that environmental constraints can be accommodated into marina developments if there is a willingness by the proponents to seek methods of sensitive management. In the words of the Port Director:

"Far from wishing to despoil the lonely places of the estuary the Company saw them as an asset and inviolable.....It was not immediately clear to the conservation industry that on the whole the Company's objectives coincided very closely with their own...."[6]

Welsh Water plc: Although not promoting marinas, Welsh Water is proposing contentious coastal development in the form of several new sewage works. Their approach to date seems to be an enlightened one: widely circulating information on their plans, seeking maximum public consultation and liaising with local green groups. This company certainly appears to be very conscious of the value of "green" PR.

Opportunities for Environmental Enhancement and Sympathetic Development:

Neyland Marina on Milford Haven: This marina was created in the late 1980's on a previously undeveloped creek on the north side of the Haven near the Cleddau Bridge, quite close to the upper reaches of the Pembrokeshire Coast National Park. The onshore development is restricted to one side of the creek where a tastefully designed row of mostly two-story flats and maisonettes, called Brunel Quay, have been built. This marina provides a good example of how sensitive development and management can ensure a successful venture.

Swansea, S. Wales: The development of Swansea's vibrant Maritime Quarter from run-down docklands has been justly acclaimed far and wide. The focal point is the five hundred and fifty berth marina in the South Dock and Tawe Basin. The scope for waterfront development is being extended with the construction of the adjacent Tawe River barrage. The legacy of derelict and contaminated land from the past industrialisation of the Lower Swansea Valley afforded the opportunity for radical environmental improvement during the redevelopment of this area. Swansea City Council has played a central role in this large-scale, long-term, imaginative project. An important factor was that the Council was the

chief land owner in the Maritime Quarter and was therefore in
a strong position to orchestrate the creation of an award-
winning built environment and also to ensure the protection of
an area$_7$ of architectural importance in the Conservation
Quarter. However, it has to be said that while the overall
design of the scheme was well conceived, the construction
standards of some of the private sector residential elements
around the marina were not suited to withstanding the exposed
maritime conditions.

Learning from the Porthmadog flats fiasco: Porthmadog
is a small town on the edge of the Snowdonia National Park in
northwest Wales. No marina has been built in the drying
harbour, but a 1970's harbour-front development of three
storey flats is relevant to this discussion. These flats have
been widely criticised for their unsympathetic design and poor
quality exterior which rapidly deteriorated. It is in
everyone's interest to ensure that similar blots on the
coastal landscape are never repeated.

6.   CONCLUSIONS

The foregoing examples illustrate both commendable and
condemnable practice in marina and coastal development. The
U.K. coastline is a very valuable national asset, therefore it
is important this resource is used responsibly. One way in
which this can be achieved is by conservationists and
developers entering into early discussion rather than assuming
entrenched positions, which often force them into protracted,
costly battles without necessarily achieving the best outcome
for the site in question. Perhaps more developers can be
convinced of the benefits of active promotion of their
company's environmentally friendly image to the point where
there is competition for the claim to be designing, building
or operating Europe's greenest marina.

REFERENCES

1.   Blain, W.R. and Webber, N.B. (Eds.). Marinas: Planning
and Feasibility, Proceedings of the International Conference
on Marinas, Southampton, UK, September 1989, Computational
Mechanics Publications, Southampton, 1989.

2.   Wallace Evans and Partners, Harbour and Marina Study
commissioned for the Wales Tourist Board, Brunel House, 2
Fitzalan Road, Cardiff, 1982.

3.   Williams, R. and Davies, F.K. A Green Paper on Marinas in
Wales. Published by the Campaign for the Protection of Rural
Wales, 1989.

4. Environmental Assessment. Joint Circular from the Department of the Environment (15/88) and the Welsh Office (23/88), HMSO, 1988.

5. McKemey, M.D. Current Requirements and Methods for Environmental Assessment of Marina Projects. In: (Eds. Blain, W.R. and Webber, N.B.), Marinas: Planning and Feasibility, Proceedings of the International Conference on Marinas, Southampton, UK, September 1989, Computational Mechanics Publications, Southampton, 1989.

6. Roberts, R. Rescuing Hayle Harbour and Estuary. Paper presented to the Royal Society for the Arts conference in London on Investing in Conservation, September 1991 (Hayle Harbour Company Ltd., The Old Customs House, Hayle Harbour, Hayle, Cornwall, TR27 4BL).

7. Edwards, J.A. Public policy, physical restructuring and economic change: the Swansea experience. In: (Eds. Hoyle, B.S., Pinder, D.A. and Husain, M.S.) Revitalising the Waterfront: International Dimensions of Dockland Redevelopment, Chapter 8. Belhaven Press, London, 1988.

Environmental Aspects in the Use of Sea Outfalls: a Sensitivity Analysis

P. Veltri, M. Maiolo

Dipartimento di Difesa del Suolo, Università della Calabria, 87040 Montalto Uffugo (CS), Italy

ABSTRACT

A dilution zone model, referring to a multiport diffuser at the end of a sea outfall, is presented and analized. Starting from the classical studies by Abraham, Cederwall, Fisher and Brooks and from the Chick law for non-conservative solute, and running through the polluted current path, all the involved processes are examined.

In order to make the model useful for practical purposes some simplified hypotheses are assumed.

In the zone model the dilution is broken up into three different dilutions : initial, subsequent and bacterial decay. The dilutions are given in terms of five dimensionless parameters, which summarize all the geometrical, hydraulic and environmental parameters.

After that a sensitivity analysis was carried out. The results are shown by means of an example in terms of dimensional parameters, by varying these ones within the ranges of normal use.

INTRODUCTION

Sewage disposal in the sea through submarine outfalls is nowadays common all over the world. International experience, see among others Charlton [1], Fisher and oth. [2], Occhipinti [3] showed that for coastal cities the most efficient and common structure consists in a submerjed pipe with a multiport diffuser, in order to achieve a higher initial dilution and to protect the surrounding environment.

The quality standards of coastal water, especially where this is used for bathing and fishing, must be guaranteed where outfalls and preliminary treatments are used. EEC Regulations establish, for example, the admitted pollution level, in terms of coliform bacteria requirement as a primary quantified parameter.

Dilution processes analysis, performed in order to obtain an optimal design of the outfall and the diffuser, can be carried out by different dispersive-diffusive models. But there are various uncertainties in the evaluation of final results, even if there are in situ analyses and experimental studies. The dilution evaluation is normally carried out not by 2-D or 3-D mathematical hydrodynamics models, but by means of synthetic or empiric formulae or, again, by means of the so- called "zone models", by which the diffusion process is broken down into different phases, and then the whole effect is recomposed in a final, overall evaluation.

However, the various uncertainties concerned with the input parameters - of geometrical, hydraulic and environmental nature - and the structure of the different zone models lead sometimes to pollution results which cause doubts for the designer and the authorities. Very often the outfall and diffuser design do not respect engineering criteria. In any case it is evident that it is necessary to get:
- a simple and intellegible numerical model;
- a fine knowledge of the local environmental parameters;
- an easy method to understand how each of the parameters modifies the diluting capability. This latter matter can be studied by means of the sensitivity analysis, which considers the effect of each parameter in turn.

ZONE MODEL

For a lot of threedimensional jets issuing horizontally into homogeneous and stagnant ambient fluid of greater density at distance x from the shoreline - measured along the current direction - and at depth Y from the sea level, let us indicate with:
C_0 the sewage concentration issuing from the treatment plant;
C_t the final sewage concentration, to be achieved near the shoreline;
$S_c = C_0/C_t$ the overall dilution.

By leaving out the so-called synthetic formulae, like that of Pommeroy, Auberte-Desirotte, Pearson, Cooley-Harris and so on, which do not admit to understand how much each parameter affects the final result and do not describe the different physical processes, let us analyze the polluted current path. A zone model, see Benfratello [4], Viviani [5],Veltri e Maiolo [6], can be defined as an ensemble of mathematical relationships, each of which is able to describe and compute an intermediate dilution process.

The overall dilution S_C can be predicted as product of the intermediate dilutions, which are obtained at the end of each process phase.

The polluted flow issuing from the small orifices of the multiport diffuser is subjected to an advective and diffusive flux and to a microbial disappearance. Referring to the definition sketches in fig.1, we may distinguish in the zone model three different dilution processes.

INITIAL DILUTION

First of all the so-called initial dilution occurs. This is done to a buoyant jet, supported by an initial momentum flux and by a density difference $\Delta \rho = \rho_S - \rho_o$ between the sea ambient density ρ_S and that of the sewage, ρ_o. In permanent conditions near the sea surface a waste field occurs. With the assumption of stagnant ambient in the lower layers it is possible to consider the dilution of a multiport diffuser with horizontal circular orifices as that of a single threedimensional jet only if the rising plumes are kept separated until they reach the surface. Initial dilution S_i can be so predicted as the dilution at the lower surface of the waste field. The rise of the buoyant effluent is con-sidered up to $Y = Y_0 - h$ in which:
Y_0 is the diffuser depth;
Y is the height of the waste field above the diffuser;
h is the thickness of the waste field.

Fisher and others [2] observed that the initial dilution ceases when the rising plume reaches the bottom of the waste field. Above this level a mixing occurs, which reduces the concentration differences among the various horizontal sections. The concentration profiles across the buoyant jets are assumed to have a Gaussian distribution (Fisher and others [2], Abraham [7], [8]). In this way the values of the concentrations along the jet axis are computed: these are 1,4 times greater than the equivalent mean ones. This process has been fully studied (Fisher and others [2], Abraham [7], [8], Rahm and Cederwall [9], Cederwall [10]), both expe-rimentally and theoretically. The results are given both by means of mathematical relationships or

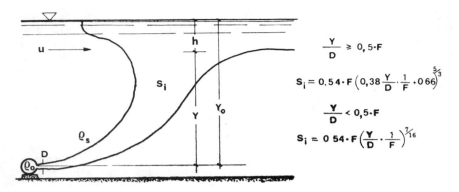

$$\frac{Y}{D} \geq 0,5 \cdot F$$

$$S_i = 0,54 \cdot F \left(0,38 \frac{Y}{D} \cdot \frac{1}{F} + 0,66\right)^{5/3}$$

$$\frac{Y}{D} < 0,5 \cdot F$$

$$S_i = 0,54 \cdot F \left(\frac{Y}{D} \cdot \frac{1}{F}\right)^{7/16}$$

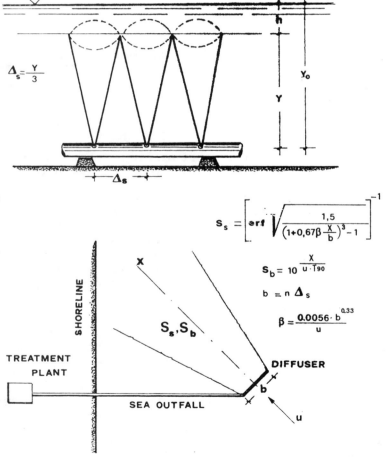

$$S_s = \left[\mathbf{erf} \sqrt{\frac{1,5}{\left(1 + 0,67\beta \frac{X}{b}\right)^3 - 1}} \right]^{-1}$$

$$S_b = 10^{\frac{X}{u \cdot T_{90}}}$$

$$b = n \Delta_s$$

$$\beta = \frac{0,0056 \cdot b^{0,33}}{u}$$

Fig. 1 Definition sketches of the dilution processes

equivalent graphical representations. It has been shown that S_i can be expressed in terms of two dimensionless parameters, Y/D and F, as follows:

$$S_i = S_i(Y/D, F) \qquad (1)$$

in which:
D is the diameter of the jet orifice;
$F = V / \sqrt{\frac{\Delta \rho}{\rho_o} g D}$ is the densimetric Froude number, being V the jet fluid velocity and g the gravity acceleration.

In the zone model the Cederwall formulae, wich have been successfully confirmed in an Italian field work (Vigliani and others [11]), are adopted:

$$S_i = 0,54 \cdot F \cdot \left(0,38 \frac{Y}{D} \cdot \frac{1}{F} + 0,66 \right)^{\frac{5}{3}} \quad for \quad \frac{Y}{D} \geq 0,5 \cdot F \qquad (1')$$

$$S_i = 0,54 \cdot F \cdot \left(\frac{Y}{D} \cdot \frac{1}{F} \right)^{\frac{7}{16}} \quad for \quad \frac{Y}{D} < 0,5 \cdot F \qquad (1'')$$

In the field of practical interest we will consider in the following only the equation (1'). In order to be applied in the case of multiport diffuser the Cederwall formula needs to get distinct single jets up to Y. Considering that for engineering purposes F ranges from 1, which is the lower value to avoid the incipient condition of primary intrusive situation, to about 20, the buoyance effect is generally prevalent. It follows that the jet length can be approximated by the jet height above the diffuser and the greatest width Δs of the circular jet can be assumed about $Y/3$ (Abraham [8]). In this case the initial length b of the superficial waste field, with thickness h, is assumed to be $b = n \cdot \Delta s$, being n the number of the diffuser ports.

SUBSEQUENT DILUTION

When the sewage-sea water mixture has approached the sewage field a further dilution process, called sub-sequent dilution, starts and continues to the shoreline. This process, as it is well known, consists in an advective-diffusive flow of the polluted plume. The 3-D or 2-D hydrodynamic models adopt the conservation of contaminant mass equation and they require, even in stationary conditions, the knowledge of the velocity vector in the overall field or that of the diffusivity coefficients. In terms of averaged values the velocities can be obtained by integrating the Navier Stokes equations.

However, for engineering purposes, a simplified
approach in the zone model has been assumed. By maintaining the Fick hypothesis, which assumes simple
proportionality between dispersive flux and
concentration gradient, we also can assume that the
subsequent dilution S_S is described by a Gaussian plume
model; in the plume a complete vertical mixing occurs
with relative rapidity as far as the dispersion time
(Holly [12]). The vertical mixing is assumed to be
negligible, the flow is plane with a depth-averaged
dispersion, the plume is advecting with constant
velocity throughout its depth. In this case it is
possible to assume a constant horizontal dispersion
coefficient. The moving plume, with initial width equal
to the diffuser lenght, can be characterized only by
the turbulent diffusivity coefficient ϵ, which can be
estimated, in the case of unbounded domain, by means
of the so-called Pearson 4/3 law. Under these assumptions
the Brooks model (Brooks [13]) gives:

$$S_s = \left[erf \sqrt{\frac{1,5}{\left(1 + 0,67 \cdot \beta \cdot \frac{x}{b}\right)^3 - 1}} \right]^{-1} \qquad (2)$$

in which:
erf(z) is the error function;
$\beta = 12 \cdot \epsilon / (u \cdot b)$ is a dimensionless parameter;
$\epsilon = 0,01 / 10^{4/3} \cdot b^{4/3}$ is the turbulent diffusivity coefficient;
x/b, ratio between the length and the initial width of
the superficial sewage field, is another dimensionless
parameter;
u is the current velocity, constant and with right angle
to the diffuser (for incidence angles close by the right
angle the diffuser projection may be assumed). As in
the case of the initial dilution, the model produces
results in the plume center which are conservative.

The currents which advect the plume can be even
indirectly valued: in particular for the Mediterranean
sea the tidal currents are negligible, the depth currents
flow sub-parallel to the shoreline, then the wind
induced currents are the only ones responsible for the
dispersion. It can be assumed (Copeland [14]) that
the superficial currents have the same wind direction
and velocity $u \approx 0.015w$, being w the wind velocity and
considering that the sea bed currents are very small.
In terms of dimensionless parameters the equation (2)
becomes:

$$S_s = S_s\left(\beta, \frac{x}{b}\right) \qquad (2')$$

BACTERIAL DECAY

The bacterial disappearance process verifies simultaneously to subsequent dilution. Physical, chemical and biological factors affect bacterial disappearance.

The polluted plume is like a reactive solute, which does not conserve itself during the path, but it reacts with the medium phase. The mass evolution is in general expressed by a balance equation, in which the reaction type, the reaction velocity and the equilibrium concentrations are taken into account. For biological decay the reaction coefficient can be put equal to 1, the second term of the reaction velocity can be put equal 0; this leads to the well known Chick law, in which the mass concentration C depends on the contact time t, equal to drift time, and on a coefficient of reaction velocity:

$$\frac{dC}{dt} = -k_e C \tag{3}$$

being Ke the bacterial decay coefficient. Equation (3) can be integrated with no respect to the subsequent dilution, in fact the hydrodinamic dispersion is not affected by decay process.

By indicating with S_b the bacterial decay, from equation (3) we can therefore write:

$$S_b = \frac{C_o}{S_i \cdot S_s \cdot C_t} = 10^{\frac{k_e \cdot t}{2.3}} \tag{4}$$

and then:

$$S_b = 10^{t/T_{90}} \tag{4'}$$

where $T_{90} = 2.3/Ke$ is the time taken for 90% decay and t is also x/u. In terms of dimensionless parameters equation (4') can be written as follows:

$$S_b = S_b(\tau) \tag{4''}$$

with $\tau = x/(u \cdot T_{90})$. In the equation (4'') compare both the environmental parameters, u and T_{90}, which are of difficult estimation and of great influence on the model result.

OVERALL DILUTION

The overall dilution Sc at the end of the process is therefore:

$$S_c = S_i \cdot S_s \cdot S_b \tag{5}$$

or, in terms of dimensionless parameters:

$$S_c = S_c\left(\frac{Y}{D}, F, \beta, \frac{x}{b}, \tau\right) \tag{5'}$$

In the subsequent dilution equation the values of erf(z) have been approximated by expanding the explicit function by means of a series with infinite convergence radius up to the third term, which shows very little errors in the range of practical interest.

Figures 2a and 2b show the diagrams of the (1') and (2) equations, which have been plotted for parameter values of practical interest. Figure 2c instead shows the diagram of S_b with varying T_{90} and t, which seems to be of easier usefulness for engineering purposes.

The use of the formulae or equivalent diagrams allows the engineer to rapidly compute the S_C value and then to design the diffuser. They also give a good tool both to compare different solutions and to understand how the diluting capability varies by varying the factors.

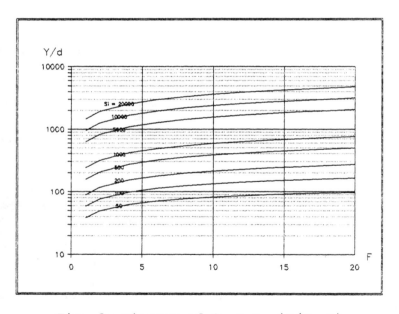

Fig. 2a Diagram of $S_i = S_i$ (Y/D, F)

SENSITIVITY ANALYSIS

In order to analyze the effects of each parameter on the overall dilution S_s the zone model has been subjected to a parametric sensitivity analysis. This can be done both by means of dimensionless or dimensional parameters: these latters present more intellegible results and therefore these will be shown in the following way.

We must consider that some of the factors which compare in S_i, S_s and S_b, as D, V and Yo, are geometrical or hydraulic factors, which are chosen by the engineer, whereas other factors, as $\Delta\rho$, u and T_{90}, are environmental parameters. Moreover it must be highlighted that some factors, as $\Delta\rho$, u, T_{90} are not unequivocally determined, but they present some distribution frequen-

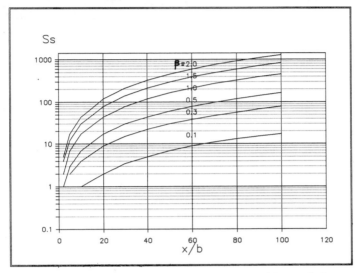

Fig. 2b Diagram of $S_s = S_s$ (β, x/b)

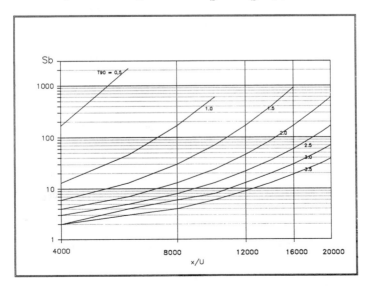

Fig. 2c Diagram of $S_b = S_b$ (x/u, T_{90})

cies. It is also necessary for them to give some expected values in terms of duration curves. Anyway if we assume $\Delta\rho=$ constant, since Y_0 and the pipe velocity Vs ($\geq$ of the scouring velocity $\approx 0,70$ m/s) have been fixed, we can reduce our sensitivity analysis to the dimensional parameters Y, D, V, u, T_{90}.

Y has been considered in the range $0,83Y_0 \div 0,92Y_0$ (Abraham [7]), with smaller values for higher S_i values, which correspond to thicker sewage fields. D has been considered varying from 0,05 m to 0,25 m. The V values are those deriving from the formula of the densimetric Froude number, with this latter in the range $1\div20$. The u values can vary in the range between 0,015 m/s and 0,20 m/s, which includes both the smaller valuable wind velocities and the higher limit velocities, above which waves are starting. T_{90} can vary between 0,5 hours and 4 hours, which represent limit values in the Mediterranean sea with conditions of cleanwater and sunshine.

We make use of the sensitivity analysis based on the factor perturbation method (Mc Cuen [15]). The results can be shown in terms of sensitivity plots (Mc Cuen [16]), if a relative sensitivity analysis is carried out. This is necessary because the values computed by means of the absolute sensitivity are not invariant to the magnitude of either S_C or each parameter. If we put: $S_c = S_c(k_1 = Y, k_2 = D, k_3 = V, k_4, = u, k_5 = T_{90})$, the sensitivity function can be expressed in terms of finite differences as follows:

$$\frac{\Delta S_c}{\Delta k_i} = \frac{[S_c(k_i + \Delta k_i, k_j; J \neq i) - S_c(k_1, k_2 ... k_5)]}{\Delta k_i} \qquad (6)$$

This measures the effect of the parametric variation ΔKi.

By using this method it is possible to analyze the model stability with respect to a reference value $S_C{}^*$, which is obtained by means of $K_1{}^*$ . . . . $K_5{}^*$ parameter values.

The stability can be shown with an example. Let us indicate with:
- C_0 the input concentration, 10^8 coli/100 ml;
- $Y_0{}_* = 29$ m;
- $S_C{}^*$ the overall dilution to be achieved, 10^6;
- $V_S = 0.70$ m/s;
- x = 1325 m.
   The parameter reference values are: $Y^* = 25$ m; $D^* = 0.10$ m; $V^* = 0.60$ m/s; $u^* = 0.05$ m/s; $T_{90}{}^* = 3.0$ hours. For

these values the sensitivity plots of the percent changes of S_C versus a percent change in each parameter in turn are shown in figg. 3a, b, c, d, e. It can be seen that for the parameters D,u and T_{90} the influence on the result is not linear, having different gradients for different parameter variations.

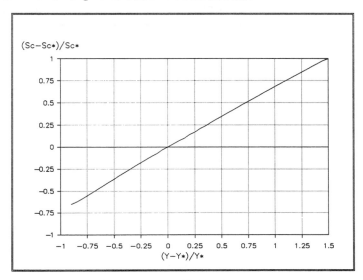

Fig. 3a Sensitivity plot of S_C versus Y

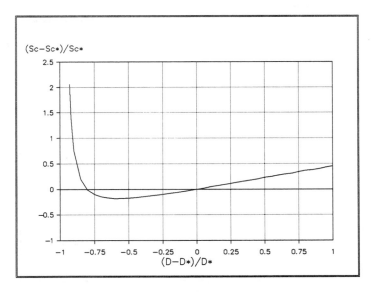

Fig. 3b Sensitivity plot of S_C versus D

In practice there are small result variations by varying the height of the waste field, so the depth can be assumed also by referring to the sea surface. In other words the influence of the thickness h is small, while on the contrary the influence of Yo, which is a given value and is linked by the sea bed gradient to the sea outfall length, is presumably high. Still

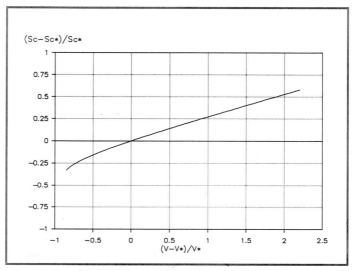

Fig. 3c Sensitivity plot of S_C versus V

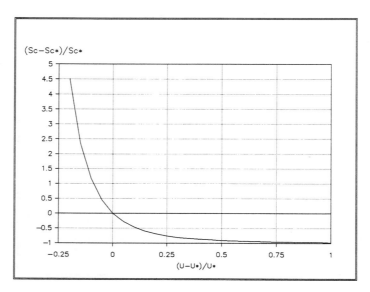

Fig. 3d Sensitivity plot of S_C versus u

less sensitive is the dependence on the orifice velocity V. The model is less stable with regard to the orifice diameter D: in fact when increasing twice D, for instance, the dilution S_C becomes about 1,5 $S_C{}^*$; while decreasing D to 60% the dilution S_C decreases only of about 15%. The sensitivity plot presents a minimum, which is due to the fact that in the zone model the parameter D compares both in S_i dilution, directly, and indirectly by the diffuser length in S_S dilution.

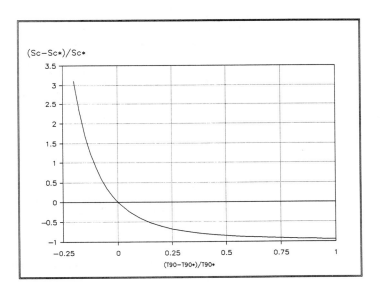

Fig. 3e Sensitivity plot of S_C versus T_{90}

Finally very high is the parametric sensitivity of both T_{90} and u. In particular plots 3d and 3e show a similar influence on the stability with little greater values for u. Anyway a clear idea on the relative incidence of each parameter can be derived by calculating the result variation for a constant percent variation of each parameter. For instance, an increase of 25% on the parameter reference values leads to the following order of relative sensitivities: u, T_{90}, Y, D, V.

CONCLUSIONS

A zone model for the dilution of biodegradable effluent by a multiport diffuser has been presented and analyzed. The overall dilution can be obtained as product of the intermediate dilutions: initial, subsequent, for bacterial decay. The overall dilution depends on five dimensionless parameters, since some simplified hypotheses are assumed. In the ranges of engineering field the corresponding diagrams of the intermediate dilutions have been plotted. Then, by making use of the relative sensitivity analysis and in particular of the parameter perturbation method, the stability of the model is highlighted. The stability is investigated in terms of dimensional parameters, which are more understandable for practical purposes than the dimensionless ones. For a study case the zone model and the sensitivity analysis have been applied. The sensitivity functions have different gradients in the range of variation of each parameter, so they depend on the reference solution of the overall dilution. Anyway it is evident how much the model result is sensitive even for small variations of the environmental parameters current velocity and time taken for 90% decay. Besides these two parameters are of doubtful valuation and therefore it is necessary to plan accurate in situ investigations. Finally, it has been shown that decreasing diffuser orifice diameter does not necessarily increase the overall dilution.

REFERENCES

[1]Charlton, J.A. Developments in Hydraulic Engineering. Chapter 3, Sea Outfalls, (Ed. Novak, P.), Vol.3, pp. 79-128, Elsevier Applied Science Publishers, London, 1985.

[2]Fisher, H.B., Imberger, J., List, E.J., Koh, R.C.Y., Brooks, N.H. Mixing in Inland and Coastal Waters, Academic Press, New York, 1979.

[3]Occhipinti, A.G. Bacterial Disappearance Experiments in Brazilian Coastal Waters, (Ed. A. I. I.), pp.429-438, Atti del Convegno sul tema: Immissione di acque reflue in mare, Ischia, Italia, 1989.

[4]Benfratello, G. Scarichi sottomarini: aspetti idraulici e fenomeni di diffusione, (Ed. A. I. I.), pp.45-77, Atti del Convegno sul tema: Immissione di acque reflue in mare, Ischia, Italia, 1989.

[5]Viviani, G. Il progetto delle condotte sottomarine: una semplificazione nell'uso dei modelli completi, Ingegneria Ambientale, Vol. 18, pp.327-339, 1989.

[6]Veltri, P., Maiolo, M. Un modello di zona per il trattamento marino dei liquami, (Ed. Frega, G.), pp.423-473, Atti del 12° Corso di Aggiornamento in tecniche per la difesa dall'inquinamento, Università della Calabria, Italia, 1991.

[7]Abraham, G. Jet Diffusion in Stagnant Ambient Fluid, Delft Hydraulics Laboratory, Publication 29, 1963.

[8]Abraham, G. Horizontal Jets in Stagnant Ambient Fluid of Other Density, Journal of the Hyd. Div., Asce, Vol. 91, pp.139-154, 1965.

[9]Rahm, S.L., Cedervall, K. Submarine Disposal of Sewage, XI Congress IAHR, Leningrad (1965).

[10]Cederwall, K. Gross Parameter Solutions of Jets and Plumes, Journal of the Hyd. Div, Asce, Vol.101, pp.489-509, 1975.

[11]Vigliani, P.G., Sclavi, B., Olivotti, R., Visconti, A., Tartaglia, G.F. Indagine sperimentale sulla diluizione iniziale dei liquami versati a mare da un diffusore, Ingegneria Sanitaria, Vol.3, pp.129-140, 1980.

[12]Holly, F.M.jr. Developments in Hydraulic Engineering. Capter 1, Sea Outfalls, (Ed. Novak, P.), Vol.3, pp.1-37, Elsevier Applied Science Publishers, London, 1985.

[13]Brooks, N.H. Diffusion of Sewage Effluent in an Ocean Current, Proc. 1th Int. Conf. on Waste Disposal in the Marine Environment, Berkeley, 1959.

[14]Copeland, C.J.M. Simulation of Shear Dispersion on an Effluent Plume caused by Wind Stress on the Sea Surface, in Water Pollution/91, (Ed. Wrobel, L.C. and Brebbia, C.A.), pp.241-248, Proceedings of Conference on Modelling, Measuring and Prediction, Southampton, UK, 1991. Elsevier Applied Science Publishers, London, 1991.

[15]Mc Cuen, R.H. The Role of Sensitivity Analysis in Hydrologic Modeling, Journal of Hidrology, n.18, pp.37-53, 1973.

[16]Mc Cuen, R.H. A Sensitivity and Error Analysis of Procedures used for Estimating Evaporation, Water Resources Bulletin, Vol.10, pp.486-497, 1974.

Water Quality in Marina Basins:
Sources of pollution and practical mitigation

M.D. McKemey
Environmental Assessment Services Ltd, 8A Cuckfield Road, Hurstpierpoint, West Sussex, BN6 9RU U.K.

ABSTRACT

A marina usually provides a base for a variety of boating activities sited around an enclosed area of water with some connection with a larger water body, river, estuary or open sea. A number of these activities can cause water pollution. Where there is a risk of pollution in a marina basin, a number of measures may be adopted to mitigate the adverse effects on water quality.

DOES IT MATTER?

The answer to this question must be yes, although there may be some divergence of view as to what constitutes a significant water quality problem and what does not. Water pollution may be considered significant from these levels:

1. Bad enough to affect the customers. 100% of those connected with marinas would agree that a fetid, stinking, septic basin which boat owners would dread falling into, full of bobbing litter will make an unsuccessful and an unacceptable marina.

2. Bad enough to affect human health. It is possible that a water body may appear aesthetically acceptable but can contain bacteria , viruses or parasites which can threaten the health of those exposed by deliberate or accidental immersion, splashes, open wounds or hand to mouth routes. Similarly, the presence of toxic substances in sufficient concentrations may also damage the health of those exposed.

3.   Bad enough to affect the ecology. The ecological impacts of water pollution extend from the enhanced growth caused by excess nutrient to the lethal effects of toxins. However, the effects of pollution on aquatic life are used as indicators of water quality, rather than the other way round. E.g. coliform counts in shell fish flesh, etc. as an indication of persistent sewage pollution.

4.   Bad enough to affect the regulators. It should be the case that before pollution has had a significant effect on aesthetics, human health or ecology it has already exceeded current regulations.   As a rule, concentrations set in regulations are below the minimum at which adverse effects have been observed. Suggested water quality criteria for a marina basin in U.K. in the form of Environmental Quality Standards are given in Table 1 below:

TABLE 1

SUGGESTED WATER QUALITY STANDARDS FOR MARINA BASINS
Based on European Community Directives et al

Parameter	Unit	Water Quality Standards					
		Inland		Estuary		Marine	
Mercury	ug Hg/l	1		0.5	D	0.3	D
Cadmium	ug Cd/l	5		5	D	2.5	D
Arsenic	mgAs/l	0.5	G	0.5	G	0.5	G
Chromium	mgCr/l	0.5	G	0.5	G	0.5	G
Copper	mgCu/l	0.5	G	0.5	G	0.5	G
Iron	mgFe/l	3	G	3	G	3	G
Lead	mgPB/l	0.5	G	0.5	G	0.5	G
Nickel	mgNi/l	0.5	G	0.5	G	0.5	G
Zinc	mgZm/l	50	G	50	G	50	G
Tributyltin	ug/l	0.02		0.002		0.002	
Triphenyltin	ug/l	0.02		0.008		0.008	
Aldrin	ug/l	0.01		0.01		0.01	
Dieldrin	ug/l	0.01		0.01		0.01	
Endrin	ug/l	0.005		0.005		0.005	
Isodrin	ug/l	0.005		0.005		0.005	
Total "drins"	ug/l	0.03		0.03		0.03	
Dichlorodiphenyltrichloroethane							
(all 4 isomers, total DDT)	ug DDT/l	0.025		0.025		0.025	
(para, para-DDT)	ug ppDDT/l	0.01		0.01		0.01	
Hexachlorocyclohexane	ug HCH/l	0.1		0/02		0.02	
Carbon tetrachloride	ug CCl$_4$/l	12		12		12	
Pentachlorophenol	ug PCP/l	2		2		2	
Hexachlorobenzene	ug HCB/l	0.03		0.03		0.03	
Hexachlorobutadiene	ug HCBD/l	0.1		0.1		0.1	
Chloroform	ug CHCl$_3$/l	12		12		12	
1,2-dichloroethane	ug EDC/l	10		10		10	
(ethylenedichloride)							
Perchloroethylene	ug PER/l	10		10		10	
(tetrachloroethylene)							
Trichlorobenzene	ug TCB/l	0.4		0.4		0.4	
(all isomers)							
Trichloroethylene	ug TRI/l	10		10		10	
Hydrocarbons	ug/l	300	G	300	G	300	G
Phenols	ugC$_6$H$_5$OH/l	50		50		50	
Surfactants ug(as lauryl sulphate)/l		300	G	300	G	300	G
Dissolved oxygen	% saturation	80-120	G	80-120	G	80-120	G
pH		6-9		6-9		6-9	
Sulphide	mgS/l	(preliminary suggestion 0.04)					
MICROBIOLOGICAL PARAMETERS							
Faecal coliforms	per 100ml	2 000		2 000		2 000	
Total coliforms	per 100ml	10 000		10 000		10 000	
Faecal streptococci	per 100ml						
Salmonella	per 1	0		0		0	
Entero viruses	PFU/10 1	0		0		0	

Key: D - Dissolved   G - Guideline

When the regulator's EQSs are exceeded or predicted to be exceeded, the marina's business (and even its existence) may be threatened.

CAUSES OF WATER POLLUTION IN MARINA BASINS

The typical sources of water pollution in marina basins include:

- marine lavatories
- fuel and lubricant spills
- anti-fouling materials
- storm drainage
- boat maintenance & repair facilities
- general litter

1. Marine lavatories

The discharge of marine lavatories into the waters of a marina basin may have aesthetic, human health, ecological and regulatory consequences. Although the waters in a marina basin are unlikely to be designated as bathing waters, the guidelines for bathing waters give something to aim for. This may not be appropriate if the surrounding waters are already below the bathing water standard. Other references suggest that the impact of marine lavatories on water quality is of significance mainly due to peak loads on the busiest occasions of the year (1). Research on water quality at marinas on the Sussex coast suggests that water quality outside the marina basin may have the greater significance most of the time. (See Table 2.)

TABLE 2

TYPICAL MICROBIOLOGICAL WATER QUALITY IN TWO SUSSEX
MARINA BASINS
(Samples taken mid-day, autumn weekend)

SITE	TOTAL COLIFORMS	E COLI	FECAL STREP
A	5/100ml	2/100ml	3/100ml
B	29000/100ml	12000/100ml	1100/100ml

Site A is a coastal basin with a large number of boats and Site B is a river mouth basin with fewer boats 2 km downstream of a sewage works.

2.   Fuel and lubricant spills

Most fuels used in marinas, e.g. petrol and derv, evaporate quickly. The visibility of a very thin layer of oil on the water surface will provide a clear visual indication of pollution in advance of any need for analysis of water samples. Small fuel and lubricant spillages at the fuel berth may drain directly into the basin unless some collection and separation device is included in the surface drainage system.

Oil leaks and breather condensate from marine engines tend to collect in bilges and will be discharged into the basin water when the bilge pump is operated at the berth. The apparent containment provided by bilges can encourage tolerance of leaky engines. Ideally a bilge water collection system with appropriate separator drain should be available within the basin, although the practical use of this type of facility by all craft would be difficult to arrange. Mini booms can be stored near a fuel berth in case of significant spillage and these work well in the calm waters of sheltered marina basins. However, although much of the spillage may be expected to evaporate away, arrangements must be included to quickly clean up the spillage retained in the boom. This also implies a need for facility to store and separate the oil and water (possibly combined with a facility for collecting the bilge water).

3.   Anti-fouling materials

Marine growth on a vessel's hull increases drag and thus decreases sailing speed. Early anti-fouling systems comprised sheathing timber hulls in copper below the water line and anti-foulings have continued to be based on substances which are toxic to aquatic organisms which try to spat or grow on the treated hull.

To be effective, the biocide in the anti-fouling needs to be released slowly and continuously during the life of the coating. Some of the most effective anti-fouling coatings have been based on organotin compounds, e.g. Tributyltin Oxide (TBTO). Organotin coatings can remain effective for up to five years. Organotins are toxic to marine organisms at low concentrations and when large areas of coating slowly release their doses into confined water bodies concentrations can increase to toxic levels some distance away from the protected hull(s). After surveys during the 1980s had revealed severe impacts on shell fish living considerable distances away from groups of moored yachts, it was decided that organotin anti-fouling was not suitable for use on small boats which were likely to be moored for long periods in concentrated groups.

The restrictions on the use of organotin anti-fouling were not extended to sea-going ships on the assumption that these are normally on the move in open waters and rarely moored for long periods in concentrated groups. Current yacht anti-fouling paints are based on, inter alia, Cuprous Oxide (Copper) in vinyl.

The risk to water quality in marinas from organotin should lessen as the existing coatings lose their potency and are replaced by other anti-fouling systems. Organotin anti-fouling compounds typically used either a soluble matrix, which had an effective life of some three years (and where some of the biocide can remain trapped within the defunct coating) or a polishing method where the coating continues to release biocide for a period of up to five years (while the coating wears completely away). The removal of old hull coatings by scraping and sanding may release residual organotins. Similarly, high pressure hosing can flush out large concentrations of the biocide. Organotin dust is also harmful to humans, the HSE gives an Occupational Exposure limit of $0.1mg/m^3$ for an 8 hour Time Weighted Average exposure (2). Dept of the Environment guidelines on the handling and disposal of organotin coatings are available. The organotin content of anti-fouling paints was typically 8 - 50% by weight. Application rates were mainly in the range of 1 litre to cover between 4 - $8m^2$.

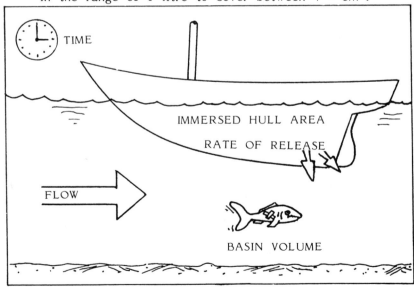

FIG. 1

Concentration of biocide in marina basin water

$$C = K.\frac{\text{Rate of release} \times \text{Area of hulls} \times \text{Time}}{\text{Basin water volume} \times \text{Rate of change}}$$

4.   Storm drainage

The easiest approach to surface drainage at a waterside site is to slope all surfaces towards the water. This approach is simple, natural and avoids puddles. However, drainage into an enclosed water body - such as a marina basin may be a threat to basin water quality as oily water from car parks, dog faeces from pedestrian areas and a range of chemicals from boat yards and workshops may be washed directly into the moorings. For example: a typical basin volume for a 300 boat marina will become significantly polluted, in terms of current water quality standards (3), if it contains the following:

50 litres of oil
2500 litres of sewage
0.5 gramme of organotin
60 kg of Copper (in solution)

Storm surface drainage may also have a significant biochemical oxygen demand and may cause a reduction in the dissolved oxygen content of the basin water.

5.   Boat maintenance and repair facilities

This is largely covered above and the least controllable route of entry for these materials will be via the drainage system. Greatest risks will be from accidental spillages and leaks. The temptation to hose spilled materials into the basin should be resisted.

Boats may be regularly pulled out for maintenance and the hulls cleaned by high pressure water jet. This is a most satisfactory method of controlling marine growth on hulls although it may provide another way for nutrients and other pollutants to be washed back into the mooring area.

Boatyard managers in the U.K. who have completed their Control of Substances Hazardous to Health (COSHH) assessments, will be aware of the hazards connected with the substances used in boat building and maintenance.

6.   General litter

The problem of litter in marina basins is well known to marina operators. Litter is more of an aesthetic rather than a water quality problem as it is not distributed throughout the water column but tends to collect at the surface. The problems of litter in marina basins may be summed up as follows:

1.   litter invariably floats

2.   it is the most visible form of pollution (bar graffiti)

3. once wind-borne litter falls into water, it rarely gets lifted out again.

4. the removal of litter is invariably labour intensive.

Occasionally floating litter can have some surprising effects. Accelerated wear of sheet piled quay walls in the windward corners of harbours has been attributed to abrasion by this type of floating debris.

HOW CAN WATER QUALITY BE CONTROLLED ?

There are a number of methods that may be considered:

1. Maximise dilution of the pollutants in the basin by increasing the basin volume. This is unlikely to be a popular method as the cost of excavation of the basin may be one of the most significant capital costs of a marina scheme. Usually the optimum scheme involves the smallest practicable basin volume for the maximum number of moorings.

2. Maximise dilution of the pollutants by maximising the water exchange between the basin and surrounding water bodies. This method enjoys surprising popularity, despite its resemblance to tidying the car by tipping the ash tray out of the window. Where there is a reasonable tidal range, it is tempting to think in terms of flushing and inducing circulatory flow in the basin by means of a cunning arrangement of the entrance channel. Some schemes use two entrances of different sizes with intention that this will act as a water "ratchet" causing the flow to mainly enter one entrance at one end and mainly leave at the other (these schemes resemble perpetual motion machines, ingenious but never quite manage to work). Water exchange in most marinas is limited and plug like. Moored boats and pontoons tend to act as baffles and limit water circulation.

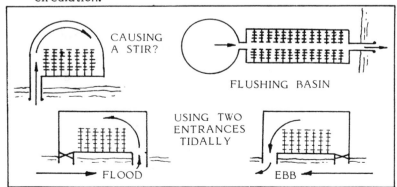

FIG. 2
Attempts to improve water exchange by basin design

With a large tidal range it is possible to induce circulation and pollution entraining current velocities in a basin but only to the detriment of safe navigation and quiescent mooring which usually are the primary objectives of the scheme. Similarly, the two entrance system may be made to work if one entrance is sealed on the ebb tide and the other on the flood. However, a pulsed throughput of water may recharge the basin sediment load on each tide and increase siltation.

3. Treating the basin water. This has been carried out with some success in enclosed dock waters where the effects of pollution have been reduced by chemical dosing and agitation (with moored tug boat propellers). Similarly, pumping basin water through ornamental fountains has been suggested as a method of replenishing dissolved oxygen levels and encouraging biological purification. For a typical 300 boat marina it would take one week to treat the entire basin volume at a rate of 500 l/s. Other options could include the removal of metals in solution by an electrical arrangement of immersed cathodes and anodes. Treating the basin water is most unlikely to be a popular option and the costs could undermine the commercial viability of any scheme. Dosing with pollution devouring micro-organisms may offer a feasible option in the future.

4. Control of polluting activities. This offers the most practicable route to the control of basin water quality. Prevention usually being less costly than a cure. The list of measures may be deduced from the typical sources of pollution listed above:

- Discourage the discharge of marina lavatories into the mooring basin. Provide pump out and slop out facilities to discharge boat sullage tanks and chemical toilets into the sewage system as an alternative.

- Minimise the risk of fuel and lubricant spills at fuel berths and have a workable system for swiftly cleaning up any that occur. Have some storage available for grossly polluted water.

- Recognise the potency of anti-fouling materials from the paint pot to final removal.

- Fit oil traps and other control devices to boat yard and parking area drainage.

- Be aware of any hazardous substances used in boat repairs and maintenance.

- Provide wind proof receptacles for litter in convenient locations.

5. Maintenance and good housekeeping. A second line of control on water basin quality is a committment to the maintenance of any pollution control measures, promptly cleaning up spillages and the collection and removal of litter from moorings and litter bins. Guidelines for the disposal of organotin and other hazardous wastes should be adhered to.

It is important to know the extent of any water quality problem in a marina basin and this may be discovered by occasional analysis of the basin waters. An original analysis for all the parameters listed in Table 1 will cost about ₤1000, subsequent analysis may concentrate on the EQSs most at risk of being exceeded.

ATTITUDE OF THE REGULATORY AUTHORITIES

The organisation most likely to be interested in marina basin water quality in England and Wales is the National Rivers Authority (NRA). The NRA is presently pre-occupied with the control of directly polluting discharges and getting to grips with its PARCOM responsibilities. Thus, water quality in marina basins remains a somewhat low priority for the NRA. However, as statutory consultee for planning applications which impinge on their areas of interest, the NRA is now looking for the inclusion of positive pollution control measures in proposed marina schemes, including attention to fuel berth siting, etc. Local planning authorities, MAFF and the Crown Estate Commisioners may also have an interest in the water quality of proposed marina basins.

IN CONCLUSION

Water quality in marina basins will increase in importance as part of overall moves towards the control of pollution in rivers and estuaries. It is necessary to consider this matter when planning a new marina and pollution control is most cost effective when designed in from concept (4). Regard should also be given to the conservation of salting and reed beds which are nature's own pollution control systems. Marina management should include a commitment to effective maintenance of pollution control measures during the operation of the moorings.

ACKNOWLEDGEMENTS

The author wishes to thank the officers of the following organisations for their assistance in preparing this overview of water quality concerns in marina mooring basins:

NRA Southern Region
The Health & Safety Executive
Department of the Environment
Southern Science Limited
Warren Spring Laboratory (DTI Environmental Helpline)
Hempel Paints
MAFF (Fisheries Laboratory, Burnham)

REFERENCES

1. Seabloom, Plews & Cox. "Influence of Boat on Marina Bacteriological Water Quality". Proc. International Conference on Marinas, September 1989.

2. "Occupational Exposure Limits 1991......for use with The Control Of Substances Hazardous to Health Regulations 1988 (EH40/91)". Health and Safety Executive, HMSO 1991.

3. Gardiner & Zabel. "United Kingdom Water Quality Standards arising from European Community Directives - an update" WRC Report PRS 2287-M/1, July 1991.

4. McKemey. "Current Requirements and Methods for Environmental Assessment for Marina Projects". Proc. International Conference on Marinas, September 1989.

The Impact of Ship Based Pollutants and their Implications in the Marina Environment

B. Beler Baykal (*), M.A. Baykal (**), A. Demir (*)
(*) Department of Environmental Engineering, Istanbul Technical University, 80626 Ayazaga, Istanbul, Turkey
(**) Taskizak Shipyard, 80480 Haskoy, Istanbul, Turkey

ABSTRACT

Bilge water and domestic wastewater constitute the two major sources of liquid pollutants in the marina environment. The former is rich in oil and the latter is rich in organics, nutrients, suspended solids, and coliforms. These pollutants have significant effects on the marina environment such as the depletion of oxygen, and the eutrophication problem. The impact of those pollutants and possible remedies are discussed. Also other environmental concerns arising from on shore services of the marinas are reviewed.

INTRODUCTION

The marine environment may be polluted by three major sources: pollutants from the air, pollutants from the land, and pollutants from the sea. Ships and boats are considered to be mobile polluters and they are the most significant source of marine pollution from the sea.

Generally ships have three different kinds of wastewater, namely ballast water, bilge water and domestic wastewater, that are quite different in character. The ballast water which is used to maintain ship stability is expected to be practically unpolluted sea water. The bilge water consists of oily wastewater that has drained from various machinery and equipment and from various cleaning processes. The domestic wastewater on the other hand, stems from kitchens and lavatories. It is rich in organic matter, nutrients, suspended solids, oil and pathogenic microorganisms. All three of these wastewaters are normally stored in separate tanks on board. As such, they do not mix with each other, they keep their characteristic integrity and may be handled separately.

Marinas are intended for berthing yachts and small craft. Frequently people live on their yachts during their stay. Marinas have to be pleasent places in which to spend some time. A standard amount of service must be provided including sanitary services like WC and showers, collection of garbage and wastewater, and laundary facilities. People living on the boats generate basically two forms of pollution: liquid and solid wastes.

The relatively clean ballast water that is characteristic of large ships is not typical for small craft that are housed in the marinas. However the bilge water and the domestic wastewater, are crucial.

The aim of this paper is to present a general overview of ship based pollution, the impact of the pollutants on the marine environment and to review the risk of environmental pollution in the marinas along with suggestions for pollution control that marinas could have. Within the scope of this work, the emphasis will go to liquid wastes.

THE MARINA ENVIRONMENT AND
THE IMPACT OF POLLUTANTS FROM SMALL CRAFT

The ecology of any inshore environment is affected by the interactions of water with the bottom in shallow waters, by rocks, beaches and shoreline structures, and by the special conditions of the semienclosed marine waters. Inshore water is often heated by the sunlight, hence inshore environments are likely to be warmer than their offshore counterparts. In shallow regions, there is little or no loss of nutrients since waves and currents transport nutrients back to the inshore water column. Thus, inshore waters tend to be significantly richer in nutrients than the offshore. In inshore environments, benthic (bottom dwelling) organisms contribute to the accumulation of metabolic products and they selectively remove nutrients and organic compounds from the water. This biological activity can have significant effects on the composition of water. Land based run offs also bring nutrients and organics as well. Thus inshore waters tend to be more productive (by as much as 1000%) than offshore waters. Typical marine conditions are not characteristic of inshore waters. In semienclosed waters, like the marinas, this holds with even greater force. In such waters, temperature is likely to be quite high, organics and nutrients accumulate in high amounts. Consequently microbial activity may become extreme, depleting the supply of dissolved oxygen and releasing products of decomposition into the water (Reid and Wood, 1976).

The bilge water and the domestic wastewater constitute the two types of wastewater that need care in handling from the point of view of marine pollution in the marinas.

The bilge water is rich in oil. Oil is one of the most significant pollutants in the marine environment and may be the most recognized pollutant originating from ships.

Being less dense than the sea water, it forms a distinct layer on the surface which hinders the diffusion of oxygen that is vital for most marine organisms, into the sea. The final result is a deficiency in the dissolved oxygen concentration and sensitive species may have to leave that marine environment in favor of areas where the oxygen concentration is high. This will mean an inbalance in that ecosystem.

Oil on the other hand, hinders the penetration of light into the depths of the sea as well. This may interfere with the photosynthetic activity of the aquatic species and again there will be problems of ecological inbalance.

Further to these problems, oil may stick to the gills of fish and inhibit their respiration activity or to the feathers of the sea birds to retard their ability to fly.

The domestic wastewater comprises of organics, nutrients, suspended solids, oil, and coliforms, each of which have significant impacts on the marine environment.

Organic matter may not always be a direct threat as it is. However, its pollutional capacity stems from the fact that most organics are biodegradable and oxygen is used up in the process of biodegredation. This may lead to an oxygen deficit, rendering the marine environment unacceptible for its natural inhabitants.

When organic matter is introduced into the receiving water, it is attacked by microorganisms that degrade it to be converted into carbon dioxide and water. Within the process of biodegradation, dissolved oxygen which is crucial for marine organisms is consumed. In extreme events, the oxygen may completely be used up. Combined with the relatively warmer character of the semienclosed inshore water which reduces the solubility of oxygen, the situation may be very severe. This is not only very harmful for the inhabitants and the ecosystem in the water but also anaerobic (zero oxygen) conditions may prevail, which will lead to the production of toxic and explosive gases and stinking. The latter is probably more important in the marinas than anywhere else because the bad smell may render them unacceptible for their purpose which includes providing people with a pleasant environment to relax.

Nutrients imply material like nitrogen and phosphorus which are essential for the growth and maintanence of organisms. They are required for sustaining life forms. However, if they are present in excessive quantities, there may be a large increase in primary production which will eventually lead to a tremendous growth in the aquatic organism population at all levels, through the process of eutrophication. This is especially a problem for relatively dormant bodies of water, like lakes and inland waters, where water circulation is comparitively insufficient. If the discharge of nutrients is not properly controlled, the basin may fill up due to the uncontrolled growth of organisms. It may get shallower and the water body may eventually turn into a marsh. This is one impact of small craft based pollution that is especially crucial for the marinas and every care should be exercised to prevent eutrophication in that environment.

Settled and suspended solids lead to various problems in the marine environment. First of all they are aesthetically unacceptible. They may eventually settle to the bottom of basins and may act to reduce the depth. The settled solids cover the natural habitat of benthic organisms and alter the ecological community. They cause turbidity, color and may also be indicative of objectionable microorganisms.

Coliforms are significant especially from the point of view of public health. The coliform group is used as an indicator of pathogenic (disease causing) microorganisms. Millions of coliforms are given off with the excreta each time. These organisms, although they may not be pathogenic themselves, are very good indicators of probable communication of water borne diseases. Coliforms are especially significant for swimmimg water quality and if swimming is not prohibited in the marina and surrounding areas, they too have to be closely controlled.

POSSIBLE PRECAUTIONS FOR CONTROLLING POLLUTION

The bilge water and the domestic wastewater are the two major sources of pollution in the marinas that are associated with small craft. They are normally stored in separate compartments, hence each one can be handled separately.

Ships use oil water separators to control pollution through oil discharge, mostly originating from their bilge water. A common type of such equipment uses a helical gear to separate oil and water as a centifuge. Oil is lifted up and leaves the separation chamber from the top while treated water leaves from the bottom. Separation is carried on till the discharge limit for oil is reached. The treated water is discharged into the sea under control. The control system starts operation when the water is being discharged and permits discharge as long as the oil content is under the specified limit. When the limit is exceeded, discharge is stopped. The oil thus separated in the process must be stored on board to be delivered to oil recepticles on shore.

Such an arrangement does not seem to be common in small craft. The general practice is to store the oily bilge water as is and deliver it to the marinas. This would dictate that the marinas have to take necessary precautions for proper collection and disposal of the oily wastewater. This can be done by an oil-water separation system within the marina or can be carried to relevant treatment facilities by scavengers as adviced by the local regulators. However, one way or the other, extreme care must be exercised not to dump any oil within the marina.

For domestic wastewater, there seem to be two choices in general. The domestic wastewater may either be treated on board, or it may be retained in holding tanks to be transferred to receiving facilities on shore (Irons, 1989). Ships generally have various treatment systems on board for their domestic wastewater. The well known activeted sludge system, a scheme of which is given in Figure 1, seems to be the most commonly preferred type of treatment. In the activated sludge process, organic wastes are degraded by microorganisms to be converted into treated wastewater, carbon dioxide, water and to newly produced microorganisms, called the excess sludge, in the presence of oxygen. Wastewater thus treated to the

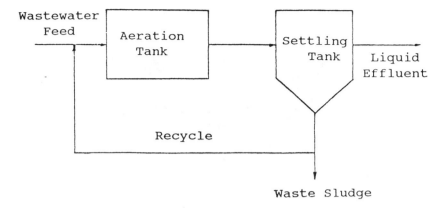

Figure 1 - A typical activated sludge scheme

required limits may be discharged overboard. The basic principle is to convert hard-to-separate organics into easily separable excess sludge. The excess sludge generated in the system has to be stored on board to be delivered to recepticles on shore to be treated further and disposed off.

On board treatment systems are not common for small craft. A feasible system to manage domestic wastewater therein seems to be the vacuum toilet system (Irons, 1989). Its main advantage is that only about 1.2 liters of water is utilized per flush due to the usage of vacuum. This figure is to be compared to an average of 10-12 liters for a regular flush, or about 20 liters for craft that are equipped with manually controlled valve system which is common in the Mediterranian and the Black Sea region. The main feature of the vacuum toilet system is to reduce the amount of water use, which in turn will reduce the quantity of wastewater produced. The wastewater is generally kept in the holding tank to be delivered to onshore recepticles. However, some provisions to prevent stinking for large periods of storage may be necessary. Some brands of vacuum toilets are also equipped with simple treatment facilities that claim to reduce pollution loads to acceptable limits.

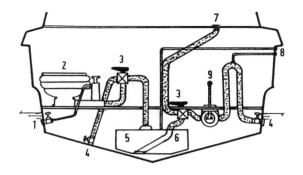

Figure 2 - A proper toilet layout for yachts.
1- seacock inlet, 2- marine toilet, 3-diverter valve,
4- seacock outlet, 5- holding tank, 6- suction line,
7-deck discharge suction, 8-air vents, 9- pump

Figure 2 shows a practical and a proper system for small craft toiletts. The system enables the user to discharge the wastewater into the sea through outlet 4 if permissible at offshore locations; or through the usage of a valve, wastewater may be directed to the holding tank from where it may either be discharged into the sea, or may be transfered to a wastewater recepticle, preferably to a wastewater treatment system ashore. Such a treatment system must be well equipped to handle the pollution load properly. It should consist of a mechanical treatment unit for removing large particles and settlable solids; a unit that would achieve organic matter removal, as a biological or a chemical treatment process; necessary provisions for the removal of nutrients; and a disinfection unit to remove pathogens. A possible layout is given in Figure 3.

Necessary provisions must be provided by the marinas to collect the wastewater brought on shore by yachts and boats. Collection equipment must be made available. Proper arrangements for treatment and disposal of domestic wastewater must be planned. This

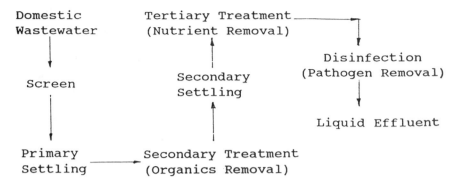

Figure 3 - A possible treatment system layout for the
          marinas

could be accomplished either in the form of a
treatment plant in the marina where the wastewater is
properly treated and the treated wastewater is
discharged on site, or by transferring it to the
sewage treatment facilities or to septic tanks which
then have to be emptied by scavangers.

ENVIRONMENTAL CONCERNS IN THE MARINAS OTHER THAN THOSE ARISING FROM YACHTS AND BOATS

Marinas are expected to provide a pleasant living
environment and a standard amount of service. Except
the pollution generated and discharged from the boats,
marinas have to watch for some on shore service
facilities to maintain a sound environment. Those
facilities include WC and showers, laundries, repair
areas, containers for solid wastes, oil, and
wastewater collected from the small craft.

WC and showers will generate sanitary sewage that
is rich in coliforms, organics and nutrients. They
have to be carefully disposed through relevant systems
like the sanitary sewers. Discharge or mixing of the
wastewater thus generated with the marina water will
bring about the same effect as discharging domestic
wastewater into the sea directly.

Wastewater from the laundries on the other hand will be rich in detergents. Detergents comprise of a surfactant, the active organic part that is responsible for cleaning the dirt, and a softener, a common form being phosphates. The surfactant may either be biodegradable (LAS) , or non-biodegradable (ABS). The former one undergoes biodegradation in the marine environment at the expense of dissolved oxygen like any organic material. The latter cannot be degraded and it accumulates in the water. If a threshold is exceeded concentrationwise, it may be harmful for some marine organisms.

Repair facilities may produce wastewater that is rich in oil and precautions must be provided to prevent any leakages into the sea.

An important function of the marinas is to provide service for collection of solid and liquid wastes from vessels. The solid refuse must be handled with care. Necessary precautions must be taken for proper storage and disposal so as to prevent any seepages or mixing into the marine environment or into the soil.

Although there might be treatment systems that treat the wastewater collected, the usual practice seems to be disposing the wastewater into sewers or septic tanks and let the local administration take care of the pollutants. If there happens to be treatment facilities in the marina, sludge produced within the treatment process and oil separated must be stored and sent to their ultimate disposal sites with care. Whatever the choice may be, proper precautions must be taken to keep these pollutants away from the marine environment in the marinas.

A typical aid for the marinas is the provision for forced water circulation. Since the marina water is a semienclosed dormant body of water, with very little opportunities of circulation, structures that will maintain water movements are frequently provided. This practice will dilute pollutants in the water enclosed in the marinas and hence reduce their negative effects in the environment, by transporting the polluted water out and relatively clean water in.

Finally, environmental considerations in site selection for the marinas are crucial. Nearby sea discharges of any type and currents which may have an effect on them must be critically examined. Entrance to the marinas must be out of the flow trajectory of sea disposals. Any pollution carried from such areas may end up being dragged into the semienclosed marina area adding a significant pollutant load from outside.

CONCLUDING REMARKS

The oily bilge water and the domestic wastewater that is rich in organics, nutrients, suspended solids, and coliforms constitute the two major sources of pollution originating from boats and yachts that are housed in the marinas. Those pollutants have significant effects on the marina environment like aesthetic detoriorations, eutrophication, depletion of oxygen, and public health problems.

Among the pollutants cited, oil seems to be the one which is most widely recognized, and it receives the greatest share in environmental considerations. However in semienclosed inshore waters like the marinas, domestic wastewater may be equally important.

Small craft generally do not have treatment facilities on board. Wastewater generated therein may either be discharged overboard or be stored in holding tanks to be delivered to recepticles on shore. There is a tendency, spurred on by actual or impending legislation, for small craft to have septic or sewage tanks instead of pumping toilets directly overboard as indicated by Nicolson (1983). The contents of the holding tank will have to be emptied into recepticles, most probably in the marinas, where they have to be managed safely. Necessary provisions for collection of liquid and solid wastes must be provided. A plausable solution seems to build a treatment system in the marina. Such a treatment system must provide means of organic matter, nutrient and pathogen removal units.

Moreover, marinas have to watch for other environmental pollutants stemming from their onshore services.

As an example of practice, treatment plants in the marinas are not common in Turkey. There are equipment for collection of the bilge water and domestic wastewater from the yachts. The former is generally transfered into holding tanks ashore, from where the oily water is carried to spots of disposal as advised by the local authorities by scavengers. The domestic wastewater collected on the other hand is either transfered into sanitary sewers or into septic tanks.

Naturally, there are precautions to be taken on board. First of all, usage of holding tanks must be enforced in the marinas. A marine toilet system like the one shown in Figure 2 will be very useful. A standard flange used at the deck discharge suction will make it easier to collect wastewater from the holding tanks through the use of a vacuum system. Also vacuum toilets should be preferred over gravity toilets. This is especially significant for small craft since they have limited space for holding tanks and restrictions of weight due to their small displacement.

Quite a large number of yachts and boats built by smaller shipyards are not classified and not all of the shipbuiders and yacht owners feel obligations for environmental protection unless they are forced to do so by regulations. Hence regulations must be set and enforced for an effective pollution control.

REFERENCES

- Irons, G., "Nature Adores a Vacuum Toilet", Shipbuilding Technology International, 189,1989.
- Nicolson, I., Small Steel Craft, Second Edition,Granada Publishing,USA,1983.
- Reid, G.K., Wood, R., Ecolgy of Inland Waters and Estuaries, Second Edition, D. Van Nostrand Co., USA, 1976.

Rehabilitation of Lake Gregory, Sri Lanka

D. Meemeduma

Sri Lanka Land Reclamation and Development Corporation, 1117 Sri Jayawardenapura Mawatha, Rajagiriya, Sri Lanka

ABSTRACT

Environmental degradation had caused the man-made Lake Gregory, where regular boating had taken place in the past [1], to be heavily silted leaving only a small water surface in the centre. Fill area capacities on the periphery of the lake limited the extent of desilting. Surface vegetation obscured the water surface exposed by dredging a part of the lake. The fill area capacities were increased by burrowing   earth required for retaining bunds of fill areas and this enabled desilting to be continued. Surface clearing of the vegetation, extension of dredging, improvements and construction of silt traps are in progress to restore Lake Gregory to its former contribution to water sports and aesthetic appeal.

1. INTRODUCTION

Lake Gregory, named after its creator Sir William Gregory, the British Governor of Ceylon (now Sri Lanka) (1872-1877), originally covered an area of about 80.9 ha. The creation of the lake enabled Sir Gregory to expand the town of Nuwara Eliya by draining the swampy areas and making more land available for urban development [2].

Nuwara Eliya town, a gently undulating valley of 3.2 square kilometers at an elevation of 1 859-1 890 m above mean sea level (msl) in a municipal area of 12.4 square kilometers, is surrounded by some of the highest mountains of Sri Lanka namely, One Tree Hill (2 100 m), Hakgala (2 126 m), Kikilimana (2 240 m), and Pidurutalagala (2 524 m).

The innumerable streams and rivulets originating in these hills flow into the main stream Nanu Oya, which flows through the Nuwara Eliya town, to feed Lake Gregory. These mountains ensure that Lake Gregory is watered during both North East and South West monsoons.

Upto recent times Nuwara Eliya, a hill town 165 km from the Capital City - Colombo of Sri Lanka, was a haven for holiday makers for its salubrious climate, neatness and scenic beauty of the environs. Foremost among the attractions of the town was Lake Gregory.In the recent past, environmental degradation of this resort city, an accidental discovery by Dr. John Davy in 1819 [1], has perhaps left only the climate to remain relatively unchanged [3]. The silted lake reflected the environmental problems encountered by the city from the lake upwards to the cleared mountain slopes and blocked streams.

Until about 1978 the lake had a boat house with boats available for hire for rowing, sailing [1]. A proposal made in 1980 regarding the management of the lake, envisaged the lake catering for wind surfing; trout and carp fishing; operation of paddle-boats, motor-boats and sailing yachts; water skiing; and an impetus for holding annual monsoon regatta [4].

2. PROPOSALS FOR DESILTING

2.1 Surveys
A study carried out in 1980 estimated .the quantity of silt to be removed to a depth of water 2.13 m below the spill level of 1873.53 m msl of the lake covering an extent of 43.3 ha, as 270 000 m3.

In 1986 a hydrographic survey conducted with soundings on a system of grids of grid distance 10 m and data presented on a contour map of 0.3 m interval (on a scale of 1:2400) revealed that in an extent of lake area of 48.2 ha, the quantity of silt to be removed to a mean level of 1872 m msl (i.e 1.53 m below spill level) to be 275 000 m3.

2.2 Proposals for disposal
Disposal of the dredged material presented the initial major problem. The proposal to pump the spoil over the lake spill to be flushed down Nanu Oya, although the most logical, easiest and cheapest, had to be abandoned as the silt would be trapped in the Kotmale reservoir downstream in which the construction of the dam had commenced by this time.

The next alternative was to dispose of the dredged material avoiding double handling and hence the proposal to deposit the dredged material on the area bordering the lake. This alternative had the limitation in the quantity of dredged material that could thus be retained. To retain as much dredged material as possible in the fill areas, it was decided to construct retaining bunds in identified areas around the lake.   A cross-sectional elevation of the retaining bund is given in Figure 1 A (inset in Figure 1).

Even for rehandling of the dredged material (i.e. taking away of the already removed silt dredged out of the lake) the dredged material had to be in a dry state as far as possible. This required draining out the water in the dredged material and therefore, the rehandling had to be carried out at a subsequent stage, whenever such rehandling becomes necessary.

The survey also indicated the possibility of depositing 208 900 m3 around the lake in the fill areas bounded by retaining bunds and the existing high ground. The areas so identified, as possible fill areas are indicated in Figure 1 and capacities of these fill areas are given in Table 1.

2.3 Economic considerations

The project had limited funds and it was necessary to make optimum use of the funds allocated among the various components of the project. On scrutiny, it became apparant that some of the proposed fill areas have to be abandoned at least in the initial stage. To determine the economically feasiable fill areas, the cost of depositing silt in the fill areas was compared with the cost of reclamation of the fill areas with burrowed material.

Let,Unit cost of dredging be D (=Rs.76 per m3)
    Unit cost of construction of retaining bund be C (=Rs.135 per m3)

Then,Unit cost of disposing silt, R
    =Unit cost of dredging +
        Cost of retaining that unit of dredged material   (1)

$$R = D + \dfrac{C \times \text{Quantity of earth required for the construction of the retaining bund}}{\text{Quantity of dredged material that could be retained in the fill area}} \qquad (2)$$

$$R = D + Cx \ \frac{\text{Bund Volume}}{\text{Fill Capacity}} \tag{3}$$

$$R = D + Cx \ \frac{1}{\text{Effectiveness Factor (E)}} \tag{4}$$

For economic feasibility, $C > R$

$$135 > 76 + \frac{135}{E} \tag{5}$$

$$E > 2.29 \ (\text{say } 2.3) \tag{6}$$

Table 1: Estimated fill area capacities (Fill Capacity) and estimated quantities of earth required for retaining bunds (Bund Volume)

Fill area code	Estimated Fill Capacity m3	Estimated Bund Volume m3	Effectiveness Factor, E
01	24 300	7 780	3.1
02	96 800	13 340	7.3
03	10 400	8 460	1.2
04	29 000	9 200	3.2
05	1 900	3 690	0.5
06	4 200	3 090	1.4
07	7 200	4 360	1.7
08	14 100	4 260	3.3
09	12 600	5 440	2.3
10	8 400	2 860	2.9
Total	208 900	62 480	

2.4 Proposals for Phase 1

With the limitations imposed on,

(a) project financing

(b) fill area capacities, and

(c) avoiding completely rehandling of dredged material

it was decided to restrict Phase I of the desilting programme to the economically feasible fill areas given in Table 2.

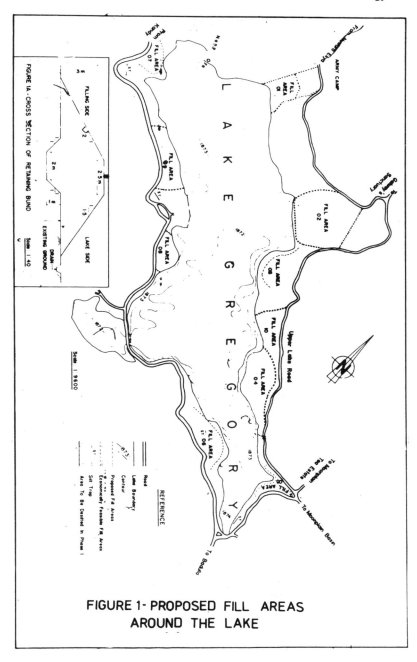

FIGURE 1- PROPOSED FILL AREAS
AROUND THE LAKE

Table 2: Details of economically feasible fill areas

Fill area code	Estimated Fill Capacity m3	Estimated Bund Volume m3	Effectiveness Factor, E
01	24 300	7 780	3.1
02	96 800	13 340	7.3
04	29 000	9 200	3.2
08	14 100	4 260	3.3
09	12 600	5 440	2.3
10	8 400	2 860	2.9
Total	185 200	42 880	

Fill area 01 was sited just opposite an Army Base Camp and therefore had to be abandoned for security reasons.

Thus Phase I of the rehabilitation of the lake had to be restricted to comprise of -

Construction of retaining bunds	40 000 m3
Dredging of lake (to 1 872±0.3 m)	125 000 m3

3. IMPLEMENTATION OF PHASE I

3.1 Construction of retaining bunds
In the search for burrow pits for burrowing of earth for the construction of retaining bunds, analysis of the soil in the identified fill areas proved to be of satisfactory quality. Analysis of soil in the fill areas with low effectiveness factor also revealed of being of satisfactory quality for the retaining bunds. This enabled the material in the feasible fill areas and the fill areas with low effectiveness factor, to be burrowed for the construction of retaining bunds and resulted in -

(a) increasing of capacity of the fill areas already identified at least by the volume of earth so extracted, and
(b) fill areas which were earlier with low effectiveness factor and uneconomical, now turning out to be economically feasible due to increase in fill capacity.

3.2 Dredging
As Phase II was not expected to materialise in the near future, the benefits accruing from the investment in Phase I had to be optimised.

Therefore the factors considered in demarcating the lake area to be desilted in Phase I were -

(a) to obtain the maximum water spread area
(b) to obtain the largest continuous stretch of water surface for water sports, and
(c) to utilise fill areas with optimum reach of dredger for productivity considerations.

These factors led to identification of the south-eastern part of the lake for desilting in Phase I.

A diesel-electric cutter suction dredger 340 kW, 300 mm diameter suction, 250 mm diameter discharge with an output rating of 500-700 m3/h at 1000 m discharge distance was deployed for dredging and the tolerance on dredging was limited to ± 0.3 m.

3.3 Deviations from plan

During heavy rainfall, overflows and flash floods caused some parts of Nuwara Eliya town to be submerged upto about 1 m of water. A canal connecting the Nanu Oya stream to the central part of the lake was proposed to enable free flow of storm water from Nanu Oya to the 'unblocked' part of the lake. The dredging plan was amended to accomodate the dredging of the peripheral canal in the silted western part of the lake.

4. RESULTS

At the end of Phase I of the rehabilitation,although 125 000 m3 of silt had been disposed of and with about 30 ha of the lake   at 1 872 m msl, a continuous stretch of water surface could not be achieved.

Aquatic weeds started to grow on the exposed water surface and covered a part of the dredged area. Portions   of the floating weed mattress around the central part of the lake broke off during the season of high wind and drifted to the dredged area. This weed mattress had to be cut through for moving the dredger out of its operating position in the south eastern part of the lake for it to be positioned for dredging the canal in the western part of the lake. The surface vegetation and the weed mattress obscured the water surface exposed by dredging. So ended hopes of establishing water sports in the stretch of water surface created by the dredging of a part of the lake. The lake at the end of Phase I showing the surface vegetation is depicted in Figure 2.

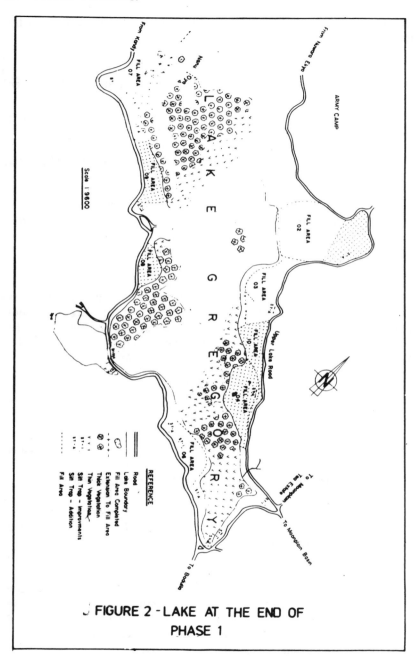

FIGURE 2 - LAKE AT THE END OF
PHASE 1

Surface clearing of the part of the lake already dredged therefore became necessary to expose the water surface already formed. In addition, it became clear that at least the floating weed mattress has to be removed so as to avoid repetition of obscuring of the water surface with torn-off parts of this mattress in the future.

The major restriction in Phase I of limited fill area capacity around the lake had now been circumvented with burrowing of earth for retaining bunds from the fill areas. The revised capacities of fill areas consequent to the burrowing, yielding an additional fill capacity of 106 800 m3 (231 800 -125 000), are given in Table 3. These fill areas at the end of Phase I are shown in Figure 2.

Table 3 : Details of fill areas after burrowing

Fill area code	Effect- iveness Facter before burrowing	Burrowed to		After burrowing		
		Fill area code	Quantity m3	New Fill Capacity m3	Bund Volume m3	New Effect- iveness Factor
02	7.3	–	–	96 800	13 340	7.3
03	1.2	02	6 726*	–	–	–
		03	8 460	25 586	8 460	3.0
04	3.2	04	9 291*	38 291	9 291	4.1
05	0.5	05	3 690	5 590	3 690	1.5
06	1.4	06	3 090	7 290	3 090	2.4
07	1.7	07	4 360	11 560	4 360	2.7
08	3.3	08	4 974*	19 074	4 974	3.8
09	2.3	09	7 103*	19 703	7 103	2.8
10	2.9	02	2 242*			
		10	2 861*	13 503	2 861	4.7
Total				231 807		
(without fill area 05 which is still uneconomical)						

Note :   As the project was mainly on desilting (i.e. dredging) exact measurements were taken only of the quantity dredged and the quantity of material used in bund construction and not of the quantities of silt deposited in the fill areas. The figures marked * are the figures for completed bunds.

5.   PROGRAMME OF PHASE II

This phase was the completion of the rehabilitation of the lake and thus it became necessary to clear the surface from vegetation, complete the desilting, and improve silt traps.

5.1 Surface clearing

Surface clearing required the removal of the vegetation in the lake. The vegetation comprised of two categories -

(a)   Thin vegetation (floating weeds) comprising of aquatic weeds which float on the water surface and included Water Hyacinth (Eichhornia crassipes), Salvinia (Salvinia molesta). It was possible to collect these weeds manually by scooping the water surface.

(b)   Thick vegetation (rooted weeds) comprising of weeds which have their roots grown into some base (mostly deposited silt) and these required shearing off from the base. These weeds included Vernonia cinerea, Crassocephalum crepidioldes, Ludewigia octovalvis, Eclipta alba, Conyza canadensis, Alternanthera sessilis, Panicum repens, Colocasia sp, Eichhornia sp [5].

Manual clearing of some stretches of the thick vegetation was possible with the balance requiring the use of machinery such as dragline, floating bulldozer and excavator.

5.2 Silt traps

To complete the project and the total investment in the rehabilitation of the lake, arresting of further silting was necessary and hence the need arose for traps across the paths that could carry silt into the lake. All streams that flow into the lake were therefore provided with silt traps.

5.3 Additional fill capacity

The extent of desilting left for Phase II was 150 000 m3 (275 000 -125 000), whereas the balance fill capacity at the end of Phase I was only 106 800 m3 (231 800-125 000). To avoid rehandling of dredged material, a further fill capacity of 43 000 m3 (150 000-106 800) had to be found.

The proposals to obtain this additional fill capacity were-

(a)   to regain fill area 01 which was abandoned in Phase I for security reasons

(b)   to increase the bund height from 3 m to 5 m in fill area 03 by increasing base width of the bund and to connect bund 03 (of fill area 03) to bund 10 (of fill area 10) and thereby increase the area and the quantity of earth burrowed for bund construction. The new fill capacity is 51 000 m3 and bund volume is 18 160 m3 giving an effectiveness factor of 2.8, and

(c)   to top up the already filled fill areas. In all the filled fill areas, the dredged material have settled and the fill levels have dropped in varying depths.

The additional fill capacity thus made available, estimated to be 59 700 m3, is summarized in Table 4.

Table 4 : Summary of additional fill capacities

Description	Fill area code	Additional Fill Capacity m3
Regain abandoned fill area	01	24 300
Increase in bund height ⌐ Extension of bund and ⌐ increase in fill area ⌐	03	26 000
Topping up of fill area		
– by 0.2 m	04	3 900
– by 0.2 m	08	1 500
– by 0.1 m	09	1 400
– by 0.2 m	10	2 600
		59 700

5.4 Proposals for Phase II

The proposals in the Phase II of the rehabilitation programme thus comprised of -

Surface clearing	268 000 m2
Dredging	150 000 m3
Construction of additional retaining bunds	30 000 m3

Silt traps - improvements   04 Nos.
          - additions   08 Nos.

6. CONCLUDING REMARKS

On completion of the rehabilitation, a lake of 48.2 ha with a minimum water depth of 1.5 m would evolve enabling water sports of a restricted scope.

Burrowing from the fill areas has enabled the fill capacity on the periphery of the lake to be doubled, avoiding rehandling of dredged material.

The fill areas on the periphery of the lake had imparted to the lake elevated extents of land 3 m above the other areas at original ground level and had cut-off the water contact of the elevated areas with the lake. Plans are already underway for some of the filled areas to be formed into childrens' parks, recreational areas and picnic spots overlooking the lake. These elevated areas had imparted undulations to the lake to harmoniously integrate with the Nuwara Eliya valley, although not as gentle as of which Nuwara Eliya itself is formed.

These fill areas have enabled the once 'environmentally alien silt' to be made use of as land fill material for the creation of recreational areas around the lake.

REFERENCES

1.   de Silva, G.P.S.H., Nuwara Eliya - the beginnings and its growth, Department of Information, Colombo, Sri Lanka, 1978

2.   Tampoe,M., The ugly face of Nuwara Eliya, Saturday Magazine, pp.1, The Island, 24/8/1991

3.   Gunaratna Associates, Environmental Study of Nuwara Eliya, pp.35-40, Report prepared for the Central Environmental Authority, Sri Lanka, 1987

4.   Government Agent, Nuwara Eliya District, Letter to Urban Development Authority, Sri Lanka, 1980

5.   Moody, K., Munroe,C.E., Lubigan, R.T., Paller Jr., E.C., Major Weeds of the Philippines, Weed Science Society of the Philippines, University of the Philippines at Los Bofios College, 1984.

SECTION 4: SITE INVESTIGATION OF WATER LEVEL CONTROL

Vilamoura Marina: Hydraulic Aspects of Lock Design

A. Sanches do Valle, A. Trigo Teixeira

Hidrotécnica Portuguesa, Consulting Engineers, Apartado 5058-1702, Lisboa Codex, Portugal

INTRODUCTION

The Vilamoura Marina is located on the Portuguese south coast, the Algarve, and started to operate in the seventies - Figure 1. The marina comprises the harbour entrance, protected from the sea by two convergent jetties and a tidal basin with a capacity for one thousand boats. This basin can accommodate boats up to 50m in length - Figure 2. To the present time it is still the largest small craft harbour complex in Portugal and the Marina is part of a modern seaside resort.

The marina was planned to be built in two stages. The second stage of the marina was recently designed and construction had already started. The second stage of the project has different characteristics from the first stage. In fact, while in the existing marina the tidal basin is connected directly to the sea and subjected to a water level fluctuation of almost 4.0m, in the new basin the water level will be kept constant. The new basin will be constructed inland and will be linked to the tidal basin through a navigation channel with a lock system.

Different factors have influenced the design of the two basins: tidal basin and locked basin to be built inland. Since the tidal range in Vilamoura is almost 4m it was decided to keep the water level constant in the locked basin reducing dredging and excavation. The global area of the basin is approximately 16.2 ha. A constant water level in the basin will provide for a better landscape design. This paper presents the main hydraulic aspects considered in the lock design. The lock capacity is assessed for several scenarios, the emptying time, the filling up time and the total lock cycle are calculated for some representative tides (Hidrotécnica Portuguesa [1]).

 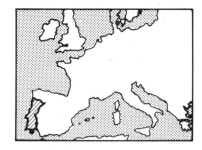

Figure 1 - Map of Europe showing the location of the Vilamoura Marina on the Portuguese coast. The marina is located along the Algarve south coast.

PLANNING AND DESIGN

The construction of a lock in a small craft harbour entrance has some advantages and disadvantages that must be assessed before the final decision to build the lock is made. In the present case the choice was made after considering the substantial saving in the overall cost of construction. This cost includes the dredging and excavation costs of the basin the cost of the retention works that usually increase with the square root of the height and the cost associated with the land uses.

On the other hand, the water level is kept constant in the basin which is a great advantage for the small craft, providing at the same time for a much better landscape design. Currents will be of very small value within the locked basin and will be associated mainly with the occurrence of floods during the winter months. The locked basin will be connected to the existing tidal basin through a short channel and the management of berthing places will be made jointly for the two basins. The manager will take advantage of this particular situation to optimize the berthing space allocation.

The lock option has also some disadvantages. The lock is expensive to construct and to maintain. The access to the locked basin will be restricted to the small craft, in this case boats having a draft up to 1.5 m. Access to the lock will be allowed only at certain times depending on the tide.

Figure 2 shows the Vilamoura tidal basin, the planned locked basin, the location of the lock, the shore line and the two jetties which protect the harbour entrance. In this picture one can see the small river entering the locked basin. This river will contribute to restoring the amount of water lost during the lock operation.

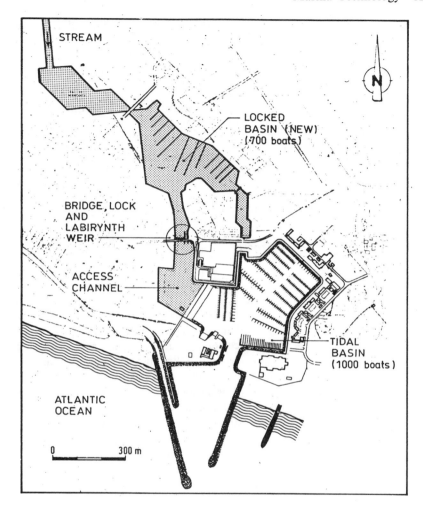

Figure 2 - General layout of Vilamoura Marina showing the position of the tidal basin, the planned locked basin, the lock, the bridge over the lock and the access channel.

Lock capacity

The selection of the lock size is very important and must take into account several conflicting factors. In the present case the water level on the locked basin is constant and above the HHWL at all times; the water level on the tidal basin has a range of almost 4.0 m; the maximum size of vessel that will have access to the locked basin and finally the water losses in each lock cycle are also factors that will influence the choice of the lock size. The number of berths planned for the locked basin will be known.

A parallel study was carried out to estimate the number of berths within the locked basin. A total of 780 berths will be available in this basin. The vessel dimensions are as shown on Table I.

Table I - Dimensions of the vessels with access to the locked basin.

Vessel Class	Dimensions		Percentual Weight (%)
	Max. Lenght (m)	Max. Width (m)	
I	6.0	2.3	22.5
II	8.0	2.7	23.5
IIA	8.0	3.1	11.5
III	10.0	3.1	15.0
IIIA	10.0	3.7	7.5
IV	12.0	3.6	13.5
IVA	12.0	4.2	6.5

The lock capacity is measured in number of boats locked per unit time and can also be expressed as the daily number of vessels locked in the peak period. The lock capacity is of great importance for the success of the new development. In the present case another factor should be taken into account: the water losses in each tidal cycle. The water level in the locked basin will be kept constant and over the summer pumping will be needed to compensate for water losses due to the lock operation, the evaporation and the seepage. In summer, when most needed the inflow from the stream will be seriously reduced.

After different studies were carried out it was concluded that the lock dimensions should be chosen between the two choices:

Table II - Lock dimensions under study.

Lock	Lenght (m)	Width (m)	Area (m2)
Lock-I	24.0	6.3	151.2
Lock-II	18.0	7.6	136.8

An analysis was made based on the average use of the lock and on the daily water losses before a final decision was reached. The average theoretical occupation of the lock was assessed considering different schemes for vessels inside the lock chamber. An occupation index was defined calculating the total area occupied by the boats. Each boat is considered as defining a rectangle in plan (length x width).

For the two hypothesis being analyzed several schemes for the lock occupation were considered as shown on Table III. All the lock occupation schemes that might be obtained by the substitution of a boat from another one of a lower class were omitted from this analysis. To all of these schemes correspond obviously a lower occupation index.

Table III - Characteristic occupation schemes for Lock-I and Lock-II.

LOCK - I
Useful area - 24m x 6.3m

Sch-me	Boats per class				Total area	
	I	II	III	IV	(m2)	(%)
1	4	0	0	2	151.2	100.0
2	6	0	0	1	135.6	89.7
3	0	3	2	0	135.6	89.7
4	1	4	1	0	140.8	93.1
5	0	6	0	0	139.2	92.1
6	8	0	0	0	120.0	79.4
				Average		**90.7**

LOCK - II
Useful area - 18m x 7.6m

Sch-me	Boats per class				Total area	
	I	II	III	IV	(m2)	(%)
1	3	0	0	2	136.2	99.6
2	0	2	2	0	112.4	82.2
3	5	0	1	0	108.0	78.9
4	0	4	0	0	92.8	67.8
5	3	2	0	0	91.4	66.8
6	5	1	0	0	98.2	71.8
7	9	0	0	0	135.0	98.7
				Average		**80.8**

According to this criterion one may conclude that the Lock-I has a higher occupation index allowing for the best use of the lock. From the point of view of the lock occupation Lock-I has another advantage. It will lock six boats of class II which is the class having more berths within the locked basin (35%).

Water losses
For every lock cycle, a lockfull of water is lost from the basin. Water consumption was another criterion taken into account to choose between Lock-I or Lock-II. Three cases were studied for the lock operation (movement of boats in one direction entering or leaving the basin): 5%, 10%, 15% of the berthing capacity (780 boats). Assuming a one day trip, these percentages correspond to the assumption of a daily movement through the lock of 10%, 20% and 30% of the total number of boats moored within the basin. In first place the water loss per square meter of locked vessel was estimated, assuming a daily operational period of 10 hours (summer months) - Table IV.

Table IV - Estimation of the daily water loss for Lock-I and Lock-II per square meter of locked vessel.

		LOCK I	LOCK II
Lock cycle (min)	(min)	21	21
Daily operational period	(h)	10	10
Number of lock cycles per operational period	(-)	28	28
Useful area of lock chamber	(m2)	151.2	136.8
Average percentual occupation	(%)	90.7	80.8
Average area of boats locked per operation	(m2)	137.1	110.5
Area of boats locked daily	(m2)	3838.	3094
Average lock lift	(m)	2	2
Total lock area	(m2)	176.4	172.5
Water loss per lock operation	(m3)	352.8	345
Daily water loss	(m3)	9878.	9660
Daily water loss per m2 of locked boat	(m3/m2)	2.6	3.1

From Table IV one may conclude that the daily water losses per square meter of vessel is smaller for Lock-I than for Lock-II. On the other hand, taking into account the fleet that will have access to the locked basin it was possible to estimate the average area occupied by each vessel, see Table V.

Table V - Estimation of the average area occupied by each vessel within the locked basin.

Class	Percentual weight	Number of boats	Total area (m2) per boat	Total area (m2) per class
I	22.5	176	15.0	2 640
II	23.5	183	23.2	4 246
IIA	11.5	90	26.4	2 376
III	15.0	117	33.0	3 861
IIIA	7.5	59	39.0	2 301
IV	13.5	105	45.6	4 788
IVA	6.5	51	52.8	2 693
		781		*22 905*

Average area occupied per boat is 29.3 m2

Bearing in mind the indexes adopted (daily water loss per square meter of locked vessel and the average area occupied per vessel) for the comparative analysis of the lock size as a function of the water losses, and assuming the three cases referred above (5, 10 and 15%) one may estimate the number of lock cycles required per day and the daily water consumption in case of Lock-I and Lock-II, see Table VI.

Table VI - Estimation of the daily water losses for Lock-I and Lock-II.

	Locked boats per day		Average area of	Lock Operations	Total Lock cycle	Daily Water losses
	(%)	Number	vessels(m2)	(-)	(h)	(m3)
	10	78	2 285	17	6.0	5 998
Lock I	20	156	4 571	34	11.9	11 995
(24.0 x 6.30 m2)	30	234	6 856	50	17.5	17 640
	10	78	2 285	21	7.4	7 245
Lock II	20	156	4 571	42	14.7	14 490
(18.0 x 7.60 m2)	30	234	6 856	62	21.7	21 390

The choice of Lock-I implies a daily water loss that is 10 to 15% less than the one corresponding to Lock-II. On the Table VII the main indexes estimated according to the criteria mentioned above for the study of Lock-I and Lock-II are summed up. The final decision was in favour of Lock-I.

Table VII - Daily water losses as a function of the number of vessels using the lock per day.

	Average occupation	Number of Occupatin Schemes	Lock Cycle	Daily water loss per m2 of boat	Daily water losses as a function of the number of vessels	
					Number of vessels (-)	Water losses (m3)
	(%)	(-)	(min)	(m3/m2)		
					78	5 998
Lock I	90.7	6	21	2.6	156	11 995
(24.0 x 6.30 m2)					234	17 640
					78	7 245
Lock II	80.8	7	21	3.1	156	14 490
(18.0 x 7.60 m2)					234	21 390

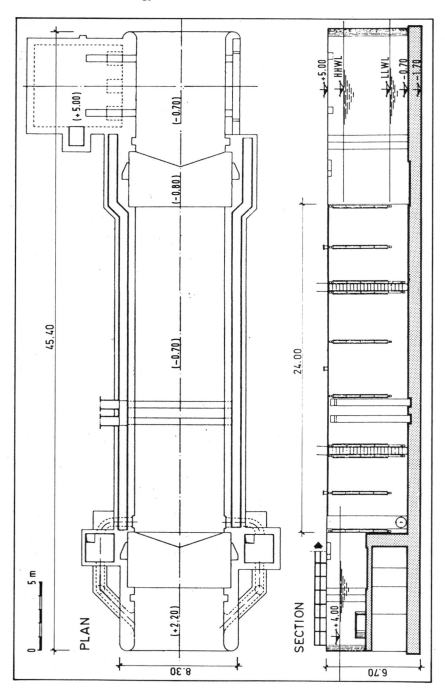

Figure 3 - lock plan and longitudinal section.

Lock characteristics

The lock will be rectangular in plan, made of reinforced concrete and will have an useful length of 24.0m and an useful width of 6.3m, see Figure 3. The upstream water level is (+4.0) and the maximum water level in the locked basin for a flood with a period return of 100 years is estimated to be (+4.5). The draft on the upper sill will be 1.8m and the draft on the lower sill will fluctuate between 0.9m and 4.5m according to the tide.

FILLING AND EMPTYING SYSTEMS

The lock will be filled by transferring water by gravity from the upper reach (locked basin) into the lock chamber until the level in the lock is the same as in the downstream reach (tidal basin). The amount of water involved is approximately the volume obtained by multiplying the area of the lock chamber by the difference between the upper and lower reach levels or the lift of the lock (PIANC [2]). For a very low lift lock (3 to 4m) a low cost filling and emptying system can be used because it is probable that the hawser stresses will be satisfactory without a long filling time. Three main principles apply when design these systems: filling and emptying times should be as short as possible, turbulence in the lock chamber and in lock approaches should be reduced to a minimum.

To classify locks according to the lift, the classification adopted by the Hydraulics Engineers of the US Army Corps of Engineers (Davies J. P. [3]) might be acceptable. Under this classification the lock is a low lift lock (lift lower than 10m). Due to the low lift of the lock the filling and emptying is made through the heads. The filling is made by short separate by-passing culverts and the emptying by using valves in the lock gates. Figure 4 depicts a section through one of the culverts used to fill up the lock.

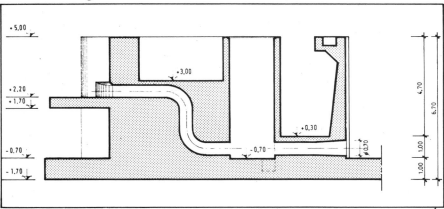

Figure 4 - Section of culvert used to fill up the lock.

The gates of the lock are of mitre type. The emptying of the lock is made through sluice valves (0.5m x 0.7m) placed in the lower part of the gates so that their tops are lower than the low water level .

Filling and emptying time
The filling up time can be estimated using Equation 1:

$$t = \frac{2A\sqrt{\theta h_0}}{n}$$

$$\theta = \frac{k}{2gS^2} + \frac{L}{(K_s SR^{2/3})^2}$$

(1)

where: t is the filling up time (s) ; A is the area of the lock chamber (m2) ; ho is the water level difference between the lock chamber and the upstream reach when the operation begins (m); n number of equal filling culverts; g is gravity (ms-2); S area of the cross section of each culvert (m2); L length of each culvert (m); Ks Strickler roughness coefficient (m1/3 s-1); R hydraulic radius (m). Table VIII shows the filling up and emptying times for different tide levels:

Table VIII - Filling up and emptying times for different water levels on the tidal basin. Water level on the locked basin is kept constant (+4.0).

Tide level	Level to Datum (m)	Lock lift (m)	Filling up time Total lift (min)	Filling up time Lift-(0.2m) (min)	Emptying time Total lift (min)	Emptying time Lift-(0.2m) (min)
LLWL - Lower low water level	0.16	3.84	7.20	5.60	6.20	4.80
MLLW - Mean Lower low water	0.90	3.10	6.50	4.80	5.60	4.20
MSL - Mean sea level	2.00	2.00	5.20	3.60	4.50	3.10
MHHW - Mean higher high water	3.00	1.00	3.70	2.00	3.20	1.80
HHWL - Higher high water level	3.75	0.25	1.80	0.20	1.60	0.20

Last columns of Table VIII refers to the time from the beginning of the operation till the difference in the water level on both sides of the gates is 0.20m. At this point lock gates can be open safely.

Emptying time
Equation 2 can be used to estimate the emptying time. This equation is valid for flow through the sluice valves on the lock gates, providing that the valves are always below the lowest water level in the tidal basin. In this equation m is the number of valves and C the contraction coefficient. The other symbols have the same meaning as in Equation 1.

$$t = 2A\lambda \frac{\sqrt{h_0}}{m}$$

$$\lambda = \frac{1}{CS\sqrt{2g}}$$

(2)

Lock cycle

Lock cycle is defined as the shortest time elapsed between two lock operations. In calculating this time three factors were considered: the time to fill up or to empty the lock, the time to open or close the gates and the manoeuvring time for vessels entering or leaving the lock. In the present situation the lock cycle was estimated assuming the movement of vessels in one direction (leaving or entering the marina). This situation is highly probable and corresponds to vessels leaving the marina during the morning and returning in the afternoon. An estimation of the time spent in each partial operation of the lock cycle is made on Table IX.

Table IX - Estimation of the time spent in each partial operation of the lock cycle.

Partial Operation of lock cycle	Time (min)
Vessels entering from tidal basin	5.00
Closing of downstream gates	1.00
Filling up of the lock	3.60
Openning of upstream gates	1.00
Vessels leaving the lock	3.00
Closing of upstream gates	1.00
Emptying of the lock	3.10
Openning of downstream gates	1.00
	18.70

REFERENCES

1-Hidrotécnica Portuguesa. Conjunto de Lagos e Canais em Vilamoura. Projecto. September 1984. (in portuguese)
2-Permanent International Association of Navigation Congress. Final Report of the International Commission for the Study of Locks. Supplement Bulletin 56. 1987.
3-Davies J.P. et al. 21st PIANC Congress, Section 1-Subject 2, Stockholm. 1965.

The River Tawe Barrage

A.W. Bleasdale

W.S. Atkins Wales (A division of W.S. Atkins Consultants Ltd.), West Glamorgan House, 12 Orchard Street, Swansea, SA1 5AD U.K.

ABSTRACT

Britain's largest single area of industrial dereliction was the Lower Swansea Valley. A cornerstone to the reclamation of the valley is the construction of a barrage across the mouth of the River Tawe. The object of the barrage is to retain water within the tidal reach and thus create an improved environment in and around the river and also a focal point for maritime activities and general amenity. The development process started in 1979 and included feasibility and model studies and a parliamentary bill followed by detailed design. The barrage comprises a lock, fish pass and weirs, the construction incorporates specific measures to minimize interference to river users and the river regime. Construction commenced in 1989 and is due for completion in Spring 1992, when the existing 380 berth marina will be supplemented by a new 200 berth marina.

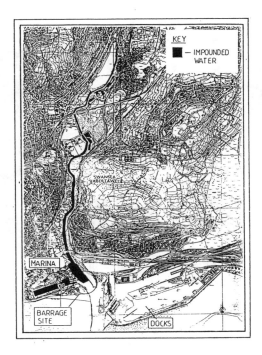

Figure 1. - Impounded Reach

HISTORY OF THE LOWER SWANSEA VALLEY

The first copper smelters came to Swansea in 1717, the attraction of the area being the abundance of coal and the fact that ore could be transported cheaply from Cornwall to Swansea by sea.

Up until the middle of the 19th Century the area developed rapidly and by then Swansea had become the metallurgical centre of the world. Almost any metal imaginable was being smelted in this area, steel, tin, iron, phosphorous, zinc, copper, lead, nickel, cobalt etc. For example, by 1913 there were 106 tinplate works alone in the Lower Swansea Valley. The result of the roasting of these ores to produce metals was that sulphur was emitted into the atmosphere and waste dumped into the river.

Figure 2. Copperworks in the Lower
Swansea Valley (1865)

But by 1920-1930 the copper and smelting industries had closed, leaving piles of bricks, slag heaps and derelict stone buildings. Even coal mining had finished in the area by this time. Unfortunately the legacy left was of desolate and unable to support any growth and a river that was an industrial sewer, in fact the area was biologically dead.

In the early 1960's the Local Authorities promoted the Lower Swansea Valley Project, involving clearance and reclamation of old works, removal of slag tips and the planting of some half a million trees. Dramatic progress has been made over the last 30 years resulting in improvement in the quality of the river. Unfortunately the mudbanks, derelict hulks, abandoned dock works and other industrial relics in and around the river were still generally unattractive.

During this time the idea of a barrage developed, this being a further step in the overall concept of improving the valley.

Figure 3. River bed and banks of the River Tawe

EARLY STUDIES

In 1978 Swansea City Council engaged W S Atkins - Wales to report on the feasibility of a barrage at the mouth of the River Tawe. A summary of the studies is as follows:-

Pollution   The effects of the tidal barrage on flow conditions and pollutant levels upstream of the structure were assessed using the computer model OXBAL to predict the hydrodynamic effects of pollutant reaction rates.

Site Investigation

An initial site investigation was carried out in the form of six shell and auger boreholes extended through rock by rotary coring, four in the river driven from a barge and two on the abutments.   The geology of the site comprises estuarine alluvium up to 4.5m thick overlying a layer of glacial till (4-12m) thick overlying sandstone bedrock.   At the abutments, estuarine alluvium is overlain by up to 12m of made ground.   A second stage investigation was carried out at both abutments and along the line of the surface water overflow.

Physical Model

A 1:55 scale model was constructed of the barrage and adjacent river channel in order to study flow patterns in the area.   Aspects under investigation included debris accumulation, stratification, navigation, and the positions of hydraulic jumps.   The testing programme showed that refinements to the design were necessary in the form of the provision of a training wall and modifications to both the spillway and the downstream bed morphology.   Tests were also carried out on various closure conditions in the river in order that construction restraints could be identified.

PARLIAMENTARY BILL

Navigation on the River Tawe was protected by the Swansea Harbour Act (1854).   Following detailed consultations with interested parties a Private Bill allowing the construction of a barrage was laid before Parliament.   Petitions were lodged by a number of bodies resulting in further discussions and amendments to the Bill or Second House Undertakings where appropriate. The Royal Assent was granted to the Swansea City Council (Tawe Barrage) Act in April 1986.

NAVIGATION

The location of the barrage is such that it is just upstream of the entrance of the docks so that major commercial shipping is unaffected.   All pleasure craft which currently moor in the marina and in the river will navigate through a lock, situated on the west side of the barrage within the navigable channel, in order to pass out to sea.

Feasibility Study
The study was to examine and report on the feasibility of constructing a barrage including investigation of alternative positions, height and construction, an estimate of costs and any constraints on implementation. Also to consider possible pollution and siltation problems and the necessity for a fish pass. It was important to ensure that flooding of property and highways should be avoided and the effect on surface water discharges upstream of the barrage had to be considered. The report was positive but indicated the need for further studies as follows:-

Environmental Impact An environmental impact study was conducted on the river regime upstream of the proposed barrier and a sedimentation study was carried out on the river estuary and the approaches to Swansea Docks. The data for the models was obtained by hydrographic, chemical and biological surveys and from a programme of water quality sampling in the river extending over a full year. This data was used to produce representative numerical models which were validated against known criteria.

Sedimentation and Erosion The effect of the proposed structure on sediment behaviour was assessed with the aid of a 2-D model of tidal flows and bed shear stresses.

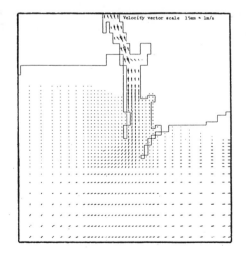

Figure 4. Plan of River Mouth Velocity Vectors

DESIGN

The barrage is of reinforced concrete construction with weir crests set at
0.75m above mean high water neap tide level. Tidal overtopping occurs on
two thirds of the days in a year thus retaining the visual interest of tidal
variation whilst maintaining a saline regime upstream of the barrage.

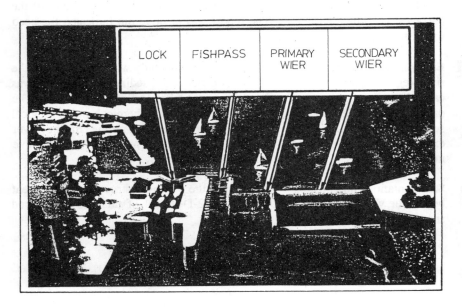

Figure 5. Display Model

The Lock
All navigation past the barrage will be controlled through the lock. In order
to size the lock a study was carried out on the boat movements through the
existing lock to the marina which currently holds 380 berths. Consideration
was given to the wide variety of type and size of craft, together with the
potential expansion upstream of the barrage and in the marina itself which
could eventually amount, in total, to the accommodation of some 800 vessels.
The interaction of pleasure and mercantile navigation was taken into account
and a downstream mooring pontoon 32m x 3m has been provided for the
Charter Fishermen.

The eventual size of the lock is 40m long x 14.5m wide with the invert
of the barrel being at -5.0m OD allowing for passage at all but the occasional
very low tides. It was decided to provide hydraulically powered double sector

gates which allow rapid through put of boating as is needed to satisfy this busy yachting centre. The gates are 12.5m wide, the inner being 7.3m high and the outer being 10.5m high with each leaf weighing approximately 40 tonnes.

The lock also incorporates a system of culverts and sluices, the operation of which improves the locking cycle time. These sluices can also be used as a flushing device to implement water change and reduce the effects of siltation and pollution. The structural elements of the lock are of substantial proportions required to resist the 10.5m variation in tidal level. The walls vary in thickness from 0.45m to 1.85m and the base is 2.5m thick founded on the gravels. At the upstream end a sheet piled cut off wall connects the structure to the rockhead thus preventing the passage of water below the structure.

Control House

The lock is operated from a control house elevated over its western side allowing clear visibility of the barrel and the approaches. The ground floor incorporates the switchgear and control panels. There is a separate room for the Electricity Board transformer and a standby generator is provided to automatically operate in the event of a power failure. The control desk is situated on the first floor and includes the control devices for the gates, sluices, penstocks and back pumping station. There is a comprehensive range of visual displays indicating water levels relative to inner and outer sills and the state of closure of penstocks etc. The gates are automatically controlled from "side sluicing"·to "centre sluicing" by means of level sensor equipment. There are also warning devices and interlocks to ensure the correct operation of the lock. The control house also incorporates the safety apparatus and visual/audio equipment needed for communicating with the yachtsmen.

Fish Pass

The fish pass is located in the centre of the barrage next to the lock as close as possible to the natural deep water channel. The purpose of the fish pass is to ensure that migratory fish can pass from the sea to the upper reaches of the Tawe where they will eventually spawn.

The design of the fish pass was subject to the approval of Ministry of Agriculture Fisheries and Food, and the Water Authority and was finalized after detailed consultations including the consideration of several alternative forms. This pass is of the "pool and traverse" type and is designed such that fish can jump through a zig zag configuration of pools but with a minimum height differential of 450mm, it is known that this minimum height differential can be easily negotiated by the fish. Each pool is of sufficient size to allow a resting area between jumps. Notches are set in each traverse so that a minimum depth of water is present even at low flows.

A back pumping station is situated on the east embankment to allow for recharging upstream of the barrage should a combination of low river flows and neap tides cause water levels to become critical. The top notch is set at such a height that the pass is preferential and there will always be flow down the fish pass even in drought conditions. A bypass culvert is incorporated in the side of the pass in order that flows can be diverted, then by an arrangement of grills and drainage devices fish become trapped in one of the pools and fish movement can be monitored.

The fish pass is also of reinforced concrete construction founded on the gravels with a sheet piled cut-off at its upstream face.

Weirs

The weirs extend the full width of the barrage remaining from the fish pass to the east abutment. They are divided into two sections, the primary weir being 19m wide and set at a sill level of 3.05m AOD and the secondary weir 45m wide and set at a sill level of 3.35m AOD. These levels were determined after consideration of opposing criteria, on the one hand to achieve the maximum retained water level and on the other hand to ensure that the backwater effect up the reach of the River Tawe leaves a satisfactory freeboard against flooding upstream of the barrage. A number of combinations of river and tidal conditions were analysed but the design event was considered to be 1 in 100 year fluvial event (490m³/sec) in combination with a mean high water spring tide (4.6m AOD).

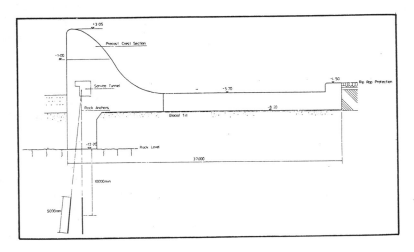

Figure 6. Section through Weir and Stilling Basin

The weirs are of Ogee shape designed to American Bureau of Reclamation criteria for hydraulic capacity. The primary weir is set at such a level that normal flows are attracted towards this area creating white water adjacent to the entry of the fish pass thus encouraging fish to find the pass. The secondary weir comes into operation to pass flood flows this being at a flow of 8.0m^3/sec which is just above the median flow.

The weirs are of reinforced concrete construction and founded on the gravels but with the addition of a concrete diaphragm wall connecting the base to the rockhead and acting as a hydraulic cut-off. In order to provide the required factor of safety against sliding it has been found necessary to incorporate a double line of rock anchors alternating one vertical to one inclined. These are terminated in an inspection gallery running through the barrage from which they can be maintained in the future and monitored by means of installed load cells. This gallery also acts as an access tunnel and allows the routing of services through the barrage from one side of the river to the other. Downstream of the weirs a concrete stilling basin with an upstand toe ensures that a hydraulic jump is created thus adequately dissipating the energy. The downstream bed morphology, protected by rip rap, ensures that flows are directed back to the natural channel in a controlled manner.

Pumping Station and Outfalls
In order to improve the water quality two storm surface water outfalls which currently discharge upstream of the barrage are diverted and extended downstream. These pipelines are of 1.8m diameter glass reinforced pipe and 2.1m diameter reinforced concrete pipe construction laid in the bed of the river. These pipelines join together and pass through a pumping station situated approximately 200mm to the north of the barrage.

The pumping station is capable of pumping 8m^3/sec by means of four propeller pumps. The design of this situation is such that at low tides flows can gravitate through the system, but when tide-locking occurs then the pumping station triggers on and overpumps the tide.

The pumping station was model tested to a scale of 1:10 in two separate models, one comprising the sump section and the other the discharge arrangement. The purpose of the sump model tests was to optimize the flow presentation to the main pumps and to develop a rational set of operating levels for the various combinations of pump duties. Regard to vorticity, pre-swirl at the pump intake and air entrainment was given throughout. There were problems of vorticity and turbulence and these were overcome by the addition of an expansion chamber, also baffles, tripper walls and benching improved the flow presentation to the pumps. A vent had to be incorporated downstream of the pumping station to dissipate a build up of air.

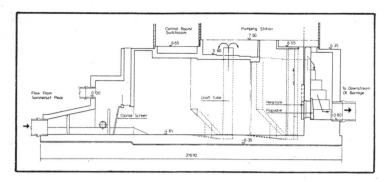

Figure 7. Section through Pumping Station

The arrangement of the pumping station is that flows enter the station at invert level -4.85m OD and are directed to the pump intakes by concrete dividing walls. The pumps are situated in 1200mm diameter draft tubes and flow is delivered accordingly to a high level area at the level of +5.6m OD. From here flow cascades down a series of steps to the seaward side of a 2.1m diameter cast iron flap valve.   An actuated penstock is provided in the upstream face of the wall supporting the flap valve in order that the station can be isolated from the seaward side for maintenance purposes.

The sub-structure to the pumping station is approximately 30m x 15m x 11m deep and of reinforced concrete construction.   This if faced by a vertical sheet piled wall driven into the gravels and tied back to the structure. The superstructure is circular and of steel framed construction clad with brickwork.

The peak electrical demand at the station is one megawatt and this is provided via two alternative routes one of which runs through the service tunnel in the barrage the other through the docks.

Constraints on Construction
The barrage has to be constructed in a particularly hostile environment. The catchment of the River Tawe has rapid run-off characteristics and river flows can rise dramatically, the mean annual flow being 230m³/sec. Also the large tidal variation in the Bristol Channel, the second greatest in the world gives rise to an extreme variation in water level of 10.5m.

Before commencement of construction it was necessary to remove the alluvium from the bed of the river and some 80,000 tonnes has been dredged from a spud pontoon dredger and transported via a split hopper barge to a licensed dumping ground in Swansea Bay.

To minimize the effect of construction on the river regime and to river users it is specified that the project be constructed in 3 distinct phases. Each phase therefore starts with the construction of a cofferdam in order to create a dry working area. For the first phase, the lock structure, the contractor elected to use a cofferdam construction of two skins of sheet piling 12.5m apart connected by two levels of ties and filled with granular material. The remaining two phases, the fish pass and primary weir, and lastly the secondary weir will be constructed within a rockfill cofferdam with a central sheet piled cut-off wall.

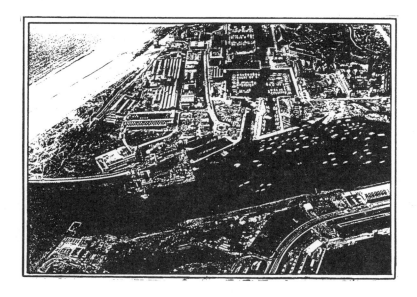

Figure 8.   Cofferdam under construction
Marina in background

Following completion of the first phase all navigation is being directed through the completed lock barrel. The lock can be used to control and assist navigation and also to regulate the river flow.

The last operation is the installation of the top 4m of the primary weir which is left off until then to provide additional hydraulic capacity through the construction period.

It is a requirement of the Tawe Barrage Act that the pipelines and pumping station are operational before impounding takes place.

CONCLUSION

On completion, the River Tawe Barrage will provide the centrepiece for the final improvement and development of the Lower Swansea Valley. The construction of a second Marina will ensure Swansea's importance as a front-line centre for maritime activities.

Design of Sector Gates for Locks

S. de Turberville, N.J. Pope, J.R. Newsom

Sir Alexander Gibb & Partners Ltd, Earley House, London Road, Reading, Berks, RG6 1BL U.K.

ABSTRACT

The choice of sector gates for marina locks is described together with the sizing of the lock and gates. Sections on design cover structural design, pivots, seals, built-in parts, sluicing and operating equipment. The paper also describes corrosion protection methods, civil requirements, safety aspects, installation, operation and maintenance and costs and lead times.

1. INTRODUCTION

Over recent years sector gated locks have become the norm in the UK for new and re-gated locks for leisure facilities. In the survey carried out for this paper, (Reference 2) 70% of the locks constructed in the last twenty years or presently being developed are sector gated, a further 12% have sector/mitre combinations whilst the remainder are mitre (6%) and other designs (12%). In the last ten years, and again with present developments included, the popularity of sector gates gives 77% for sector gated locks and 15% for sector/mitre combinations.

The appeal of the sector gated lock can be readily seen from its advantages and ease of use to both the marina operator and his customers. However, it is the most expensive of the common lock gate installations and has only been able to be adopted as a result of the large capital sums spent on modern marina developments and the earning power of that capital investment.

With such a large capital outlay, and with the consequences to the marina users - that is the revenue providers - if the lock is unsatisfactory, it is clear that the design of the lock and gates is critical to the success of a locked marina.

This paper draws attention to some of the areas which a designer must attend to in order to provide a safe, functional design which maximises the added value to the client.

2. CHOICE OF SECTOR TYPE GATES

Choosing a sector gated lock starts with determining the reasons for needing water level control within the marina. This may result from environmental, aesthetic or siltation reasons, as well as the more obvious level and flood control purposes. Varying degrees of level control can be achieved by fixed sills, rising sills, flap gates, mitre gates and sector gates (listed in general order of increased control, cost and sophistication). Reference 1, which can be taken as a precursor to this paper, gives a detailed appreciation of the reasons for water level control and the means of achieving it and this complementary paper gives details of the sector gated lock solution. A typical layout of a sector gated lock is given in Fig. 1.

F

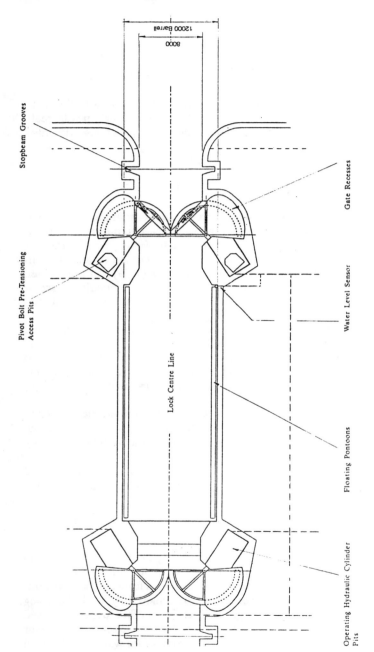

Plan of typical sector gated lock   Fig. 1

The main advantages of sector gates and sector gated locks are:

*   Suitable for heads from either direction.
*   Ability to close against flow, resulting in a maximised free-flow period.
*   Integral sluicing avoiding need for separate sluices.
*   Low operating forces due to hydrostatic loads acting radially through the pivot.
*   Reduced filling time.

The only disadvantages are:

*   Cost of the gates.
*   Possible increased civil costs as a result of the recesses for the gates.

A correctly designed sector gated lock can achieve all of the following, unless other factors make individual items impossible or inappropriate:

*   Access in and out of the marina 24 hours a day at any state of the tide.
*   A free-flow period, when tides allow, at which time locking delays are avoided.
*   Minimisation of salt water ingress into a fresh water basin if required.
*   Protection of marina from external flood levels and unwanted drawdown.
*   Visual and amenity interest.
*   Restriction on siltation.
*   Convenient and safe operation.

A variation of the sector gate with its curved skin plate is the delta gate which has a flat skinplate forming an isosceles triangle with the arms. This can have certain advantages concerning sluicing and associated forces but has not found general acceptance. A problem with it is that the bottom seals sweep across the sill which can cause problems of tearing and debris wedging under the seal. This type of design, which was used most notably at Brighton Marina, is detailed in Reference 5. Virtually all matters dealt with in this paper can be taken as being applicable to delta gates.

A further possibility is a lock with one pair of sector gates and one pair of mitre gates.

3. SIZING AND ORIENTATION

The choice of lock size is the difficult and subjective analysis of the following factors and requirements, which can be partly conflicting:

* Size of largest vessel to be locked
* Number of vessels per locking
* Size of site
* Water availability and usage

* Running cost
* Locking times
* Construction cost

The relationship between lock size and vessel through-put is not linear. To indicate this two indicative cases are given below. The smaller lock is say 8m x 30m in the barrel and the larger one 50% bigger in both directions):

	Smaller Lock	Larger Lock
Open gates	1	1
Admit and moor vessels	4	6
Fill lock	3	7
Exit vessels	3	4
Close gates	1	1
Empty lock	3	6
	15 mins	25 mins
Average capacity/locking	8 vessels	20 vessels
Throughput/hour	32 vessels/hour	48 vessels/hour

Against this increase in throughput has to be set a considerable increase in capital and running costs. In addition, during periods of average or low demands, vessels will be delayed by the longer filling time for the larger lock. A single vessel which might take 5 minutes in the smaller lock could take twice as long in the larger and also use much more water.

The lock filling time is dependent upon the height of the gates (and the head difference) to a much larger degree than the width. This is because the turbulence caused during sluicing has to be limited by restricting the flow rate and flow velocity around the edge of the gates to figures appropriate to the size of lock. See Section 10.

If a lock is to be part of a large marina development with on-shore residential, commercial and recreational facilities, it may well be that the increase in throughput for the larger lock is sufficient to justify the disproportionate lock costs. Conversely, it may be difficult to justify the construction of a new larger lock for an existing marina as an increase in berthing charges that could result would be unlikely to generate sufficient revenue for the project. If the marina was not fully utilised as a result of lock "bottlenecks" the reverse might well apply.

A compromise arrangement for the sizing of locks is to use a wide barrel (i.e. the barrel is significantly wider than the lock gates). This increases lock capacity whilst leaving gate and associated civil costs unaffected. Against this the civil costs of the barrel would rise, the manoeuvring in and out of a wide barrel lock is more difficult, and mooring and manoeuvring in the corners is unattractive. The increase in capacity is therefore less than if the gates are the full width of the lock, particularly if the beams of the vessels are about half the width of the lock. However, the cost advantage could well be overriding. This is particularly the case in locks where new gates are being installed in an old wide commercial lock, for example in South Dock Marina in London's Docklands.

The survey (Reference 2) indicates average lengths and widths for sector gated locks of 36m and 9m respectively. These figures were derived ignoring locks designed to accommodate commercial vessels in addition to recreational, and those where lock widths have been significantly affected by existing wide barrels. As expected the larger locks are associated with more prestigious developments as well as large berthing capacities, but it is worth noting that even the extensive Port Solent development has adopted the wide barrel compromise arrangement.

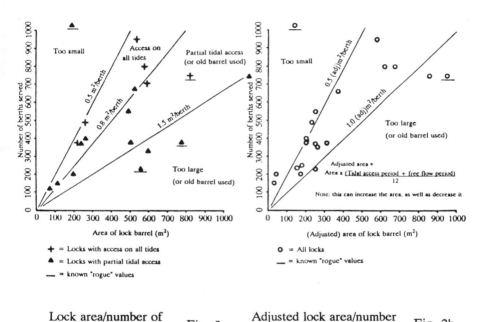

Lock area/number of berths relationship   Fig. 2a   Adjusted lock area/number of berths relationship   Fig. 2b

A guide to the sizing of locks is given in Fig 2a. This relates the area of a lock to the number of berths catered for and distinguishes between those where access is available on all tides and those which will have a limit either side of high tide. This guide indicates that unless the lock area/berth ratio is greater than $0.5m^2$/berth the lock will probably be too small. Where tides limit access this ratio should be at least 0.8 m^2/berth. A ratio of 1.5 m^2/berth would probably be excessive unless tide access was very poor or an existing wide barrel had to be utilised in conjunction with a minimum realistic length.

Clearly other factors such as free-flow period, average vessel size and typical usage patterns will influence this, but the statistics for existing locks fall within these criteria with few exceptions other than known "rogue" cases.

A further method of establishing a guide to lock sizing is to adjust the actual area by a factor taking account of tidal access and any free-flow period. An empirical formula has been used in Fig 2b where the adjusted lock area (which can be greater than as well as less than the actual) equals:

actual area x   (Access period per tide + free flow period per tide)
                                        12

This indicates that the lock (adjusted) area/ berth ratio should fall between the previous minimum size limit of 0.5 m²/berth and a new maximum of 1.0 m²/berth.

These approaches are only intended as a guide and, particularly for larger locks, a more exhaustive analysis should be carried out. See Reference 3.

Concerning the orientation of the gates, the norm is to have the pivots on the inside of the lock and the curved surfaces facing outwards. There are three main advantages to this; firstly when the lock is pumped dry for maintenance and inspection of gate and civil works, the pivots, seals, arms and general structure can all be accessed. Secondly although collision with the gate arms is possible with any orientation, the greatest vessel speeds are associated with the arrival at the lock, from either direction, rather than in entry into or exit from it. Thirdly the loads due to wave action are reduced. The only significant disadvantage is that lock volume and hence water usage is increased with no improvement in vessel capacity.

The survey (Reference 2) located only a few sector gated locks (as opposed to a sector and mitre combination), in which the arrangement was other than the above.

4. STRUCTURAL DESIGN

Over recent years there have been two basic approaches to the design of sector lock gates, with advantages and disadvantages associated with each.

Design Approach "A"

a.   The gate skin plate is supported by two main horizontal beams coincident with the arms. These are spanned by vertical stiffeners.

b.   The gate leaves are supported by two arms which are at approximately 50 degrees to each other when viewed in plan.

Design Approach "B"

a.   The gate skin plate is supported by three main vertical members, two at the outside edges and one at the gate leaf centre joint, with intermediate vertical stiffeners and plated horizontal stiffeners (egg box construction).

b.   The gate leaves are supported by three arms which are displaced at approximately 45 degrees to each other when viewed in plan and are coincident with the main vertical members.

The most important areas of the gate structure to maintain flat, square and rigid are the gate leaf sealing faces, and hence it follows that design approach "A" does not fulfil the task as well as design approach "B".

It is preferable to avoid the need for wheels on the bottom of gate leaves and this can be helped by having adjustable cross bracing to compensate for short or long term deflection.

In a marine environment it is essential that the gates are designed with generous factors of safety in order to withstand unexpected occurances. For example, the outer sector lock gates at Hull marina were struck by a dredger shortly after going into service. These gates were of the "egg box" construction and designed with good factors of safety and as a result the gates withstood the impact and needed only a small section of the gate leaf to be cut out and replaced. However, if the gate had been as design type "A" a greater area of the leaf would have been affected and may have totally collapsed. A normal minimum plate thickness of 10mm is recommended and this helps achieve a general robustness.

Regardless of design type it is now quite common for a finite element analysis of the entire structure to be undertaken in order that deflections and stress concentration levels can be fully and properly analysed. This type of analysis allows deflections to be accurately quantified - which is particularly important in avoiding leakage under load, and similarly stresses - which is important for critical areas such as arm/leaf junctions.

When manual calculations are used, indeterminate aspects of the structure can lead to a large degree of overdesign to ensure the required rigidity and strength are achieved. For all but small and basic gates, the savings in weight coupled with the increased confidence in the design should justify the cost of finite element analysis.

The method of determining acceptable loads can in theory either be by limit state analysis or the permissible stress approach although the latter is probably still the most common and suitable and has many adherents. No specific gate design standards are produced by British Standard (BS) and hence it is common to use a related steel structure code such as that for bridges; BS 153 (withdrawn but useful) and BS 5400. A German standard, DIN 19704, which is particular to hydraulic steel structures, is also in common use. There is a Japanese standard, (Reference 8), which approximates to a British Standard Code of Practice format, which includes sections on gates, but has to be used with caution.

5. MAIN PIVOT DESIGN

It should be noted that every hydrostatic, dynamic, and operational impact loading involves this vital component. Even with a modestly proportioned pair of sector gates, say 8m span, 8m deep the lower pivot of each leaf will need to resist loads in the order of 200 tonnes from hydrostatic forces alone. Other types of gates will have this load spread more evenly throughout the structure, for example onto the quoin blocks in the case of mitre gates.

For medium and large installations such as that proposed at the Cardiff Barrage, the water loads on the gates are very high and the pivot attachments to the concrete abutment structure can be by the use of bolts tensioned after installation. The bolts are located in grease filled ducts to permit future inspection/replacement. They are designed such as to facilitate re-tensioning, and access to the ends of the pretensioned rods remote from the pivots are by way of local chambers with access arrangement.

To avoid fatigue, it is important that pivot retaining bolts (as well as others) clamp parts together with a tension greater than any external force tending to separate them. The bolt then remains under constant stress and is immune from fatigue; if the initial tension is too low, varying loads act on the bolt and it will ultimately fail. Reliability therefore depends upon correct initial tension and is ensured by specifying and controlling the tightening torque.

The bolt ducts are provided with a means of allowing old grease to extrude under pressure from the trunnion end of the ducts. By this means any tendency for water to seep into the end of the ducts are countered by progressive replacement of the grease during maintenance procedures. Provision is made for the extruded grease to be collected in order to avoid water polution.

The pivots are normally of welded construction or cast steel. Means should be provided for adjusting the position of the pivots in each direction after erection. The pivot pins are of stainless steel, stainless steel clad or finished with hard chromium plating if suitable for use with the bearing.

The bearings in the pivot ends of the gate arms, and the thrust washers, are usually of the self-lubricating type. Care must be taken in choice of self-lubricating material, and graphite in the lubricant should be avoided. This is because firstly it is very cathodic and in the presence of sea water can cause electrolytic corrosion, and secondly in the case of stainless steel pins it can cause fretting corrosion.

6. SEAL DESIGN

There are a number of different seal sections and materials/grades which can be used for sector lock gates, the more successful ones are shown in Fig. 3.

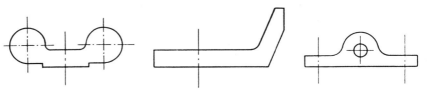

"Double Music Note"   "Hockey Stick"   "Centre Bulb"

Various types of seals   Fig. 3

The double music note seal is the most widely used for both vertical centre and side seals as well as horizontal lower seals. This type of seal is of the "on-seating" type which is pushed onto the sealing face harder as the water pressure rises and consequently requires little pre-compression to affect an excellent seal. A well set up music note seal of the correct material and hardness, typically neoprene rubber of 60/70 shore hardness, can be expected to last upwards of ten years with only minor routine repairs.

The hockey stick seal is sometimes used for lower gate horizontal seals, as a cheaper method of sealing, but has the disadvantage that vertical to horizontal seal transitions are overcomplicated. Seal leakage at the gate corners is common with this method.

The centre bulb seal can be used for vertical and horizontal seals but as it is of the "off-seating" type higher pre-compressions are necessary to obtain a good degree of sealing.   Due to the need for pre-compression, operating loads increase and the expected seal life decreases.

Both music note and centre bulb seals can be supplied with either PTFE or brass inserts which are designed to reduce the operating loads. This reduces friction and provides, in the case of the brass inserts, a harder wearing face.   However, unless the quality of the water is particularly abrasive or the possible minor savings in power consumption are considered of particular benefit it is normally better to opt for the relatively inexpensive simple neoprene rubber sections without inserts. The main reason for this is that once a seal insert starts to become detached from the main seal section it is usually necessary to take the gates out of service to fit complete new seals. Maintenance of neoprene seals is easier because only the damaged areas which are generally at the water line need to be replaced.

Seal support, adjustment and protection is of great importance if long and trouble free life is to be expected from the seals. With all types of rubber seal it is imperative that they are supported and fixed correctly in order that they resist buckling under or tearing when the gates are operated under full head.

The seals must be supported by rigid steel backing plates which should be capable of being adjusted by at least $\pm$ 15mm in all planes. They should also have the ability to tilt the seal, thus applying more pressure to one bulb of a double music note seal than the other if necessary. It is also desirable for the seal support to articulate in order that irregularities over the seal length can be accommodated, as no two lengths of rubber seal are identical.

If the seals are designed correctly, it should be possible to carry out all routine adjustments from within the lock, without having to install stopbeams and take the lock out of service.

In areas where floating debris is expected it is advisable to incorporate at the design stage a system of seal protection to prevent, as much as possible, any floating debris either becoming trapped between the seal and stainless steel embedded sealing face or hitting the seals as the debris is drawn between the side seals during sluicing. This can be achieved by shielding vertical side seals or, for persistent debris problems, by providing a bubble barrier at the end of the lock.

7. DESIGN OF PARTS BUILT INTO CONCRETE

For larger and traditional locks the gate sealing, and sometimes bearing, faces were of fair faced stone blockwork built to very high accuracy by specialised craftsmen. More recently, and for marina lock applications, it is normal in the UK to use steel built-in parts.

The requirements for these are that they should be accurately located, robust and resistent to corrosion. This is achieved by the type of arrangement shown in Fig 4. The main features of this are:

*   Adjustment in all planes to achieve accurate location. A tolerance of better than $\pm$ 2mm is normally required.

*   Rigidity during second stage concreting. Plate thicknesses would normally be a minimum of 10mm and anchor bolts of 12mm diameter, or preferably 16mm, would be located at say 500mm centres.

*   Corrosion resistance is obtained by using a suitable grade of austenitic stainless steel for exposed surfaces. The stainless steel can either be solid plate or clad. If clad the minimum stainless steel thickness should be 3mm. Ideally no mild steel should be exposed to the water or have less than 75mm cover or less than a 75mm water path to it. Where exposed mild steel is unavoidable it must be fully painted (see Section 8) and must not encroach upon sealing surfaces. Steel parts embedded in concrete but within 75mm of the water (in terms of water path or cover) should be blast cleaned and primed, again in accordance with Section 8.

*   Provision of a smooth surface on sealing faces to achieve good sealing with low seal wear. The as-rolled surface of the stainless steel is sufficient, but care has to be taken during design and manufacturer to avoid welding distortion. If this occurs the faces have to be machined which is expensive and can reduce the surface finish.

The use of Terrazzo faced slabs for sealing faces has been tried and could well be worth considering, for example for the bottom sill of delta gates (see Section 2).

8. CORROSION PROTECTION

The combination of steel construction with sea water dictates careful attention to the risks of corrosion and how they can be reduced or eliminated.

Although the main attention will be on surface protection, care must also be taken to avoid bi-metallic corrosion. A typical gate installation will include steel, stainless steel, zinc, brasses/bronzes, chromium, etc and all points where they come in contact, or in close proximity in submerged areas, have to be considered. Either electrical separation or a change of material has to be adopted.

Regarding surface protection, the submerged or partially submerged items are normally the most vulnerable to corrosion and the most costly to re-paint or repair. It should be noted that the splash zones of the gates, rather than the permanently submerged areas, are the worst. For all these areas use of an epoxy coal tar paint is almost universal and with diligent application a paint from a reputable manufacturer can give a system lasting 15 or more years without major work. There can, however, still be problems. The surface must be adequately prepared by blast cleaning. The norm has, until recently, been SA 2½ to Swedish Standard SIS 05-59-00. However, there is an increasing trend towards designers and suppliers requiring SA3 but it is questionable if this is realistic in practice. Ambient and substrate temperature, overcoating times and humidity/dew point limitations must be observed, and for painting at site in a marine environment this can be a problem. Remedial painting of damaged areas must be properly executed with the full system and not a single brush applied coat to give the appearance of a repair.

A typical and proven coal tar epoxy system is:

* 50 microns (in one coat) of zinc rich epoxy, zinc phosphate, zinc silicate or other high quality zinc rich primer.
* 400 microns (in two coats) of high build, two-pack, isocyanate cured epoxy coal tar.

For areas which are not submerged at any time or are not in a splash zone, such as handrailing, ladders, covers, frames and external control pedestals and cabinets etc, a more decorative finish is required.

A typical system for this is:

* 50 microns (in one coat) of primer as above.
* 200 microns (in two coats) of two-pack, high build epoxy or polyurethane coating. Chlorinated rubber or micaceous iron oxide may be an acceptable and cheaper solution, although the former should not be used in hot climates.

In certain circumstances it may be thought worthwhile to supplement the underwater surface coating with a suitable cathodic protection system.

The use of galvanising, without further protection, in a marine environment should be adopted with great care for any parts. If it is to be painted it must be coated with a mordant wash or similar beforehand.

British Standard 5493 (Code of Practice for protective coating of iron and steel structures against corrosion) gives a very high rating to sprayed metal finishes. These are, however, expensive and uncommon.

9. CIVIL ENGINEERING REQUIREMENTS

Three areas of the civil works have particular influence over the gates and can and do cause problems. These are accuracy, second stage concrete placement and protection.

As indicated in Section 7, the accuracy required for the mechanical works is measured in millimetres. A normal accuracy for civil works would be in centimetres and this can cause problems of alignment and interference. Problem areas are all around the sealing frames, particularly in the side seal/sluicing arrangement and possibly at the pivots for the hydraulic cylinders.

A further problem regarding accuracy is the location of the first stage anchors in the box-outs. Ideally these need to be located to within approximately $\pm$ 10mm to achieve the maximum strength resulting from the adjusting bolts aligning with the anchors welded to the backs of the plates.

The design of the built-in parts and their anchorage/adjustment method has to be heavily influenced by the rough treatment, large forces and vibration that occur during concrete placement and compaction. Even so, damage can still occur if the concreting gang do not take some care.

The final main problem is that of damage caused by local civil construction works to the built-in parts, gates and operating equipment after erection and where applicable, after painting.

10. SLUICING METHODS

As mentioned in Section 2, for a sector gate the hydrostatic forces act radially through the gate pivots. This theoretically results in no net opening or closing torque.

In the absence of such torque the gates can be opened against a differential water head on either side of the gates, (as well as being able to close against flow) thus eliminating in many cases the need for lock filling and emptying culverts or sluices/paddles on the leaves themselves.

This is achieved by the use of the gap between the side seals and the built-in frames that naturally occurs when a gate revolves a few degrees from the fully shut position. The size of gap is limited by the maximum flow rate and velocity that should occur and by the unbalanced hydrostatic forces that result from the nib on the gate (see Fig 4). The flow passes into the specially shaped gate recesses (see Fig 1) and continues for the duration of the water level equalisation period, but is normally supplemented for the last 1m or so when further rotation allows the centre of the gates to be opened (see Fig 5). Such sluicing may give a total filling or emptying time of say 3 or 4 minutes for many locks. It should also be noted that yachtsmen are not keen to remain in a lock for lengthy periods awaiting access, and often wish to leave and enter at low water which aggravates the problem.

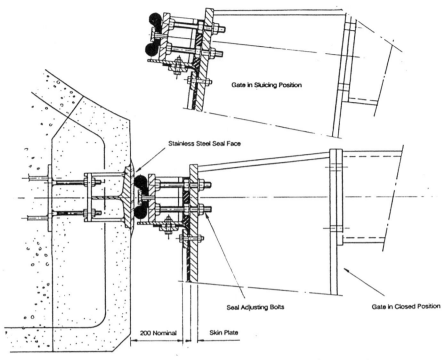

Fig. 4

Section through vertical seal assembly (showing nominal 200mm sluice position)

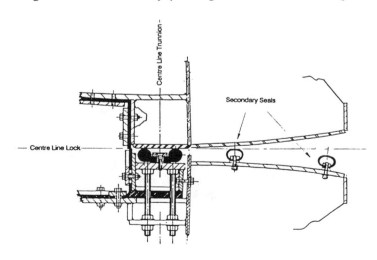

Section through centre seal assembly with gates closed   Fig. 5

Limitation, control and dissipation of the energy of the water is critical to the satisfactory design of the lock. The normal method of side sluicing, which is preferable to centre sluicing, takes advantage of the natural energy dissipation caused by the passage of flow around the specially shaped recesses, followed by a further dissipation caused by the two flows impacting each other across the lock.

Model tests should be considered, particularly for locks with a large head differences. Sluicing flow and energy dissipation can be studied and the sluicing arrangements can then be designed to reliably achieve optimum sluice times and minimum water disturbance. Such model tests have been carried out for the design of sluicing of the proposed Cardiff Bay Barrage sector gated locks.

11. METHODS OF OPERATION

Earlier approaches to the operation of mitre and sector gates were often by means of chains or steel wire ropes. An elegant and ingenious system was in common use until quite recently whereby the chain or rope was directly wrapped around the periphery of each leaf and operated by winches from the abutments. This method can suffer from the need for frequent maintenance in order to cater for rope stretch, etc. the avoidance of which in hydraulic cylinder applications gives them their main advantage.

Of the sector gated locks located in the survey (Reference 2), approximately 80-90% of the recent installations use hydraulic cylinders whilst the remainder use ropes.

The detailed design of the operating equipment needs to satisfy the following main requirements:

* Control and operation should be accurate, rapid, reliable and convenient.
* It should be robust in construction.
* It should include protection from overload.
* The machinery should be accessible for maintenance and repair.
* It should be located to avoid dockside obstruction or danger.
* It should be weatherproof and protected from possible impact damage.
* Synchronisation of the leaves.

The principle guidelines and recommendations for design are taken from the German Standards DIN 19704 and 19705 (see also Section 4). These standards are used by a number of European countries including the UK. The operating machinery must be designed for the loads listed in the standard, appropriate to the individual site. These are:

* Dead weight	* Friction
* Hydrodynamic Pressures	* Pressure due to ice
* Wind Forces	* Closing and opening pressures
* Impact	* Temperature effects

Calculations are by the method of allowable stresses and pressures and cover three distinct loading conditions; standard operating case, special operating case, and exceptional conditions. Allowable stresses for operating devices are limited to percentages of the material yield point, these are: 40% in the standard case, 50% in the special case and 80% in the exceptional case.

Modern oil hydraulic systems together with state of the art electronic control lend themselves to sector gate operation and adequately satisfy the main requirements, listed above.   Well engineered hydraulically operated cylinders working at approximately 140 bar are provided for operating each leaf either from a central hydraulic power pack or by using proprietory integral pump/cylinder arrangements. There can be considerable advantages in using this latter type of factory built semi-standard unit, such as a reduction in hydraulic pipework and ease of maintenance.   A spare "standby" unit of identical construction can be held on site for replacement should this become necessary.

As with any hydraulic drive, cleanliness of the fluid is essential. Oil filtration systems have advanced considerably over recent years together with component and pipework technology, such that designing and installing a sound, economic and reliable system is quite possible given adequate knowledge and care.

Another type of proprietary actuator which may gain favour is that which uses an internal screw thread to extend a ram in an assembly that looks and acts in a similar way to an hydraulic cylinder.

Subject to certain factors, traffic can often be handled in high capacity without recourse to any locking operations during a "free-flow" period which is popular with yachtsmen, but which must be undertaken with care by lockmaster and boat owner alike. It is important that the lock is equipped with traffic signalling of an internationally recognised configuration when such operations are undertaken, to avoid vessels travelling in opposite directions within the lock. Operations are generally concluded when barrel flows exceed approximately 2 knots unless limited by other factors such as safe maintenance of basin water levels. The survey (Reference 2) indicated that half of the locks had a free-flow period, at least on some tides.

In recent years advances in electronic control have seen improvements in all fields of engineering, and marina locks are no exception. Many of the new installations are equipped with programmable logic controllers (PLCs). These, together with modern centrally located consoles in purpose made buildings provide reliable, compact and virtually maintenance free control facilities which are easy and safe to use and adaptable to any development in the control philosophy.

If the lockmaster's office does not directly overlook the lock or if more than one lock is to be controlled from a single control building, as in the case of the proposed Cardiff Bay Barrage, it is advisable to incorporate close circuit television systems in order to allow full monitoring of the locks.

For any size and arrangements of lock it is crucial that the operator has a clear view of the inside of the lock and of the approaches.

12. SAFETY IN DESIGN

Safety aspects of the installation must be a key consideration during design and construction.

If a marina is accessible to the general public, as most are, then the position and design of barriers and handrailing must prevent contact with the sector lock gates during operation. This can generally be achieved by the use of semi-automatic gates across walkways, controlled directly by the lockmaster. The design of the gate top walkways and handrailing must be carefully detailed together with gaps and potential crushing or scissoring points resulting from movement of a gate and its accessories relative to the concrete and relative to the opposite leaf.

To protect the arms of sector gates from impact by vessels, and to reduce risk of related damage to vessels in the lock, a wooden grillage is sometimes provided on the arms of the gates.

A number of sector lock gates include wire rope operated limit switch attached to the leading edge of each pair of gates, to prevent the gates closing onto and crushing vessels.

As sector lock gates are capable of being operated against a full head of water, it is not inconceivable that both pairs of gates could be opened together (outside of any controlled "free-flow" regime), thus possibly draining a marina. The gates can be interlocked electrically to prevent such an event.

13. GATE INSTALLATION

The gate installation sequence begins with a number of steel plates and fabricated assemblies being built into the first stage concrete works. These have to be suitably supported and anchored down to prevent movement during pouring of the concrete. These items include the main load bearing pivot foundations, sealing frame foundation plates and other miscellaneous plates. Once the concrete has had time to cure and the site is cleared of all debris, the gate installation can commence. For an average size lock this can take 10 to 14 weeks.

Due to transportation restrictions, the sector gates would normally arrive at site in a number of sections in a finish painted condition. All welding carried out on site has to be closely controlled to obtain satisfactory results.

The sections would be assembled as close as possible to the lock in order that the lifting-in operation, using a large mobile crane, is made as simple as possible.

Although an average sized gate assembly could weigh between 20 to 30 tonnes the size of crane required to lift it may need to have a capacity of 100 to 150 tonnes, due to the outreaches required.

Crane access may be required for maintenance during the life of the lock, not just during construction, and hence crane access to the lock side, coupled with space and paving strength for the crane and outrigger must all be catered for in the design of the lock surrounds.

Whilst the gate leaf assembly is being carried out, all of the sealing frames and pivot brackets would be accurately set-up and secondarily grouted. Once the pivot brackets had been set the upper and lower arm assemblies complete with cross bracing would be attached to the pivot brackets and adjusted square and level. The gates would then be lowered into position, complete with all seals and seal carriers, and attached to the arms.

It would only remain for the operating equipment to be attached and set up to allow the gates to be operated.

Following this, time is spent setting the seals and carrying out dry testing prior to the wet testing. By this means only minor adjustments should be necessary during the final tests.

Any damaged areas of paint should be made good, prior to wet testing, in line with the paint system manufacturer's recommendations.

14. COMPARITIVE COSTS AND LEAD TIMES

As previously stated, sector lock gates are not the cheapest solution for providing a locked basin. The following table indicates the relative costs of installed M&E equipment for various gate types, for an opening of 9m wide x 8m high. The indices will vary for locks which are much larger or smaller than this. The sector gated lock is given an index of 100 as the basis for the comparison. At 1992 values the index of 100 would approximate to £500,000.

		Index
2 No:	Flap Gates	65
2 Pairs:	Mitre Gates	75
2 Pairs:	Delta Gates	95
2 Pairs:	Sector Gates	100

The approximate lead times for a typically sized lock, depending on scope, complexity of control equipment and civil construction programme, would normally be as follows:

Design and approval to start of manufacture	10-12 weeks
Manufacture and deliver to site, fully painted	26-30 weeks
Install and commission	12-14 weeks

giving a total period of approximately one year from award to handover.

15. OPERATION AND ROUTINE MAINTENANCE

It has become more common in recent years for marinas to have quite grand lock control buildings, for example on two storeys with the first floor affording views of the lock barrel and lock approaches. However, marinas should also normally have the facility to operate the gates from lockside, which can prove to be very useful when carrying out any gate maintenance as well as being a safe control point for busy periods or critical locking operations.

The amount of maintenance necessary to keep sector lock gates fully and reliably operational depends largely on the type of operating equipment, build quality and quantities of floating debris. The items requiring routine maintenance are seals, operating equipment, gate pivot bearings, corrosion and cathodic protection systems.

It is advisable that all gate pivot bearings are suitably piped to common battery plates in order that greasing of bearings can be carried out from lockside level.

Consideration must be given to the possible future need to remove a complete gate leaf and it is therefore necessary to have a sufficient number of stopbeams, stored as close to the marina as possible, with which to isolate at least one end of the lock. Crane access to the site for gate removal is sometimes overlooked with very costly consequences. (See section 13).

The costs of any period when the lock is out of service, (outages) can be considerable both in terms of direct cost and loss of good will; or with repeated problems loss of customers.

The aim should be to have no outages during daylight hours during the spring/summer/autumn period with planned daytime outages occuring in the winter. With a well designed and operated lock given thorough and planned maintenance this should be quite achievable. A very occasional accidental out of service period would then be tolerated by marina users.

16. CONCLUSIONS

The introduction draws attention to the critical nature of a marina lock to the success, enjoyment and viability of a marina, and the subsequent sections deal briefly with some of the technical areas worthy of detailed study.

The lock gates and associated items must not be procured by a short performance specification nor must they be designed and manufactured by a company experienced only in general steel work fabrication or similar.

To achieve a good end product requires the teamwork of consulting engineers and manufacturers both highly experienced in this specialised field. The survey (Reference 2) revealed few very bad installations (an exception being one that had apparently been provided in a manner particularly at odds with the above recommendations) but it highlighted a number of areas in which the owner or operator, through no fault of his own, had justifiable criticisms of the installation. This emphasises the need to use the right expertise for the work and to continually improve the standard of design and installation achieved.

Reference List

1. Water Level Control in Marinas and Harbours by S de Turberville and P D Hunter, presented at the International Conference of Marinas, Southampton, UK, September 1989 and published in "Marinas: Planning and Feasibility" published by Computational Mechanics Publications.

2. A survey carried out by the authors covering approximately sixty locked or gated facilities for leisure craft. The help given by the operators, owners or engineers is gratefully acknowledged.

3. Some aspects of pleasure navigation in the Netherlands. The Capacity of Locks and Waterways by Dr R Filarski, Ing CG den Hartog and Ing P de Ridder presented at the 26th International Navigation Congress in June 1985.

4. Sell's Marina Guide, published by Sell's Publications Ltd.

5. Marinas and Small Craft Harbours, Proceedings of a symposium held at the University of Southampton in April 1972 and published by Southampton University Press.

6. PIANC - Final Report of the International Commission for the study of locks.

7. Marinas - a working guide to their development and design.   D.W. Adie, published by the Architectural Press Ltd.

8. Technical Standard for Gates and Penstocks, published by the Japanese Hydraulic Gate and Penstock Association.

SECTION 5: WAVE MODELLING AND ANALYSIS

An Efficient and Cost Effective Wave Absorber for Marinas

W.W. Jamieson, G.R. Mogridge
Coastal Zone Engineering, National Research Council of Canada, Montreal Road, Ottawa, Ontario, K1A OR6 Canada

ABSTRACT

This paper describes the development of an upright wave absorber to be used in a marina where vertical sheet pile walls cause wave conditions at the berthing facilities to be unacceptable. The situation required the wave absorbers to be reasonably efficient, to be inexpensive, and to occupy as little space as possible. A wave absorber was designed which consists of rows of tires supported in a steel frame, using the principle of progressive porosity. A section of the wave absorber was tested at full scale in the laboratory, and was then installed in the marina for observations of its behaviour in waves and survival through the ice conditions of winter. Laboratory test results are presented in terms of wave reflection coefficients as functions of wave height and wave steepness for the design wave length and water depth conditions of the marina.

INTRODUCTION

Dawson's Marina on Lake Simcoe, Ontario, covers an area of approximately 200 m by 500 m and has slips for up to 400 pleasure craft. Wind waves entering the marina and incident on vertical sheet pile walls are up to 30 cm high with a period of 1.7 s. Boat wakes are less severe. It is generally preferable that wave heights inside a marina at wave periods of about 2 s, should not exceed 15 cm more than once a year (Northwest Hydraulic Consultants [1], Abraham [2], Cox [3]). When waves reflect off vertical walls, standing waves of up to twice the incident wave heights can occur and disturb moored boats. Thus, Dawson's Marina requested designs for wave absorbers to be placed against the sheet pile walls. The water depth at the site is 1.8 m, and the bottom is very soft so that it cannot support any structure or stone slope that might normally be used to control wave reflections. Also, there is heavy marine growth in late summer, which could clog small openings in porous structures. The absorber was required to be inexpensive and easily constructed by marina staff, and light enough to be handled by a front-end loader with a lifting capacity of 500 kg.

Breakwaters using various types of perforated or porous structures have been studied on numerous occasions. Jarlan [4] was the first to propose a perforated breakwater consisting of a porous wall, a chamber, and a solid rear wall (see also Marks and Jarlan [5]). Considerable research on similar structures has since been conducted by Nagai and Kakuno [6], Sawaragi and Iwato [7], Onishi and Nagai [8], and others. These breakwaters with perforated or slit walls, and sometimes a perforated floor in the chamber, are all reasonably effective in absorbing wave energy near the design wave length. Reflection coefficients were found by Onishi and Nagai [8] to be lowest at a ratio of breakwater length on wave length (ℓ/L) of approximately 0.13 to 0.18. For this range of ℓ/L, reflection coefficients as low as about 30% were measured for the slit-type breakwater. For the box-type breakwater with a perforated floor, the measured coefficients were reduced to about 10% at similar relative absorber lengths. However, in both cases reflections increased rapidly at other values of ℓ/L. Kondo [9] studied a perforated breakwater with two perforated walls in front of a solid wall. The lowest reflection coefficients of 15% to 20% were measured for a relative breakwater length of 0.25.

Perforated quay walls incorporating a slope within the inner chamber have been described by Matteotti [10]. Reflection coefficients at a relative absorber length of 0.2 were approximately 40% to 50%.

Porous concrete wall elements called "igloos" have been used for wave absorption inside harbours, as described by Shiraishi, Palmer and Okamoto [11]. This type of structure is quite effective, with measured reflection coefficients as low as 18% to 35% for a limited range of wave conditions, and an ℓ/L ratio of between 0.14 and 0.20. A similar type of breakwater structure made up of concrete units called "warocks" was described by Ijima, Tanaka and Okuzono [12]. It was concluded that the best wave absorption occurred when the total absorber length was 0.18 times the wave length. Reflection coefficients were measured as low as 10% to 20%.

Other techniques have been used for wave absorption such as rows of closely spaced piles described by Truitt and Herbich [13], Dalrymple, Seo and Martin [14], and Herbich and Douglas [15]. These studies did not test the rows of piles at some distance in front of a vertical wall, but measured wave transmission in a long wave flume. The test data indicates that two rows of closely spaced piles must be used before wave transmission is significantly reduced. These tests imply that the system would not be effective as a wave absorber with a small relative length.

Gardner, Townend and Fleming [16] tested porous screen breakwaters consisting of vertical rectangular slots. Three horizontally-slotted vertical screens were used by Weckman, Bigham and Dixon [17].

Somewhat more unusual types of energy-absorbing breakwaters that have appeared in the literature are buoyant flaps (Sollitt, Lee, McDougal, and Perry [18]), buoyant spheres (Agerton, Savage and Stotz [19]), and sea balloons (Uwatoko, Ijima, Ushifusa and Kojima [20]). However, none of these ideas seem to be suitable for the present marina application.

The design of the proposed absorber is based on earlier work carried out at the National Research Council of Canada (NRCC) on perforated expanded metal wave absorbers for laboratory facilities (Jamieson and Mansard [21]). Expanded metal sheets are suitable for laboratory absorbers, but are unlikely to be practical for field conditions. However, based on the same principles, a progressive absorber has been designed for marinas, which occupies little space, is reasonably efficient, is easily fabricated and installed, and is constructed of inexpensive scrap automobile tires. Tires have been used extensively for the construction of floating breakwaters (Bishop et al. [22], [23]), but never, to the authors' knowledge, for wave energy absorption by fixed structures inside marinas.

A narrow test section of the proposed tire absorber was tested at full scale in the laboratory, and was then installed at Dawson's Marina.

DESCRIPTION OF THE TIRE WAVE ABSORBER

A drawing of the wave absorber is shown in Figure 1. It consists of a welded steel frame (constructed using 7.5 cm channels and 3.75 cm angles) holding tires in vertical rows such that the first row of stacked tires has a relatively high porosity, The next row consists of two layers offset so as to considerably reduce the porosity. The absorber shown is one module, which can be duplicated as many times as necessary to form a wide absorber. Because the modules are 80 cm wide, they would be placed at 90 cm centres to form a continuous absorber, thus leaving a small gap of 10 cm between units.

The side elevation (Figure 1) shows that the module is approximately 75 cm from the front to the back of the rear row of tires. There is a 15 cm space between rows of tires inside the module and also between the back row of tires and the rear wall, which would be the marina quay wall. The tires are held firmly in place by 2.5 cm square steel tubes so that they do not move under wave attack. At the base of the module is a steel cage filled with rocks to hold the absorber in place under its own weight. The module is mounted by hanging it from a 4.22 cm diameter steel pipe fastened to the wall as shown in Figure 1. The weight of the absorber keeps it stationary against the wall. When mounted, the front of the module is 90 cm from the face of the wall. This dimension is referred to as the length, ℓ, of the absorber.

The total weight of the tire wave absorber module in air is approximately 473 kg. The weight of the steel used in the construction is 144 kg, and the weight of the rock in the base is 229 kg.

EXPERIMENTAL SETUP

The tire wave absorber was tested in the laboratory in a flume 26 m long by 2.3 m deep and 0.9 m wide. An array of five capacitive-wire wave gauges were located approximately 9 m in front of an hydraulically driven wavemaker.

On-line computer control for wave generation and data sampling, and the subsequent reflection analysis were carried out using the GEDAP (Generalised Experiment control, Data acquisition and Analysis Package) software system developed by NRCC (Crookshank [24], Miles [25], Miles and Funke [26]). Both regular and irregular waves were generated, although in this paper only the regular wave data results are presented for brevity. The time series of the partial standing wave system was measured using the array of five wave gauges. Depending on the period of the regular waves, a pre-processing algorithm was used to select the three optimally spaced gauges required for the subsequent reflection analysis. The data was sampled at 10 Hz for 180 s from the start-up of the wavemaker. Analysis was carried out on the last 30 s of the time series, which allowed time for reflections and re-reflections in the flume to stabilize.

The reflection coefficients (C_r), defined as the reflected wave height on the incident wave height (H_r/H_i), were determined using a least squares method of analysis (Mansard and Funke [27]), which requires simultaneous measurements of the waves at three known positions in a line parallel to the direction of wave propagation. The amplitudes and phase relationships required for the reflection analysis were determined by fitting monochromatic wave trains to the waves measured at each of the three gauge locations. This fitting was performed by optimizing the frequency, phase and amplitude in a least squares sense. This particular method of analysis for regular waves is much superior to the conventional technique of nodes and antinodes, which is generally prone to inconsistencies caused by non-linearities. The analysis technique has been validated extensively using numerically simulated data with different degrees of noise levels (Mansard, Sand and Funke [28], Isaacson [29]).

EXPERIMENTAL RESULTS

The experimental results of the tire wave absorber tests for regular waves of periods T = 1.00 s, 1.50 s, 1.74 s and 1.96 s are shown in Figures 2 to 5, respectively. Wave reflection coefficients are plotted against incident wave steepness, with an upper coordinate showing the incident wave height.

Jamieson and Mansard [21] recommended that for high efficiency laboratory wave absorbers, the relative length of the absorber (ℓ/L), should be greater than 0.35 and preferably 1.0. However, in the present case, it is necessary for the absorber to be short because of space limitations, so that for the design condition of T = 1.74 s, the relative length ℓ/L is only 0.194. Therefore, the absorber is entirely within the area between the wall and the first node at 0.25L. For an absorber to be relatively efficient in this location where there is considerable vertical fluid motion, it must dissipate significant energy from the vertical component of the fluid flow. Figure 6 shows the fluid turbulence that results in energy dissipation at the tire absorber. Relatively thin perforated walls or rows of piles do not achieve this effect as well as the present system. Nagai and Kakuno [6] also showed improved performance of a slit-type breakwater by taking this into consideration and providing the chamber with a horizontal perforated floor.

As would be expected from an absorber of this type, the lowest reflection coefficients in the range of 5% to 35% occur for the wave periods of 1.00 s, 1.50 s and 1.74 s corresponding to relative absorber lengths of 0.577, 0.259 and 0.194, respectively. For the design wave conditions (Figure 4), that is, T = 1.74 s, the range of wave reflection is approximately 12% to 35%, which is excellent performance for such a short absorber. In fact, if the data for wave heights less than 15 cm is ignored as being unimportant, then the range of reflection coefficients is from about 20% dipping to 12% for the steepest waves at a height of about 38 cm.

Figure 5 shows that for longer waves, T = 1.96 s (ℓ/L = 0.156), the reflection coefficients quickly become quite high, although even these values of 22% to 48% are not too bad for this type of absorber. However, for longer periods it would be preferable to increase the length of the absorber; a relative length of 0.156 is obviously too low.

FIELD INSTALLATION OF THE TIRE WAVE ABSORBER

A single module of the tire absorber was installed at Dawson's Marina in the fall of 1990, to test stability in waves and durability. The photographs in Figures 7 and 8 show the absorber being installed, and also hanging in place on a 4.22 cm diameter steel pipe welded to the sheet pile wall. The absorber was left in place over the winter to determine if movement of ice in the spring break-up caused any damage to the module or its removal from the sheet pile wall. To date there has been no problem with the absorber in waves and it also remained unaffected by ice conditions in the winter and spring. Because the inside of the marina is reasonably protected, there were no large wind-driven ice movements to damage the absorber. The marina operator also suggested that perhaps because of the black tires absorbing heat in the sunshine, they never seemed to freeze into the ice sheet. However, because no problems occurred during the first winter doesn't mean that damage will not occur in future winters. If there appears to be some risk of ice damage, the absorbers can be removed quickly and easily in late fall and stored on the dock for the duration of the winter season.

CONCLUSIONS

A new type of marina wave absorber has been designed and tested. It has the following advantages:

a) It is highly effective with reflection coefficients at design conditions of 12% to 20%;

b) It can be installed where very little space is available, and if preferred it can be covered to increase the dock area and improve its appearance;

c) It is relatively inexpensive, because it uses scrap automobile tires and can be constructed by unskilled labour and installed using relatively light equipment such as a fork-lift;

d) It can be easily removed for the winter season where there may be risk of damage by moving ice sheets;

e) It has large porous openings that do not clog with marine growth; and

f) It can be used where a deep soft mud bottom would make construction difficult for other types of absorbers.

ACKNOWLEDGEMENT

The authors wish to thank Dawson's Marina for their cooperation in the development of this new type of wave absorber.

REFERENCES

1. Northwest Hydraulic Consultants. Study to Determine Acceptable Wave Climate in Small Craft Harbours, Report prepared for Small Craft Harbours Branch, Fisheries and Oceans Canada, March 1980.

2. Abraham, C.H. A New Australian Standard for Marinas, Proceedings of the First International Conference World Marina '91, ASCE, Long Beach CA, pp. 596-605, September 1991.

3. Cox, J.C. Breakwater Attenuation Criteria and Specification for Marina Basins. In Marinas: Design and Operation (Ed. Blain, W.R. and Webber, N.B.), Proceedings of the International Conference on Marinas, Southampton, UK, September 1989, Computational Mechanics Publications, Southampton and Boston, pp. 139-155, 1989.

4. Jarlan, G.E. A Perforated Vertical Wall Breakwater, The Dock and Harbour Authority, Vol. XLI, No. 486, April 1961.

5. Marks, W. and Jarlan, G.E. Experimental Studies on a Fixed Perforated Breakwater, Proceedings of the 11th Coastal Engineering Conference, London, September 1968.

6. Nagai, S. and Kakuno, S. Slit-Type Breakwater: Box-Type Wave Absorber, Proceedings of the 15th Coastal Engineering Conference, Honolulu, Hawaii, pp. 2697-2716, July 1976.

7. Sawaragi, T. and Iwato, K. Irregular Wave Attenuation Due to a Vertical Breakwater with Air Chamber. ASCE Speciality Conference on Coastal Structures 79, Alexandria, Virginia, pp. 29-47, March 1979.

8. Onishi, H. and Nagai, S. Breakwaters and Sea Walls with a Slitted Box-Type Absorber. ASCE Speciality Conference on Coastal Structures 79, Alexandria, Virginia, pp. 9-28, March 1979.

9. Kondo, H. Analysis of Breakwaters Having Two Porous Walls. ASCE Speciality Conference on Coastal Structures 79, Alexandria, Virginia, pp. 962-977, March 1979.

10. Matteotti, G. The Reflection Coefficient of a Wave Dissipating Quay Wall, The Dock and Harbour Authority, Vol. 71, No. 825, pp. 285-291, February 1991.

11. Shiraishi, N., Palmer, R.Q. and Okamoto, H. Quay Wall with Wave Absorber "Igloo", Proceedings of the 15th Coastal Engineering

Conference, Honolulu, Hawaii, pp. 2677-2696, July 1976.

12. Ijima, T., Tanaka, E. and Okuzono, H. Permeable Seawall with Reservoir and the Use of "Warock", Proceedings of the 15th Coastal Engineering Conference, Honolulu, Hawaii, pp. 2623-2642, July 1976.

13. Truitt, C.L. and Herbich, J.B. Transmission of Random Waves through Pile Breakwaters, Proceedings of the 20th Coastal Engineering Conference, Taipei, Taiwan, pp. 2303-2313, November 1986.

14. Dalrymple, R.A., Seo, S.N. and Martin, P.A. Water Wave Scattering by Rows of Circular Cylinders, Proceedings of the 21st Coastal Engineering Conference, Costa del Sol-Malaga, Spain, pp. 2216-2228, June 1988.

15. Herbich, J.B. and Douglas, B. Wave Transmission through a Double-Row Pile Breakwater, Proceedings of the 21st Coastal Engineering Conference, Costa del Sol-Malaga, Spain, pp. 2229-2241, June 1988.

16. Gardner, J.D., Townend, I.H. and Fleming, C.A. The Design of a Slotted Vertical Screen Breakwater, Proceedings of the 20th Coastal Engineering Conference, Taipei, Taiwan, pp. 1881-1893, November 1986.

17. Weckman, A.M., Bigham, G.N. and Dixon, R.O. Reflection Characteristics of a Wave-Absorbing Pier, Proceedings ASCE Specialty Conference Coastal Structures '83, Arlington, Virginia, pp. 953-960, March 1983.

18. Sollitt, C.K., Lee, C.P., McDougal, W.G. and Perry, T.J. Mechanically Coupled Buoyant Flaps: Theory and Experiment, Proceedings of the 20th Coastal Engineering Conference, Taipei, Taiwan, pp. 2445-2459, November 1986.

19. Agerton, D.J., Savage, G.H. and Stotz, K.C. Design, Analysis and Field Test of a Dynamic Floating Breakwater, Proceedings of the 15th Coastal Engineering Conference, Honolulu, Hawaii, pp. 2792-2809, July 1976.

20. Uwatoko, T., Ijima, T., Ushifusa, Y. and Kajima, H. Wave Interception by Sea-Balloon Breakwater, Proceedings of the 20th Coastal Engineering Conference, Taipei, Taiwan, pp. 2353-2367, November 1986.

21. Jamieson, W.W. and Mansard, E.P.D. An Efficient Upright Absorber, Proceedings ASCE Speciality Conference on Coastal Hydrodynamics, Newark, Delaware, pp. 124-139, June/July 1987.

22. Bishop, C.T., DeYoung, B., Harms, V.W. and Ross, N.W. Guidelines for the Effective Use of Floating Tire Breakwaters, Information Bulletin 197, A Cornell Cooperative Extension Publication, Cornell University, Ithaca NY, 20 pp., 1983.

23. Bishop, C.T., Broderick, L.L. and Davidson, D.D. (Ed.). Proceedings of the Floating Tire Breakwater Workshop 8-9 November 1984, U.S. Army Corps of Engineers, Technical Report CERC-85-9, 131 pp., November 1985.

24. Crookshank, N.L. Experiment Control and Data Acquisition Systems at the Hydraulics Laboratory of the National Research Council Canada, Proceedings IAHR Workshop on Instrumentation for Hydraulics Laboratories, Burlington, Ontario, pp. 309-323, August 1989.

25. Miles, M.D. The GEDAP Data Analysis Software Package, National Research Council of Canada, Hydraulics Laboratory Technical Report TR-HY-030, August 1990.

26. Miles, M.D. and Funke, E.R. The GEDAP Software Package for Hydraulics Laboratory Data Analysis, Proceedings IAHR Workshop on Instrumentation for Hydraulics Laboratories, Burlington, Ontario, pp. 325-339, August 1989.

27. Mansard, E.P.D. and Funke, E.R. The Measurement of Incident and Reflected Spectra Using Least Squares Method, Proceedings of the 17th Coastal Engineering Conference, Sydney, Australia, pp. 154-172, 1980.

28. Mansard, E.P.D., Sand, S.E. and Funke, E.R. Reflection Analysis of Non-Linear Regular Waves, National Research Council of Canada, Hydraulics Laboratory Technical Report TR-HY-011, 1985.

29. Isaacson, M. Measurement of Regular Wave Reflection, ASCE Journal of Waterway, Port, Coastal and Ocean Engineering, Vol. 117, No. 6, pp. 553-569, November/December 1991.

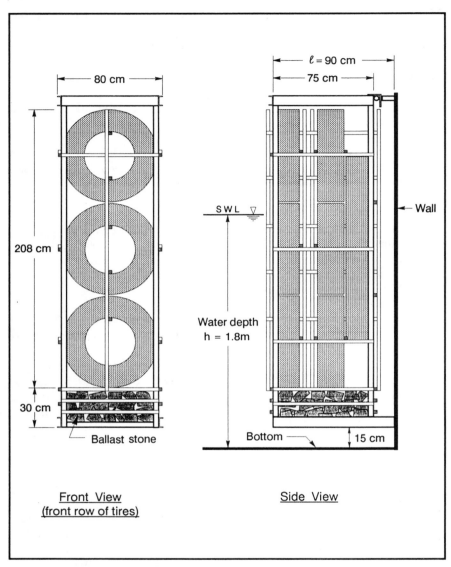

Figure 1. Tire wave absorber.

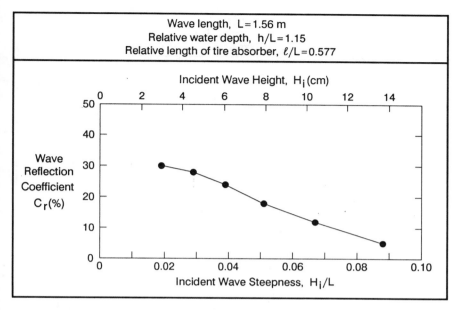

Figure 2. Wave reflection results (wave period, T = 1.00 s).

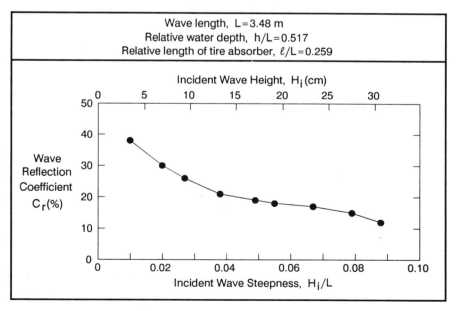

Figure 3. Wave reflection results (wave period, T = 1.50 s).

G

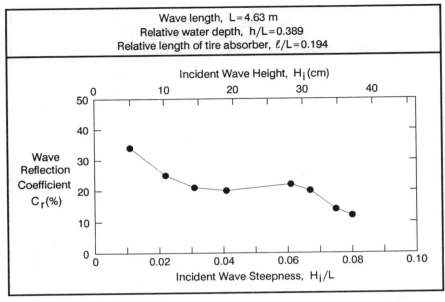

Figure 4. Wave reflection results (wave period, T = 1.74 s).

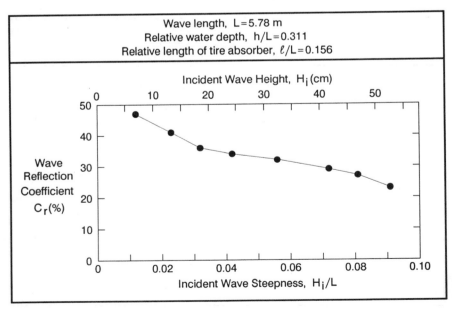

Figure 5. Wave reflection results (wave period, T = 1.96 s).

Figure 6. Fluid turbulence at tire wave absorber.

Figure 7. Tire wave absorber being lowered into position.

Figure 8. Field installation of tire wave absorber module.

Wave Propagation Over Sloping Bottom and Submerged Breakwaters

T. Sabuncu, Ö. Gören

Istanbul Technical University, Department of Naval Architecture, 80626 Ayazaga, Istanbul, Turkey

ABSTRACT

An approximate method is presented for the prediction of wave propagation over impermeable submerged breakwaters as well as over sloping sea bottom. A monochromatic small amplitude 2-D gravity wave of frequency ω propagating on the surface of an irrotational and inviscid fluid is considered. Submerged breakwaters or sloping sea bottom is approximately represented by a stepwise geometry so that the fluid region over those areas are devided into rectangular subregions extending from the corresponding bottom to the surface of the fluid. In each region the potential solutions are expanded into series of eigen functions satisfying the Laplace equation and boundary conditions associated with the corresponding sea bottom and the free surface. These solutions are then matched on the common vertical boundaries of the neighbouring subregions such that the continuity of mass flux and pressure are secured throughout the fluid. In order to reach the limiting solution the above analysis is successively repeated by "interval halving". This procedure is terminated when the resultant numerical error is made smaller than a permissable value. Based on the present analysis some examples of sloping underwater geometry and submerged breakwaters are examined with regard to their wave transmission and reflection characteristics.

INTRODUCTION

Waves propagating over various two dimensional underwater geometries have been widely studied. Wave reflection from a sloping beach and wave transmission over submerged breakwaters are said to be the important practical aspects of this general problem. There are different approaches to the problem of wave propagation over variable depth ; the earliest solutions of the linear reflection problem are found in [1], [2], [3], and followed by other solutions of [4], [5], which are all purely analytical. On the other hand, one can cite the numerical studies on submerged breakwaters such as [6] and [7].

The difficulty of this boundary value problem (among other difficulties of modeling breaking waves, dissipation and other physical phenomenon) arises due to the uniform or nonuniform slopes of underwater geometry. In order to handle this problem an approximate method, which employs both analytical and numerical methods, is developed for the prediction of wave propagation over impermeable submerged breakwaters as well as over sloping sea bottom. In the present method, the sloping sea bottom is approximately represented by a stepwise geometry so that the fluid region over those areas are devided into rectangular subregions. This idea described herein is parallel to that of [8]. Moreover, the present method seeks the resulting solution by making successive ' 'interval halvings' ' until the desired convergence is attained. Although the method is based on successive approximations, it is simple and acceptable for engineering purposes. As examples of numerical study, special geometries of sloping sea bottom and submerged breakwaters are considered and numerical results associated with reflection and transmission coefficients are given. The energy balance of the wave system is also checked and found to confirm the numerical computations.

THE METHOD OF SOLUTION

A monochromatic small amplitude 2-D gravity wave of frequency ω propagating on the surface of an irrotational and inviscid fluid is considered. A sample of underwater geometry is shown in Figure 1 with its two dimensional coordinate system (x,z) fixed at the still water plane. The other geometry of a submerged breakwater which is also studied in this paper is given within the Figure 3. The x coordinate is taken to be positive in the direction of wave propagation. The water depth below the still water level is taken as h_0 and the local slope angle of the underwater obstacle can be determined by the distance b and the depth difference $h_0 - h_N$ in the present case. According to the method given here the slope of the inclined surface is not necessarily constant.

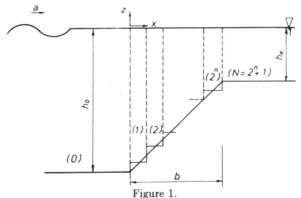

Figure 1.

Assuming the flow to be inviscid with irrotational motion, the fluid motion can be described in terms of the velocity potential $\Phi(x, x, t)$ satisfying Laplace equation of the form

$$\frac{\partial^2 \Phi}{\partial x^2} + \frac{\partial^2 \Phi}{\partial z^2} = 0 \tag{1}$$

Then one may complete the boundary value problem by stating the following conditions:

$$\left(\frac{\partial^2}{\partial t^2} + g\frac{\partial}{\partial z}\right)\Phi = 0 \qquad \text{on} \qquad z = 0 \tag{2}$$

$$\frac{\partial \Phi}{\partial n} = 0 \quad \text{on the fluid bottom} \tag{3}$$

and the radiation condition at $x = \pm\infty$ which implies the waves propagating in landward and seaward directions. The vector $\mathbf{n}$ is the normal vector pointing to the fluid bottom.

The incident wave potential can be written as

$$\Phi_i = -\mathrm{Re}\,\frac{ag}{\omega}\frac{\cosh\,k_{0,0}(z+h_0)}{\cosh\,k_{0,0}h_0}\,\exp\,ik_{0,0}x\,\exp-i\omega t \tag{4}$$

Here, i is the imaginary unit, t the time variable, g the gravitational acceleration and a is the amplitude of the incident wave. $k_{0,0}$ is the shallow water wave number of the region with uniform depth of h_0. Accordingly the velocity potential is written in the form of

$$\Phi(x,z,t) = \mathrm{Re}\,\frac{ag}{\omega}\varphi(x,z)\exp-i\omega t \tag{5}$$

where $\varphi(x,z)$ is the spatial part.

In the present method, the fluid region is devided into subregions in which the velocity potential is expanded into series of eigen functions satisfying Laplace equation and related boundary conditions. The fluid region over the inclined bottom surface is treated as a special case. This inclined surface is approximated by a stepwise geometry, so that it is possible to work with rectangular subregions in which the eigen function expansions are easily utilized.

Beginning with the region which has a uniform depth of h_0, the spatial nondimensional velocity potential is written as

$$\varphi_0(x,z) = -\frac{\cosh\,k_{0,0}(z+h_0)}{\cosh\,k_{0,0}h_0}\,\exp\,[ik_{0,0}x]+$$

$$A_{0,0}\exp\,[-ik_{0,0}x]Z_{0,0}(z) + \sum_{m=1}^{\infty}A_{0,m}\exp\,k_{0,m}xZ_{0,m}(z) \tag{6}$$

The second term with the remaining terms represents the reflected waves. $A_{0,m}$ is the complex unknown coefficient. Here the wave numbers $k_{j,m}$ and ω are related through the equations

$$k_{j,0}\tanh(k_{j,0}h_j) = \frac{\omega^2}{g}\quad(m=0)$$

$$k_{j,m}\tan(k_{j,m}h_j) = -\frac{\omega^2}{g}\quad(m\geq 1) \tag{7}$$

where j denotes the j th region with uniform depth of h_j. The eigen functions are then given by

$$Z_{j,m}(z) = N_{j,0}^{-1/2}\cosh\,k_{j,0}(z+h_j)\quad(m=0)$$

$$Z_{j,m}(z) = N_{j,m}^{-1/2}\cos\,k_{j,m}(z+h_j)\quad(m\geq 1) \tag{8}$$

with normalization factors ;

$$N_{j,0} = \frac{1}{2}\left[1+\frac{\sinh\,(2k_{j,0}h_j)}{2k_{j,0}h_j}\right]\quad(m=0)$$

$$N_{j,m} = \frac{1}{2}\left[1+\frac{\sin\,(2k_{j,m}h_j)}{2k_{j,m}h_j}\right]\quad(m\geq 1) \tag{9}$$

Utilizing the orthogonality properties of eigen functions $Z_{j,m}(z)$, valid in the subregion bounded by $-h_j \leq z \leq 0$, gives

$$\int_{-h_0}^{0} \frac{\partial \varphi(0,z)}{\partial x} Z_{0,0}(z)\, dz = -k_{0,0}h_0 A_{0,0} - ik_{0,0}h_0 \frac{N_{0,0}^{1/2}}{\cosh k_{0,0}h_0}$$

$$\int_{-h_0}^{0} \frac{\partial \varphi(0,z)}{\partial x} Z_{o,m}(z)\, dz = k_{0,m}h_0 A_{0,m} \qquad (10)$$

In a similar manner, the eigen function expansion of the velocity potential for the region nearest the land bounded by $-h_N \leq z \leq 0$ can be given as

$$\varphi_N(x,z) = \alpha_{N,0} \exp\left[ik_{N,0}x\right]Z_{N,0}(z) + \sum_{m=1}^{\infty} \alpha_{N,m} \exp\left[-k_{N,m}(x-b)\right]Z_{N,m}(z) \qquad (11)$$

which obviously represents transmitted waves with $\alpha_{N,m}$ complex unknown coefficients, which are expressed by ;

$$\alpha_{N,0} \exp ik_{N,0}b = \frac{1}{h_N} \int_{-h_N}^{0} \varphi_N(b,z)Z_{N,0}(z)\, dz, \quad (m=0)$$

$$\alpha_{N,m} = \frac{1}{h_N} \int_{-h_N}^{0} \varphi_N(b,z)Z_{N,m}(z)\, dz, \quad (m \geq 1) \qquad (12)$$

As mentioned before, the inclined sea bottom is approximately represented by a stepwise geometry so that the fluid region over the sloping bottom is devided into recrangular subregions extending from the corresponding bottom to the surface of the fluid. The approximation is begun with $n = 0$ implying that there is one $(2^n = 1)$ rectangular region with a base of $\Delta x = b$ depth of $h_1 = |\frac{(h_0 - h_N)}{2} - h_0|$. As a consequence of the zeroth approximation, the inclined surface is roughly approximated just by a step above which a rectangular subregion is defined by $0 \leq x \leq b$ and $-h_1 \leq z \leq 0$. In this subregion the spatial velocity potential can be expanded in a similar way as given by [9] ;

$$\varphi_1(x,z) = \alpha_{1,0} \exp\left[ik_{1,0}x\right]Z_{1,0}(z) + \sum_{m=1}^{\infty} \alpha_{1,m} \exp\left[-k_{1,m}x\right]Z_{1,m}(z) +$$

$$+ A_{1,0} \exp\left[-ik_{1,0}x\right]Z_{1,0}(z) + \sum_{m=1}^{\infty} A_{1,m} \exp\left[k_{1,m}(x-\Delta x)\right]Z_{1,m}(z) \qquad (13)$$

in which the first series of terms represent transmitted waves and the last series of terms represent reflected waves within $0 \leq x \leq \Delta x$. Hence, again by using orthogonality properties of $Z_{1,m}(z)$

$$\frac{1}{h_1} \int_{-h_1}^{0} \varphi_1(0,z)Z_{1,0}(z)\, dz = \alpha_{1,0} + A_{1,0} \quad ,(m=0)$$

$$\frac{1}{h_1} \int_{-h_1}^{0} \varphi_1(0,z)Z_{1,m}(z)\, dz = \alpha_{1,m} + A_{1,m} \exp -k_{1,m}\Delta x \quad ,(m \geq 1) \qquad (14)$$

and

$$\frac{1}{h_1} \int_{-h_1}^{0} \frac{\partial \varphi_1(0,z)}{\partial x} Z_{1,0}(z)\, dz = ik_{1,0}h_1 \exp\left[ik_{1,0}\Delta x\right]\alpha_{1,0} -$$

$$- ik_{1,0}h_1 \exp -ik_{1,0}\Delta x A_{1,0}$$

$$\frac{1}{h_1} \int_{-h_1}^{0} \frac{\partial \varphi_1(0,z)}{\partial x} Z_{1,m}(z)\, dz = -k_{1,m}h_1 \exp\left[-k_{1,m}\Delta x\right]\alpha_{1,m} + k_{1,m}h_1 A_{1,m} \qquad (15)$$

are obtained by the use of equation (13).

Now by using matching conditions, which assure the continuity of the mass flux and the pressure at the common boundaries,

$$\varphi_0(0, z) = \varphi_1(0, z) \quad ; -h_1 \leq z \leq 0 \tag{16}$$

$$\frac{\partial \varphi_0(0, z)}{\partial x} = \frac{\partial \varphi_1(0, z)}{\partial x} \quad ; -h_1 \leq z \leq 0 \tag{17}$$

$$\varphi_1(\Delta x, z) = \varphi_N(\Delta x, z) \quad ; -h_N \leq z \leq 0 \tag{18}$$

$$\frac{\partial \varphi_1(\Delta x, z)}{\partial x} = \frac{\partial \varphi_N(\Delta x, z)}{\partial x} \quad ; -h_N \leq z \leq 0 \tag{19}$$

in potential solutions of equations (14), (10), (12) and (15), evaluating the necessary integrals one can arrive at the following simultaneous equations ;

$$ik_{0,0}h_0 A_{0,0} + ik_{1,0}h_1 I_{0,0;1,0}\alpha_{1,0} - \sum_{k=1}^{\infty} k_{1,k}h_1 I_{0,0;1,k}\alpha_{1,k}$$

$$- ik_{1,0}h_1 I_{0,0;1,0}A_{1,0} + \sum_{k=1}^{\infty} k_{1,k}h_1 \exp\left[-k_{1,k}\Delta x\right]I_{0,0;1,k}A_{1,k} =$$

$$= -ik_{0,0}h_0 \frac{N_{0,0}^{1/2}}{\cosh k_{0,0}h_0} \quad ,$$

$$- k_{0,m}h_0 A_{0,m} + ik_{1,0}h_1 I_{0,m;1,0}\alpha_{1,0} - \sum_{k=1}^{\infty} k_{1,k}h_1 I_{0,m;1,k}\alpha_{1,k} -$$

$$- ik_{1,0}h_1 I_{0,m;1,0}A_{1,0} + \sum_{k=1}^{\infty} k_{1,k}h_1 \exp\left[-k_{1,k}\Delta x\right]I_{0,m;1,k}A_{1,k} = 0 \tag{20}$$

$$I_{0,0;1,0}A_{0,0} + \sum_{k=1}^{\infty} I_{0,k;1,0}A_{0,k} - \alpha_{1,0} - A_{1,0} =$$

$$= \frac{N_{0,0}^{1/2}}{\cosh k_{0,0}h_0}I_{0,0;1,0}, (m = 0)$$

$$I_{0,0;1,m}A_{0,0} + \sum_{k=1}^{\infty} I_{o,k;1,m}A_{0,k} - \alpha_{1,m} - \exp\left[-k_{1,m}\Delta x\right]A_{1,m} =$$

$$= \frac{N_{0,0}^{1/2}}{\cosh k_{0,0}h_0}I_{0,0;1,m}, (m \geq 1) \tag{21}$$

By introducing n, denoting the n th approximation, for the sake of generality the last set of equations become

$$- ik_{2^n,0}h_{2^n} \exp\left[i2^n k_{2^n,0}\Delta x\right]\alpha_{2^n,0} + ik_{2^n,0}h_{2^n} \exp\left[-i2^n k_{2^n,0}\Delta x\right]A_{2^n,0} +$$

$$+ ik_{N,0}h_N \exp\left[i2^n k_{N,0}\Delta x\right]I_{2^n,0;N,0}\alpha_{N,0} - \sum_{k=1}^{\infty} k_{N,k}h_N I_{2^n,0;N,k}\alpha_{N,k} = 0,$$

$$k_{2^n,m}h_{2^n} \exp\left[-k_{2^n,m}\Delta x\right]\alpha_{2^n,0} - k_{2^n,m}h_{2^n}A_{2^n,m} + ik_{N,0}h_N \exp\left[i2^n k_{N,0}\Delta x\right] \cdot$$

$$\cdot I_{2^n,m;N,0}\alpha_{N,0} - \sum_{k=1}^{\infty} k_{N,k}h_N I_{2^n,m;N,k}\alpha_{N,k} = 0 \tag{22}$$

and

$$\exp\left[i2^n k2^n, 0\Delta x\right]I_{2^n,0;N,0}\alpha_{2^n,0} + \sum_{k=1}^{\infty}\exp\left[-k_{2^n,k}\Delta x\right]I_{2^n,k;N,0}\alpha_{2^n,k} +$$

$$\exp\left[-i2^n k_{2^n,0}\Delta x\right]I_{2^n,0;N,0}A_{2^n,0} + \sum_{k=1}^{\infty}I_{2^n,k;N,0}A_{2^n,k} - \exp\left[i2^n k_{N,0}\Delta x\right]\alpha_{N,0} = 0$$

$$\exp\left[i2^n k_{2^n,0}\Delta x\right]I_{2^n,0;N,m}\alpha_{2^n,0} + \sum_{k=1}^{\infty}\exp\left[-k_{2^n,k}\Delta x\right]I_{2^n,k;N,m}\alpha_{2^n,k} +$$

$$\exp\left[i2^n k_{2^n,0}\Delta x\right]I_{2^n,0;N,m}\alpha_{2^n,0} + \sum_{k=1}^{\infty}\exp\left[-k_{2^n,k}\Delta x\right]I_{2^n,k;N,m}\alpha_{2^n,k} +$$

$$\exp\left[-i2^n k_{2^n,0}\Delta x\right]I_{2^n,0;N,m}A_{2^n,0} + \sum_{k=1}^{\infty}I_{2^n,k;N,m}A_{2^n,k}\alpha_{N,m} = 0 \tag{23}$$

where

$$I_{i,k;j,l} = \frac{1}{h_j}\int_{-h_j}^{0}Z_{i,k}(z)Z_{j,l}(z)\,dz \tag{24}$$

Note that $\Delta x = b$ and $n = 2^n + 1$, where $n = 0$ at this stage. The system of equations (20), (21), (22) and (23) is the result of the zeroth approximation of the sloping sea bottom. In order to reach the limiting solution of this approximation procedure, the above analysis is successively repeated by halving the interval. The procedure beginning with $n = 0$ is then continued by taking $n = 1, 2, 3, \cdots$. Normally the higher n is taken, the better resolution or approximaton of the sloping sea bottom is obtained. Thus, there are 2^n subregions formed over the sloping underwater surface for a specific value of n. The subregions over the inclined bottom surface, which has a constant slope in the present study, are equally spaced with $\Delta x = b/2^n$ having water depths of

$$h_j = \left| \frac{h_0 - h_N}{b} \frac{(2j-1)\Delta x}{2} - h_0 \right|.$$

Accordingly, in the j th subregion over the sloping bottom, where $j = 1, 2, 3, \cdots, 2^n$, the spatial velocity potential can be given in the light of the previous paragraphs as ;

$$\varphi_j(x,z) = \alpha_{j,0}\exp\left[ik_{j,0}x\right]Z_{j,0}(z) + \sum_{m=1}^{\infty}\alpha_{j,m}\exp\left[-k_{j,m}(x-(j-1)\Delta x)\right]Z_{j,m}(z) +$$

$$+ A_{j,0}\exp\left[-ik_{j,0}x\right]Z_{j,0}(z) + \sum_{m=1}^{\infty}A_{j,m}\exp\left[k_{j,m}(x-j\Delta x)\right]Z_{j,m}(z) \tag{25}$$

Again the solutions for each subregion $j = 1, 2, 3, \cdots, 2^n$ are then matched on the common vertical boundaries of the neighbouring subregions such that continuity of the mass flux and the pressure are secured throughout the fluid. Thus the resulting system of equations at the n th approximation for the case of sloping bottom appears to be ;

(Equation(20))

(Equation(21))

....

....

$$- ik_{j,0}h_j\exp\left[ik_{j,0}j\Delta x\right]\alpha_{j,0} + ik_{j,0}h_j\exp\left[-ik_{j,0}j\Delta x\right]A_{j,0} + ik_{j+1,0}h_{j+1}.$$

$$\cdot \exp\left[ik_{j,0}j\Delta x\right]I_{j,0;j+1,0}\alpha_{j+1,0} - \sum_{k=1}^{\infty}k_{j+1,k}h_{j+1}I_{j,0;j+1,k}\alpha_{j+1,k}-$$

$$- ik_{j+1,0}h_{j+1}\exp\left[-ik_{j+1,0}j\Delta x\right]I_{j,0;j+1,0}A_{j+1,0} + \sum_{k=1}^{\infty}k_{j+1,k}h_{j+1}\cdot$$

$$\cdot \exp\left[-k_{j+1,k}\Delta x\right]I_{j,0;j+1,k}A_{j+1,k} = 0,$$

$$k_{j,m}h_j\exp\left[-k_{j,m}\Delta x\right]\alpha_{j,0} - k_{j,m}h_j\,Aj,\,m + ik_{j+1,0}h_{j+1}\exp\left[ik_{j+1,0}j\Delta x\right]\cdot$$

$$\cdot I_{j,m;j+1,0}\alpha_{j+1,0} - \sum_{k=1}^{\infty}k_{j+1,k}h_{j+1}I_{j,m;j+1,k}\alpha_{j+1,k} - ik_{j+1,0}\cdot$$

$$h_{j+1}\exp\left[-ik_{j+1,0}j\Delta x\right]I_{j,m;j+1,0}A_{j+1,0} + \sum_{k=1}^{\infty}k_{j+1,k}h_{j+1}\cdot$$

$$\cdot \exp\left[-k_{j+1,k}\Delta x\right]I_{j,m;j+1,k}A_{j+1,k} = 0,$$

$$\exp\left[ik_{j,0}j\Delta x\right]I_{j,0;j+1,0}\alpha_{j,0} + \sum_{k=1}^{\infty}\exp\left[-k_{j,k}\Delta x\right]I_{j,k;j+1,0}\alpha_{j,k}-$$

$$- \exp\left[-ik_{j,0}\Delta x\right]I_{j,0;j+1,0}A_{j,0} + \sum_{k=1}^{\infty}I_{j,k;j+1,0}A_{j,k} - \exp\left[-ik_{j,0}j\Delta x\right]\alpha_{j+1,0}-$$

$$- \exp\left[-ik_{j+1,0}j\Delta x\right]A_{j+1,0} = 0,$$

$$\exp\left[ik_{j,0}j\Delta x\right]I_{j,0;j+1,m}\alpha_{j,0} + \sum_{k=1}^{\infty}\exp\left[-k_{j,k}\Delta x\right]I_{j,k;j+1,m}\alpha_{j,k}+$$

$$+ \exp\left[ik_{j,0}j\Delta x\right]I_{j,0;j+1,m}A_{j,0} + \sum_{k=1}^{\infty}I_{j,k;j+1,m}A_{j,k} - \alpha_{j+1,m}-$$

$$- \exp\left[-k_{j+1,m}\Delta x\right]A_{j+1,m} = 0$$

....

....

(Equation (22))

(Equation (23))

$$(26)$$

This procedure of successive halving of the interval is terminated when the resulting numerical error in reflection and transmission coefficients is made smaller than ϵ which is taken 10^{-3} in our numerical study. The present method is not confined to handle only inclined underwater surfaces with constant slopes. The method is also capable of analysing the effect of a wide class of underwater geometries by a proper definition of Δx and h_j's which are the basic parameters of the approximation. The numerical aspects of the solution of the system (26) are discussed later in this paper.

The system (26) gives the solution for a sloping bottom. In case of a submerged breakwater such as shown in Figure 3, the system (26) remains to be the same except the last two equations (22) and (23), and two more additional equations are required to meet the continuity conditions at $x = b + t$. Therefore the dimensions of the resulting complex matrix in this case are $(2^{n+1} + 4)$ by $(2^{n+1} + 4)$, instead of $(2^{n+1} + 2)$ by $(2^{n+1} + 2)$ in case of a sloping bottom.

REFLECTION AND TRANSMISSION COEFFICIENTS

Although the local reflected and transmitted waves take place in every subregion, the amplitude of transmitted waves and reflected waves in question are to be investigated in the far field. Therefore transmitted waves can be calculated by using the potential solution of equation (11) and the reflected waves by using equation (6).

Since the free surface elevation is given as ;

$$\zeta = -\frac{1}{g}\frac{\partial \Phi}{\partial t} \quad ; \text{at } z=0 \tag{27}$$

in the linear theory, one can obtain the transmitted waves, by considering equation (27) and taking the limit while $x \to \infty$ as

$$\zeta_T = a\sqrt{\alpha_{N,0}\alpha_{N,0}^{\star}}Z_{N,0}(0)\sin(k_{N,0}x - \omega t + \sigma) \tag{28}$$

where $\sigma = \arctan(\text{Im}\alpha_{N,0}/\text{Re}\alpha_{N,0}) + \pi$ is the phase angle and $(\star)$ shows complex conjugate. Thus the transmission coefficient is

$$\frac{a_T}{a} = \sqrt{\alpha_{N,0}\alpha_{N,0}^{\star}}Z_{N,0}(0) \tag{29}$$

One can also obtain the reflected waves when x tends to $-\infty$ as

$$\zeta_R = a\sqrt{A_{0,0}A_{0,0}^{\star}}Z_{0,0}(0)\sin(k_{0,0}x + \omega t + \tau) \tag{30}$$

Here the phase angle τ is given as

$$\tau = \arctan\frac{\text{Re}A_{0,0}}{\text{Im}A_{0,0}} - \frac{\pi}{2},$$

and $a\sqrt{A_{0,0}A_{0,0}^{\star}}Z_{0,0}(0)$ is the amplitude of reflected waves.

Wave reflection and transmission over submerged obstacles are generally analyzed considering the wave energy balance. Since the depth is variable in our case, it would be better to use the rate of energy flux as the sum of kinetic and potential energies per unit width :

$$\overline{E} = \frac{1}{2}\rho g a^2 c_g$$

Here $\overline{E}$ denotes the time-averaged rate of energy flux based on linear theory and c_g is the group velocity and ρ is the specific mass of the fluid. By taking two control surfaces: one is at $-\infty$ the other is at $+\infty$, the energy balance equation gives

$$(\frac{a_T}{a})^2\frac{c_g(+\infty)}{c_g(-\infty)} + (\frac{a_R}{a})^2 = 1 \tag{31}$$

During the numerical studies the transmission and reflection coefficients are analyzed taking the energy balance equation into consideration, and found to confirm the numerical computations.

NUMERICAL STUDY AND RESULTS

A computer programme was written which enables to define the elements of the linear system of equations (26) for successive n values, e.g. n=0,1,2,... . The coefficient matrix as may be seen from equation (26), is a kind of band matrix having a stepwise character. The dimensions of the matrix is determined by the number of terms used in the eigen function expansions of the velocity potential and the number n which denotes the order of approximation. Before obtaining the ultimate numerical results for reflection and transmission coefficients, some numerical tests were done in order to find the truncation limits in the series expansions. The tests showed that KMAX=4, as a truncation limit in the series expansions, is adequate for a sensitivity of 10^{-4} in transmission and reflection coefficients for constant n. Thus the number of rows or columns in the coefficient matrix associated with the configurations depicted in Figure 1 and Figure 3 are obtained as $(2^{n+1} + 2) \cdot \text{KMAX}$ and $(2^{n+1} + 4) \cdot \text{KMAX}$, respectively.

The most of the computer time is devoted to the solution of matrix equations. Because of some properties of the coefficient matrix, Gaussian elimination with maximum pivot strategy is first preferred and this in turn significantly increases the computer time. Then LU-Decomposition is decided which dramatically reduces computer time as compared to that of Gaussian elimination with maximum pivot strategy. Jacobi's and Gauss-Seidel's iterative methods are not effective due to convergence problems.

On the other hand, the convergence of the approximating procedure is shown numerically in Table 1.

Table 1.

n	a_T/a	a_R/a
0	1.045587	0.039104
1	1.043777	0.070530
2	1.042759	0.083106
3	1.042404	0.086914
4	1.042298	0.087869
5	1.042339	0.087378

$h_N/h_0 = 0.5$, $b/h_0 = 1.0$, $\text{KMAX} = 4$, $(\omega^2/g)h_0 = 1.0$

The results in Table 1 show that the 4 th order approximation gives results correct to the 3 significant digits of transmission and reflection coefficients , and this can be regarded as adequate for engineering works.

Having been satisfied with the results of the test data, the numerical results of the wave transmission and reflection coefficients for a sample configuration are given in Figure 2. In Figure 3, the transmission and reflection coefficients are given for a submerged breakwater with two different relative depths of crest submergence (h_N/h_0).

Unfortunately, it could not be possible to gain access to the necessary experimental data regarding the present configuration, therefore it was not possible to compare the results with those of experimental ones. But it can be said that the results are consistent with the energy balance equation within the frame of the linear theory.

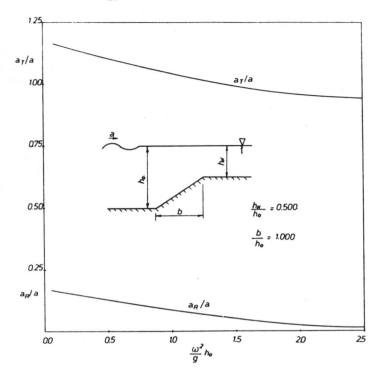

Figure 2

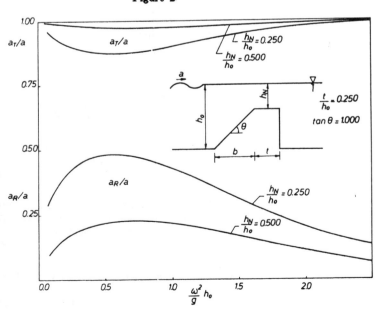

Figure 3

CONCLUDING REMARKS

An approximate method which employs both analytical and numerical methods is presented to examine the wave reflection and transmission over sloping sea bottom as well as over submerged breakwaters. The present method based on successive approximations reaches the solution by a numerical limiting process. The convergence, depending on the numerical error which is acceptable for engineering works, is said to be rapid enough to validate the method. The method presented can also be used as a basis for calculating the reflection and transmission characteristics of a wide class of underwater geometries. On the other hand, a verification study is required to demonstrate the agreement of the theory with experiment.

REFERENCES

1. Lamb, H., *Hydrodynamics*, Cambridge University Press, Cambridge, 1932.

2. Stoker, J.J., *Water Waves*, Interscience Publishers, NewYork, 1957.

3. Keller, J.B., Surface waves on water of nonuniform depth, J. Fluid Mech., Vol.4, pp.607-614, 1958.

4. Tuck, E.O., *Waves on Water of Variable Depth*, Lecture Notes in Physics, Ed. by D.G.Provis and R.Radok, Vol. 64, Springer-Verlag, pp.9-20, 1976.

5. Miles, J., Wave reflection from a gently sloping beach, J. Fluid Mech., Vol. 214, pp.59-66, 1990.

6. Kobayashi, N. and Wurtanjo, A., Wave reflection and runup on smooth slopes, Proc. Coastal Hydrodynamics, ASCE, pp.548-563, 1987.

7. Kobayashi, N. and Wurtanjo, A., Wave transmission over submerged breakwaters, J. Wtrway, Port, Coast. and Ocean Eng'ng, Vol. 115, No.5, pp.662-680, 1989.

8. Kokkinowrachos, K., Mavrakos, S. and Asorakos, S., Behaviour of vertical bodies of revolution in waves, Ocean Eng'ng, Vol.13, No. 6, pp.505-538, 1986.

9. Massel, S.R., Harmonic generation by waves propagating over a submerged step, Coastal Eng'ng, Vol.7, pp.357-380, 1983.

Wave Diffraction Patterns Around Isolated Offshore Structures

M. Huygens, R. Verhoeven
Hydraulics Laboratory, University of Ghent,
Sint-Pietersnieuwstraat 41-9000 Ghent, Belgium

ABSTRACT

The described numerical model "DIFFRAK" computes the diffraction-reflection deformation around isolated constructions by using the boundary element method to solve the two-dimensional Helmholtz-equation. The discretisation of only the boundaries of the breakwater configuration, by using this numerical technique, is a great advantage for this kind of practical problem. Starting from a reference example to prove the accuracy of the program, two specific combinations are worked out in detail.

INTRODUCTION

Coastal engineering encompasses a variety of problems of practical importance. The knowledge of the wave climate in a specified region near the coastal zone is of great importance to some projects. The sea waves and their enormous quantities of energy have an important influence on the nearshore morphology. A sound protection of inshore beaches or coastlines against attacking waves can be realised by one or a combination of offshore breakwaters. Their dimension and their implementation determine the wave climate around this breakwater configuration and from there on their effectiveness of shore protection. Another interesting application in this domain is the diffraction pattern around a combination of circular piles. This configuration can be seen as the foundation for any offshore platform, so the knowledge of the wave climate can be very helpful in the design and maintenance of such constructions.

For practical applications, the data on wave characteristics are collected at discrete spots where the wave height, the wave period and the propagation direction are measured continuously. As these characteristics are stochastic variables, the wave data are examined statistically. Out of the total wave spectra one can extract a linear monochromatic wave field for design purpose. Usually the spots where wave data are collected do not coincide with the regions in which wave characteristics are needed. A transfer model can figure the propagation of waves from the "offshore" points to the "inshore" zones of interest. During its travel, wave appearance changes due to several phenomena like energy dissipation, wave breaking, refraction, shoaling, longshore currents, diffraction or reflection.

Using linear Airy wave theory one can model refraction, shoaling, diffraction and reflection. The 2D Mild-slope equation (Berkhoff [1]) describes the combined effects of these phenomena on time harmonic, monochromatic and linear surface water waves. The assumption of constant or infinite water depth reduces the Mild-slope equation to the 2D Helmholtz equation, that takes into account the effect of diffraction and reflection. Diffraction is the process in which wave energy is transported orthogonally to the wave propagation direction.

This paper deals with a numerical method to solve the 2D Helmholtz equation, using a boundary element method (BEM) with constant elements. From the calculated two dimensional wave potential one can figure several wave characteristics.

NUMERICAL BACK GROUND OF THIS MODEL

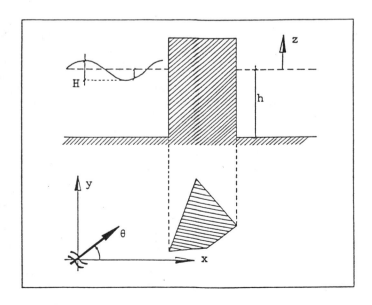

Figure 1

With reference to figure 1, the boundary-value problem defining the fluid motion may be set up as follows. The fluid is assumed to be incompressible and inviscid, while the flow is irrotational. Therefore, the fluid motion of the coastal water can be described by a complex velocity potential Φ (x,y,z,t) which satisfies the Laplace equation within the fluid domain. Considering only time harmonic Airy water waves the time parameter t can be removed from the formulation. Since the problem is linear, the complex three dimensional velocity potential is given as the superposition of a known incident wave potential and an unknown scattered wave potential. Assuming a small H/L ratio for deeper water or a small H/h ratio for shallow water, the free water surface condition is reformulated and the initial wave potentials are rewritten as a product of a two dimensional wave potential φ (x,y) and a known function of the water depth h and the vertical coördinate z.

$$\Phi(x,y,z,t) = \psi(x,y,z) \cdot e^{-iwt}$$

with $\psi(x,y,z) = D(h,z) \cdot \varphi(x,y)$

where * $D(h,z) = \dfrac{\cosh[k(h+z)]}{\cosh[kh]}$

* $\varphi(x,y) = \varphi_i + \varphi_s$

The two dimensional scattered wave potential φ_s satisfies the so-called Mild-Slope equation (Berkhoff [1]), which describes the combined effects of refraction, shoaling, diffraction and reflection around any marine construction :

$$\nabla(cc_g\nabla\varphi_s) + \omega^2 \frac{c_g}{c} \varphi_s = 0 \qquad (1)$$

where $c = \dfrac{\omega}{k}$ = wave phase velocity [m/s]
   (k = wave number)

   c_g = wave group velocity [m/s]

Since the numerical solution of this computer model only applies to problems where reflection and diffraction of linear, monochromatic and time harmonic water waves are caused by any vertical structure, the water depth h appears as a constant in the expression above. The general Mild Slope equation reduces to the two dimensional Helmholtz equation :

$$\nabla^2 \varphi_s + k^2 \varphi_s = 0 \qquad (2)$$

with the following boundary conditions :

$$\frac{\partial \varphi_s}{\partial n} = -\frac{\partial \varphi_i}{\partial n} \qquad (3)$$

$$\lim_{r \to \infty} \sqrt{r} \left[\frac{\partial}{\partial r} - \hat{\imath} k \right] \varphi_s = 0 \qquad (4)$$

(3)   is the fully reflection condition at the surface of the rigid boundary of the obstacle.

(4)   is the so-called Sommerfeld radiation condition which guarantees a proper decrease of the scattered wave potentials for large arguments i.e. φ_s will be stationary at infinity. It is a sufficient condition to ensure the mathematical uniqueness of the numerical solution (Mei Chang [2]).

Once these two dimensional scattered wave potentials φ_s are known, in superposition with the known incident wave potentials φ_i , the total three dimensional wave potential can be reconstructred. From there, all physical wave variables of interest can be calculated :

-   the wave diffraction coefficient Kdif as the ratio between incident and scattered wave height;

$$Kdif = \frac{|\varphi_s + \varphi_i|}{|\varphi_i|}$$

-   the instant water surface elevation η;

$$\eta = Re \left\{ \hat{\imath} \; \omega/g . \Psi . e^{-i\omega t} \right\}_{z=0}$$

-   the wave phase shift F at the free surface;

$$F = arctg \left\{ \frac{Re \; (\Psi)}{Im \; (\Psi)} \right\}_{z=0}$$

-   the linearised hydrodynamic pressure p.

$$p = -\rho g z + \omega \rho \; [Re(\Psi) \; sin\omega t - Im(\Psi) \; cos\omega t]$$

THE BOUNDARY ELEMENT METHOD AS A SOLUTION METHOD

The Boundary Element Method (BEM) is more than just a numerical solution technique. The traditional partial differential equation of a system is first transformed into a boundary integral equation, using Greene's second identity. Doing so, no additional assumptions or approximations are imposed on the formulation.. In this case the two dimensional Helmholtz equation governs the physical process. Introducing the Greene's function of the problem, applying Greene's second identity and integrating over the solution domain S, one obtains :

$$-\varphi_s \, (r') = \int_{\partial S} \left[\frac{\partial \varphi_s}{\partial n} \, G - \varphi_s \, \frac{\partial G}{\partial n} \right] ds \qquad (5)$$

So, the scattered wave potential in a field point r' is calculated as the total disturbance induced by two source distributions (with unit strength and caused by Greene's functions), each with a strength distribution (caused by the potential functions), all along the domain boundary ∂S (Brebbia [3]). The Greene's function suitable for the two dimensional Helmholtz equation and satisfying the Sommerfeld radiation boundary condition is given as :

$$G = - \frac{\hat{\imath}}{4} \, H_0^{(1)} \, (kR)$$

where $H_0^{(1)}$ is the Hankel function of the first kind and order zero.
To perform the integration numerically, the boundary is split up into H straight line segments ("Elements"). The unknown values of φ_s are taken in the middle points of these elements ("Knodes") and are assumed to hold a constant value over the element.

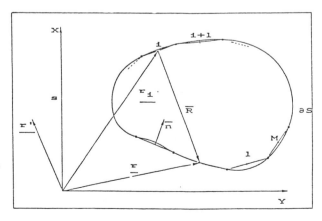

Figure 2 : Boundary Element Method

A set of M linear equations is built by placing the field point at the M different knodes on the boundary. Once the unknown scattered wave potentials in the M knodes on the boundary are calculated by solving the set of M equations, the scattered wave potential can be figured, as a line integral, in any point of the solution domain. The complete analytic formulation of this problem is worked out in [3] and [7].

The Boundary Element Method is an efficient solution method. First, because it reduces effectively the computational dimension of the problem by one without introducing any approximation : an initial two dimensional partial differential equation is reformulated to a one-dimensional line integral. Secondly, because infinite solution domains are treated easily. With the Finite Element or Finite Difference solution methods one has to discretize this complete region in gridpoints or elements while with the Boundary Element Method only the boundaries of the structures itself need to be divided in elements. So with isolated breakwaters; you only have to discretize the boundaries of these constructions, to get a proper idea about the wave potential in any point of interest. Since it makes no approximations to the governing equation, this solution method gives accurate results. In fact, the calculations in the surrounding domain satisfy exactly the Helmholtz equation. Approximations are only made at the breakwater construction itself, but they can be made as accurate as desired by taking more or modified boundary elements.

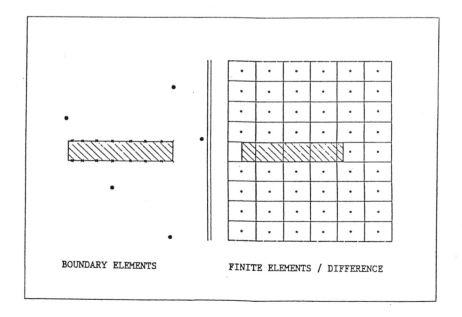

BOUNDARY ELEMENTS   FINITE ELEMENTS / DIFFERENCE

Figure 3 : Distinction between the boundary element method and finite element / difference method as solution technique

A BASIC REFERENCE EXAMPLES

In order to verify the mathematical model of two-dimensional wave diffraction by using boundary elements, an example has been designed for a wave deformation around a detached breakwater with a length of 100 m. Comparisons have been made between the analytic solutions given by Stassnic et al. (by using Mathieu functions) or by Goda (V.W. Harms [4] and Chongzhun & Gonghu [5]). It can be easily seen from figure 4-8 that the agreement with these results seems to be very satisfactory. The incident linear, monochromatic wave is characterized by the following parameters :

-   Incident Wave Height   H = 1.0 m
-   Incident Wave Length   L = 40.0 m
-   Constant Water Depth   h = 10.0 m
-   Incident Wave Direction θ = 90°-60°

In figure 4 the contour lines of equal wave diffraction coefficients Kdif around the offshore breakwater are shown for perpendicular incidence (θ = 90°). The wave phase evolution in the shadow zone of the detached breakwater is given in figure 5, while a three dimensional projection of the instant water surface elevation in the wave shelter area is seen in figure 6. The tendency of these numerical results reflects the physical process in which the wave energy distribution in the shadow zone is the superposition of the energy of the incident wave entering the shelter area diffractively from the two side tops of the breakwater. For an oblique incident wave type (θ = 60°) the agreement is again very good as shown in figure 7 (wave diffraction pattern) and figure 8 (wave phase shift).

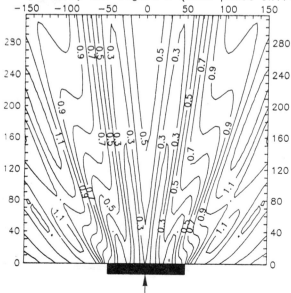

Figure 4 : Wave Diffraction Pattern   Perpendicular Incidence

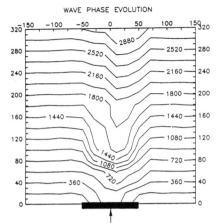

Figure 5 : Wave Phase Evolution   Perpendicular Incidence

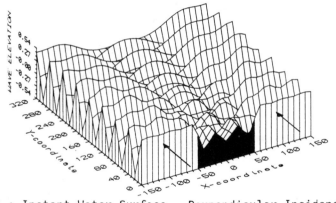

Figure 6 : Instant Water Surface   Perpendicular Incidence

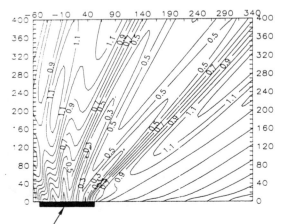

Figure 7 : Wave Diffraction Pattern   Oblique Incidence

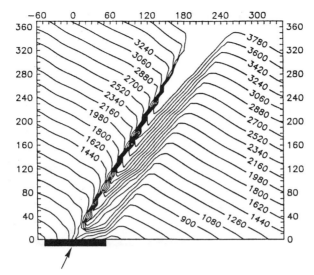

Figure 8 : Wave Phase Shift   Oblique Incidence

COMBINATION OF ISOLATED STRUCTURES

The first configuration deals with the diffraction scattering
behind a series of aligned offshore breakwaters. The example
that is worked out here computes the wave diffraction patterns
around 10 aligned detached breakwaters each with a finite
length of 25 m, placed at some discrete distance S from each
other. By this calculation an optimum design for a ·sound
beach protection or marina development that is situated in a
200 m x 100 m zone on the lee side of these aligned breakwaters, can be worked out in detail.
This structure is designed to stop the wave action between the
offshore breakwater and the beach, by sand deposition in the
shadow zone.
The incident wave form is given as :

-    Wave Height H = 1 m
-    Wave Length L = 50 m
-    Water Depth   = 11.31 m
-    Incident Wave Direction θ = 90°
     (perpendicular to the breakwaters)

By placing the 10 detached breakwaters at different interval
distances, this numerical model gives the engineer an optimum
design for an adequate protection of the shadow (working)
zone. The following figures show some general diffraction
patterns in the shadow zone (200 m x 100 m) for four discrete
interval distances S.

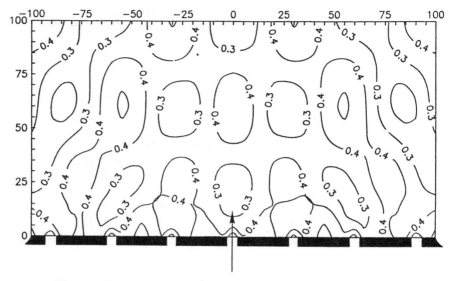

Figure 9: Wave Diffraction Pattern - S = 5.0 m

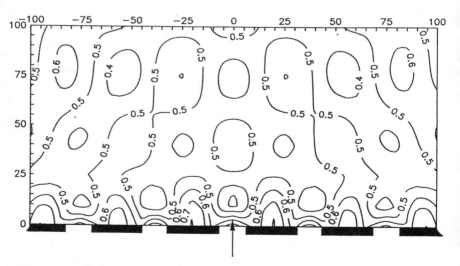

Figure 10: Wave Diffraction Pattern - S = 12.5 m

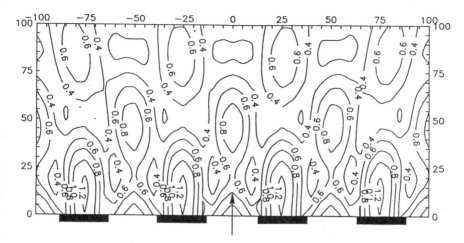

Figure 11: Wave Diffraction Pattern - S = 25.0 m

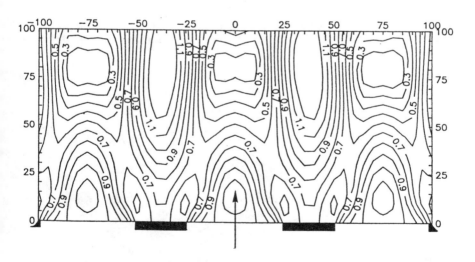

Figure 12: Wave Diffraction Pattern - S = 50.0 m

By comparing this calculated result with the engineering demands an optimum design of the placement of this offshore breakwaters can be worked out in detail. The variation
of the diffraction coefficient Kdif along the central line x =
0 m for each inter distance S is given in figure 13.

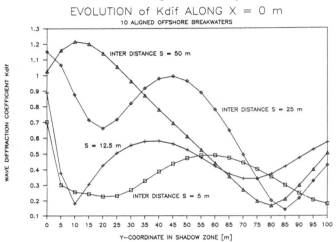

Figure 13: Kdif Evolution along Center Line

To get a clear 3D-view on the instant water surface elevation
over the gridzone behind the aligned breakwaters two representative figures are shown below.

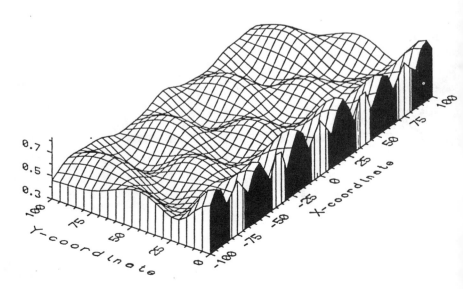

Figure 14 : Instant Water Profile - Inter Distance S = 12.5 m

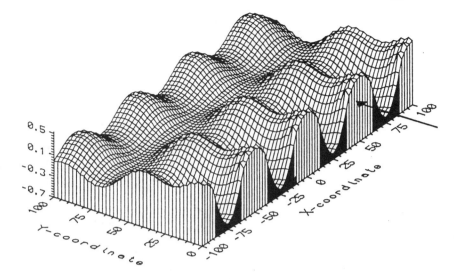

Figure 15 : Instant Water Profile - Inter Distance S = 25.0 m

The last example considers a four cylinder configuration. The four cylindric piles are placed in a square form a center distance S from each other as seen in figure 16. The configuration is characterized by the dimensionless parameter

BETA = $\dfrac{2 \times Pile\ Radius\ a}{Center\ Distance\ S}$ or by the center length T.

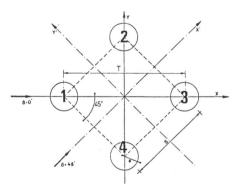

Figure 16: Four Cylinder Combination

The incident wave form is given by the following data :
- Incident Wave Height   H = 1 m
- Wave Number   k = 0.04
- Water Depth   h = 200 m
- Pile Radius   a = 25 m
- Incident Wave Direction θ = 0°-45°

A general wave diffraction pattern around the four cylinder configurations is generated for a center distance S = 100 m. In figure 17 a general view on the wave pattern is given for perpendicular (θ = 0°) incidence, while figure 18 shows a more detailed picture of the diffracted wave heights between the four cylindrical piles for an oblique incident wave (θ = 45°).

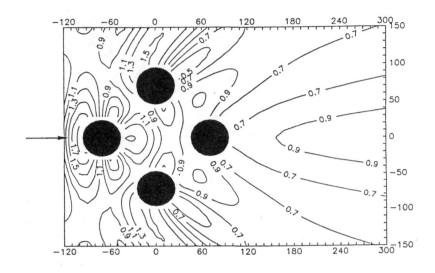

Figure 17

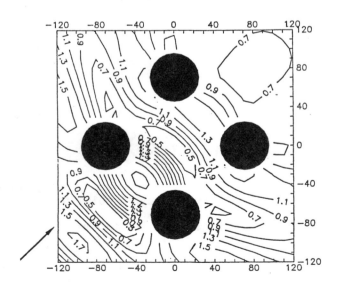

Figure 18

The relative position of the piles is of great influence to the diffraction and reflection behaviour of waves around such a cylinder combination. The longitudinal variation of the wave diffraction coefficient Kdif between pile I and III for $\theta = 0°$ is reconstructed in the following figures. The interaction between the cylindric piles modifies the evolution of Kdif by the multiple scattering between the pile constructions. Especially for the wave force calculations the presence of neighboring cylinders is quite important.

The effect of multiple scattering between the piles is shown in figure 19 and 20 by the evolution of the wave diffraction coefficient Kdif over the center line y = 0 between the piles I and III and over y = T/4 for perpendicular incident waves.

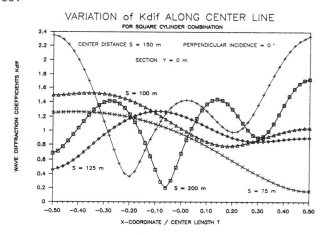

Figure 19: Evolution of Kdif in section y = 0 m

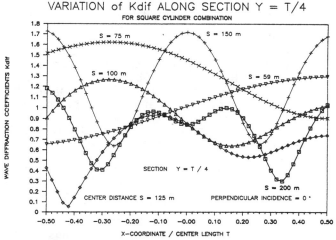

Figure 20: Evolution of Kdif in section y = T/4

NOTATIONS

$\hat{i}^2 = -1$	:	imaginary unit	
h	:	constant water depth	[m]
H	:	wave height	[m]
n	:	instant wave surface elevation	[m]
k	:	wave length	[m]
ω	:	pulsation of the wave = $2\pi / T$	
T	:	wave period	[s]
g	:	acceleration due to gravity	$[m/s^2]$
k	:	wave number; derived from the linear dispersion relation	[1/m]
θ	:	propagation direction of the incident wave field	[degr]
S	:	solution domain	
∂	:	finite border of the solution domain, i.e. the cross section of the rigid body	
ρ	:	density of the fluid	$[kg/m^3]$
Re ( )	:	real part of a complex function	
Im ( )	:	imaginary part of a complex function	
$H_i{}^{(1)}$	:	Hankel function of the first kind and i'th order	
r	:	vector pointing at source point on boundary	
r′	:	vector pointing at field point	
R	:	vector pointing from field to source point	
n	:	outward normal to the rigid boundary	
$R_n = \dfrac{\partial R}{\partial n}$	:	normal derivative of vector R	

REFERENCES

1. Berkhoff J.C.W., Mathematical Models for simple harmonic linear water waves, diffraction-refraction, Delft Hydraulics Laboratory, Publication nr. 163, april 1976.
2. Mei Chiang C., The applied dynamics of ocean surface waves, John Wiley & Sons, 1983.
3. Brebbia C.A. and Dominguez J., Boundary Element: an introductory course, Computational Mechanics Publications, McGraw-Hill, Avon, 1989.
4. Volker W. Harms, Diffraction of water waves by isolated structures, Journal of Waterways, port, coastal and ocean division, may 1979, p.131-147.
5. Chongzhun G. and Gonghu D., Mathematical Model for Wave Diffraction by a detached Breakwater, China Ocean Engineering, vol. 1, nr. 4, 1987.
6. Seghers W. & Van den Wijngaerde J., Voortplanting en vervorming van lineaire harmonische graviteitsgolven : numerieke simulatie met randelementenmethode, Ghent Hydraulics Laboratory, 1989.
7. Huygens M. and Seghers W., Boundary Element Solution for Coastal Diffraction Problems, Proceedings Third International Conference on Coastal and Port Engineering in Developing Countries, Mombassa, Kenya, 16-20 September 1991.

Wind Wave Forecasts For Extended Spans of Time

M. Pršić, Ž. Pauše
University of Zagreb, Faculty of Civil Engineering
Zagreb, Kačićeva 26, 41000 Zagreb, Croatia

ABSTRACT

The aim of the paper is to present wave forecasts for periods of time ranging between several years and a hundred years, and yielding numerical results for practical engineering needs. Such results are produced by local long-term and extreme value wave prediction, i.e. by prediction for extended spans of time. As an orientation when selecting the method, wave forecasts, as well as the existing methods for extended spans of time, are reviewed taking into consideration the source of data and the required result. A number of methods for the prediction of representative wave parameters, such as the significant wave height H_s, the maximum wave height H_{max}, the expected maximum wave height $\bar{H}_N$, the maximum wave height H_N^{mod} with highest probability among the N highest waves, the average wave period $\bar{T}_o$ the maximum wave period $T_{o,max}$, as well as the methods for the prediction of individual wave heights H, are presented.

1 WAVE FORECASTS REVIEW

A wave forecast is generally defined as the procedure for the evaluation of the representative wave profile parameters (height and period) or the wave spectre of real waves. A short-term forecast refers to a certain stationary sea state lasting from some 10 minutes to an hour, while a forecast for an extended span of time refers to rare occurrences taking place only during a long period of years. A forecast for an extended span of time is based on results of a large number of short-term forecasts, and, practically, is realized for periods ranging from several years to a century. In engineering problems associated with real waves, probability forecasts and time wave forecasts are of interest. Namely, the probabilities of reaching or exceeding given magnitudes of a particular representative parameter are determined by means of probability forecasts, while the magnitude and time of occurrence of a particular wave parameter are determined by means of time forecasts. According to geographical location, wave forecasts may be global, regional, or local.

Since the issue of this paper are forecasts for extended spans of time of wind waves (of highest specific energy) as the most interesting from the engineering aspect, these may be strictly defined applying the above general definitions as local probability forecasts

H

of representative wave profile parameters for wind sea waves for extended spans of time
having periods ranging between several seconds and several tens of sec.

2   WAVE DATABASES

Wave forecasts for extended spans of time, defined in Section 1, are carried out by
statistically processing the random process of surface wind sea waves at a fixed
geographical point within a time period of one to several years. The first step is to form
a wave sample. Samples may be formed from available databases of wave parameters
observations (Tab. 2::I) which, in principle, consist of information on wave heights,
periods and direction. Another way to form data for statistical processing is indirectly,
from meteorological information on the wind (Table 2::I).

TYPE OF OBSERVATION	POSSIBLE CONTENTS OF WIND WAVE DATABASES	
	from wave observation	from wind observations
sporadic long-term visual observation from cargo ships [36-40]	H_v, T_v	H_s, T_s
regular long-term visual observation from light-houses, meteorological ships and oil platforms	H_v, T_v	H_s, T_s
long-term instrumental recording of sea surface elevation every 3 hours or continuous wind recording	wave records, $H, T_o, H_{max}, T_{o,max}, T_{Hmax}$ $H_s, \overline{T}_o, \overline{H}_N, H_N^{mod}$	H_s, T_s

Table 2::I   Possible contents of wind wave databases according to the type of wave
or wind observation

3   REVIEW OF PROBABILITY DISTRIBUTIONS OF REPRESENTATIVE
WAVE PROFILE PARAMETERS

Only wave profile parameters shall be considered, i.e. the elevation of the physical sea
surface, since these are the only existing measurements available. The general theoretical
model for the presentation of the physical sea surface elevation in time on a given
location is a stochastic process defined by a random function $\hat{\eta}(t)$ $\forall$ $t \epsilon < -\infty, +\infty >$.
Statistical characteristics cannot be defined for such a complex process, therefore three
simpler models for which this is feasible and for which samples can be formed by
measurements at sea have been introduced. These are: I-Short-term model of physical
sea surface elevation, II-Long-term model of physical sea surface elevation, and III-
Extreme values model of sea surface elevation.

3.1   Short-term Wave Model and Corresponding Probability Distributions of Wave
Profile Parameters

A short-term model of physical sea surface elevation on a certain location, of an order of magnitude of some 10 minutes to an hour, is defined by the random function $\hat{\eta}(t)$ $\forall$ $t \epsilon$ [0,10 min (1 h)]. The short-term model, or sea state, may be considered as stationary and ergodic stochastic process with zero center. Related to it, the short-term initial random variables $\hat{\eta}$ -physical sea surface elevation, $\hat{H}$-individual wave heights, $\hat{T}_o$-individual zero crossing period), and random vector $(\hat{H}, \hat{T}_o)$, and random variables of wave parameter extremes $(\hat{H}_{max},$ and $\hat{T}_{o,max})$ may be considered. Each one has its short-term probability distribution [1,2,3,4,5,6,9,11,12,13].

3.2 Long-term Wave Model and Corresponding Probability Distributions of Wave Profile Parameters

Long-term model I is derived from the model of physical sea surface elevation at a given geographical point. It is defined by a series of stationary sea states; i.e. a stochastic process $\hat{\eta}(t)$ $\forall$ $t \epsilon$ [0,10 min] which, during many years period $\widetilde{T}$ from one to several years, is repeated N_{LT} times at regular intervals $ri = 3$ hours (Fig.3.2::1), so that $\widetilde{T} = N_{LT} * ri$. In relation to it, the series of random variables of particular long-term wave parameters may also be considered, such as: $\hat{H}_i$, $\hat{H}_{s,i}$, $\hat{H}_{max,i}$, $\hat{H}_{N,i}^{mod}$, $\overline{T}_{o,i}$, $\hat{T}_{o,max,i}$ and the similar. They have corresponding long-term probability distributions. According to this model, statistical processing includes all random variable values of a particular wave parameter within a period of several years (for example, all 11680 significant wave heights recorded during the 4-year 10-minute instrumental observations with registrations at 3-hour intervals), meaning that long-term "initial distribution" of probability is considered. Not a single distribution type for any wave parameter has been

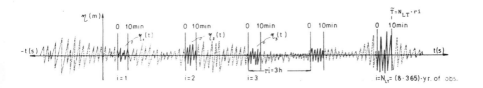

Fig. 3.2::1 N_{LT} realisations of the random process of physical sea surface elevation $\hat{\eta}(t)$ according to the long-term model

lawfully established, but, empirically, for the just mentioned random variables and based on the goodness of fit, it has been established that Fischer-Tippet's types of distribution I, II, and III (Gumbel's, Frechet's, and Weibull's) define well long-term wave parameters distributions [8,12,14]. The maximum of a particular parameter, for example the significant wave height, is defined as the highest value among N_{LT} realisations of independent identically distributed random variables $(\hat{H}_i, \hat{H}_{s,i}$ and others; $i = 1, 2,...,N_{LT})$ having the same distribution function. Unlike the short-term wave models, long-term models do not fully meet the above assumptions due to seasonal dependence and trends. If a given series of random variables, say $\hat{H}_{s,i}$; $i = 1, 2,...,N_{LT}$ is broken into subseries of "d" length (e.g. $d = 10$), and a maximum in each subseries is registered, a new series of maximums $\hat{H}_{s,j}$; $j = 1, 2,. .....N_{LT}/d$; is formed which may be considered to fulfil the conditions of independence and equal distribution. This will enable theoretical derivation of the distribution of extremes (from the assumed initial distribution of the considered wave parameter) which will also be of Fischer-Tippet's type. In engineering practice d is often 1, as it was done in example 5.1.4.

It has been shown that the log-normal distribution also fits well [8]. Statements by different authors, saying that small and medium wave heights are well distributed according to the log-normal law [21,48] and very large ones according to the Weibull's [7,14,17,48] and the Gumbel's [22,14], may serve as an orientation. Wave periods follow well the log-normal distribution [6, 48]. Frechet's distribution does not fit well the data and is rarely used. Many authors, however, have also established these distributions in different order for some points of the world sea, so it is recommendable to test all four distributions.

In engineering practice, long-term wave parameter distributions are used to determine the extreme values of wave parameters when a sample of approximately 30 years, necessary for the analysis of annual extremes, is not available. A complete wave sample produced by long-term instrumental or visual wave observations in the form of a contingency table of bivariate distribution for wave heights and wave periods is under consideration. Marginal distributions are applied for the analysis of random variables of wave heights or periods, and the entire contingency table for random vectors.

3.3　Extreme Value Wave Model and Corresponding Probability Distributions of Wave Profile Parameters

Related to the process of physical sea surface elevation at a given point for the period of one year $\hat{\eta}(t)$ $\forall$ tϵ $[\![0, 1yr]\!]$ (in the climate cycle of one year), any one of the random variables of representative wave profile parameters may be considered: $\hat{H}_s, \hat{H}_v,$ $\hat{H}_{max}, \hat{H}_N{}^{mod}, \bar{T}_o, \hat{T}_{o,max}$ etc. marking an annual extreme. The initial probability distribution of assumingly Fischer-Tippet type corresponds to it, resulting in that the corresponding distribution of extremes for an N_{LT}-year period is of the same type. The same stands for the extreme of a period somewhat shorter than one year. Depending on the time period for which the extreme wave parameter is considered, the following models are distinguished:

- the annual extreme value probability distribution model;
- the probability distribution of extreme values exceeding a given "threshold".

The validity of the assumption that one of the F-T type distributions is chosen as the initial distribution lies in the fact that the same distributions appear as the distributions of extremes for wave parameters in the previous long-term model II. The most frequently used is Weibull's (F-T type III) and Gumbel's (F-T type I) distribution [8,12,28,33]. For the extreme distribution of wave heights Thom [35] has established, from a sample produced of forecasts on meteorological data about winds, that it best fits Frechet's distribution (F-T type II). Also frequent is the use of the log-normal distribution, first applied by Jasper [21] and later much studied by others [8]. Its application is based on

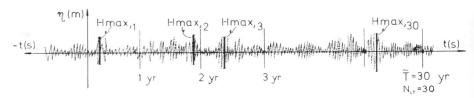

Fig.3.3::1　　NLT realisation of the random process of the physical sea surface elevation $\hat{\eta}(t)$ according to the annual extreme value model

the empirical knowledge of good fit. This model is used when a sample of approximately 30 years is available [12]. For such extended periods there are today no regular instrumental or visual wave observations (every 3 hours), and such long series may only be obtained from sporadic observations from ships or forecasts on meteorological data on winds. In cases of such samples of extremes, only random variables of wave parameters (mostly wave heights) are analysed, and not their random vectors. Most frequently processed random variables of annual extremes are the significant and visually observed wave heights.

4   SAMPLES OF RANDOM VECTORS AND RANDOM VARIABLES OF
         REPRESENTATIVE WAVE PROFILE PARAMETERS FOR LONG
         TERMS AND EXTREME VALUES

For the quantitative definition of parameters of the wave parameters probability distributions, samples should be formed from wave databases. There are two sampling methods: the total sample method, for the formation of distributions according to the long-term model, and the peak value method, for the formation of distributions of wave parameters extremes. Due to the fact that the principal inaccuracies in wave forecasts result from samples which are too small, it is essential to have available a sample of sufficient volume.

4.1   Total Samples - Samples of Long-term Random Vectors and Random Variables
         of Representative Wave Parameters

Such samples may be formed from databases created by regular (every 3 hours) observations of waves or winds during a period of 1 to several years. Here, the sample comprises all recorded, or from the wind derived, information on waves (approx. 2920 records annually). The problem with such a sample is that the requirement for statistical independence of the sample is not met. For (two-dimensional) random vectors of wave parameters (Section 3.2), the sample is in the form of a contingency table (scatter diagram); such as, for example, in Table 4.1::I for the random vector $(\hat{H}_s, \overline{T}_0)$. For the random variables (Section 3.2), such samples have a marginal empirical distribution of a random vector. The reliability of wave forecasts and the necessary volume of sample has been analysed in [26,12,46]. The most detailed such analysis, carried out in [47], has shown that for a good evaluation of wave parameters within return periods of 20, 50, 100 years, the necessary sample volume is 3-7, 8-14, 13-25 years, given a 3-hour observation interval. Since there are yet no such long wave observations, the solution lies in the formation of wave parameter samples from information on the wind.

4.2   Peak Values Samples - Samples of Random Variables of Extreme   Values of
         Representative Wave Parameters

This type of sample may be formed from long observation only, by selecting only peak values during individual storms. By analysing the reliability of distribution parameters evaluation in hydrology, a rule of thumb has been derived saying that a sample should be less than 1/3 of the predicted return period. It has been shown that for approx. 100-year forecasts of extremes a 30-year sample will suffice. According to [12], this range is also acceptable for wave extremes forecasts, and, according to [28], 25 to 50 years is necessary. To date, only sporadic visual wave and wind observations from cargo ships are that long. A sample may be formed as an annual maximum series or as partial duration series. The annual extremes sample includes only one, the largest value among all recorded, or from wind derived, data on waves during each of some 30 odd years.

MARGINAL DISTRIBUTION OF H_s				AVERAGE WAVE PERIODS $\bar{T}_{o,j}$ [s]									CONDIT. DISTR.
No $\underline{i}$	SIGN WAV HEIG $H_{s,i}$ [m] $\underline{j}$	EXCEED. PROBABILITY $P(\hat{H}_{s,i})$ [‰]	ABSOL. FREKV. OF $H_{s,i}$ z_i	20	18	16	14	12	10	8	6	4	$(\bar{T}_o|H_s)$
1	19	0.075	1	1									20.0
2	17	0.225	2	1			1						17.0
3	15	0.601	5	1			2	2					14.4
4	13	4.658	54		3			51					12.3
5	11	15.475	144					121	23				11.7
6	9	54.537	520		3		8	160	342	7			10.7
7	7	151.442	1,290		2		14	124	546	604			9.3
8	5	352.389	2,675					71	385	1943	276		8.2
9	3	743.915	5,212				5	86	438	1517	3140	26	7.0
10	1	999.925	3,408					13	133	531	1800	931	4.9
MARGINAL DISTRIBUTION OF $\bar{T}_o$	ABS. FREKV. z_j			3	8	0	30	628	1,867	4,602	5,216	957	13,311
	EXCEED.PROBAB. $(\bar{T}_o \geq \bar{T}_{o,j})$ [‰]			0.2	0.8	0.8	3.1	50.3	190.5	536.2	928.0	999.9	
	No. $\underline{j}$			1	2	3	4	5	6	7	8	9	

Table 4.1::1   Empirical two-dimensional distribution of wave random vector $[\hat{H}_s, \bar{T}_o]$, from all directions, from the center of North Atlantic (Grid Point 149; 55.9°N, 26.6°W) based on 9.5-year (1959 - 1969) continuous instrumental observation approx. every 6 hours [42]

According to Section 3.3, analyses of extremes are carried out for random variables of different wave parameters, so their samples will then bee in the form of a series of 30 odd annual extremes of the particular wave profile parameter (see Table 4.2::I).

i	1	2	3	4	5	6	7	8	9	10	11	12	13	14	15
H_{si}	13.9	13.4	12.2	12	11.6	11.4	10.9	10.7	10.6	10.4	10.2	10	9.7	9.5	9.2
z_i	1	1	1	2	2	2	1	2	2	3	1	1	1	2	3

Table 4.2::I Annual extreme value random variable sample for the northern North Sea from all directions based on 25-year (1949-1974) continuous wind observation

A partial duration series is formed of wave parameters at peaks of particular violent storms throughout observation periods somewhat shorter than 30 years. The sample includes all data exceeding a given threshold characterised by a typical violent storm so that a number of data can be recorded in a year.

5. LONG-TERM AND EXTREME VALUE PROBABILITY DISTRIBUTION LAWS FOR RANDOM VARIABLES OF REPRESENTATIVE WAVE HEIGHTS

5.1 Long-Term and Extreme Value Probability Distribution Laws Fitting to Samples of Random Variables of Representative Wave Heights

The fitting procedure is identical for long-term random variables - Model II and for random variables of extremes - Model III. Most frequently analysed random variables are: $\hat{H}_s$, $\hat{H}_v$, $\hat{H}_{max}$, $\hat{H}_N^{mod}$.

5.1.1 Empirical Probability Distributions

Clearly, according to Section 3.4, the probability distribution of wave parameters for a given location may be quantitatively determined from empirical data - a sample of the respective random variable. Engineering practice requires the prediction of the wave parameter value to which, for functional or safety reasons, a given probability of reaching or exceeding is associated, as well as the prediction of a value, or such a wave parameter, which will be reached or exceeded once in a return period of RP years. Therefore the empirical distribution of a wave parameter is also formed as the exceeding probability distribution. With this in mind, the sample is arranged in descending order, and the exceedence probability is evaluated by means of the plotting formulae, e.g. for wave heights - H (Table 4.1::I):

$$Weibull......P(\hat{H} \geq H_i) = \frac{m_i}{N_{LT}+1} \quad ; \quad m_i = \sum_1^i z_i \qquad (5.1.1::1)$$

$$Hazen........P(\hat{H} \geq H_i) = \frac{2m_i-1}{2N_{LT}} \qquad (5.1.1::2)$$

where: $\hat{H}$-wave height random variable, i-number of element in the sample arranged in descending order, H_i is the i-th wave height in the sample arranged in descending order, N_{LT}-random variable sample volume, m_i-cumulative total of absolute frequencies z_i of wave heights which are equal or larger than H_i. There is a series of similar expressions by other authors, surveyed in [43]. A more general expression is:

$$P(\hat{H} \geq H_i) = \frac{m_i - A}{N_{LT} + B} \quad \text{...(5.1.1::3)}$$

(see [8,26,44,51]). For the type I Fischer-Tippet's distribution, A=0.44 and B=0.12. For the type III_L F-T distribution, A=0.3+0,18/α and B=0.21+0,32/α. But, since α - the parameter of shape of III_L distribution is not previously known, iteration has to be applied, which is rather tedious. Consequently, the most frequently used expression is the simplest one, (5.1.1::1) [27].

5.1.2   Fitting Methods

One practical application of empirical distributions is the prediction of extreme wave parameters, i.e. those which have very low probability of reaching or exceedence (or very large return periods) associated to them. However, this is not feasible by means of the above presented empirical distributions, since the extreme values of the random variable of a considered wave parameter are located at the tail of the distribution, which is probably not covered by numerical data from the empirical distribution table (due to a relatively short sample from which it was obtained). To be able to forecast extremes, fitting of a suitable law of distribution is carried out, either graphically or analytically, to an empirical distribution. Once the law of distribution is known, its "extrapolation" into the domain of very low probabilities at the tail of the distribution becomes possible; namely, the prediction of the extreme values of the random variable of a considered wave parameter. Four fitting methods have been developed:
- the method of moments,
- the maximum likelihood method,
- the least squares method, and
- the probability paper method (the subjective graphical fitting).
Each of the four methods may yield a somewhat different result.

Fitting the Law of Distribution to the Empirical Distribution Applying the Method of Moments

This is the analytical method with which, for the known expressions of the law of distribution, parameters from the available sample of the random variable of the considered wave parameter are evaluated. In this context, the general expressions for the four cited laws of distribution have to be shown (Table 5.1.2::I), with parameters defined through the means, mean squares, mean cubes, and statistical moments. The procedure is exceptionally well presented in [8].

Fitting the Law of Distribution in the Linear Form to the Empirical Distribution Applying the Least Squares Method and Probability Paper Method

The transformation of the scales for distribution functions according to Table 5.1.2::II results in that the log-normal and Fischer-Tippet's I, II, and III distributions have a straight line graph. In the case of the log-normal, Gumbel's, Frechet's and Weibull's two-parameter distributions, each having two parameters, the scales are previously known, whilst in the case of Weibull's three-parameter distribution, the parameter ϵ

NAME	RANGE	PROBABILITY DISTRIBUTION FUNCTION FOR THE WAVE HEIGHT RANDOM VARIABLE , $\check{H}$ $P(\check{H})=P(\check{H}\geq H)=$
Log-normal	$0<H<\infty$ $-\infty<\theta<\infty$ $0<\alpha<\infty$	$\frac{1}{\sqrt{2\pi}}\int_H^{+\infty}\frac{1}{\alpha H}e^{-\frac{1}{2}\left[\frac{\ln H-\theta}{\alpha}\right]^2}dH=\frac{1}{\sqrt{2\pi}}\int_H^{+\infty}\frac{a}{H}e^{-\frac{1}{2}[a(\ln H)+b]^2}dH$
F.T. type I Gumbel's	$-\infty<H<\infty$ $-\infty<\epsilon<\infty$ $0<\theta<\infty$	$1-e^{-e^{-\left(\frac{H-\epsilon}{\theta}\right)}}=1-e^{-e^{-(aH+b)}}$
F.T.type II Frechet's	$0<H<\infty$ $0<\theta<\infty$ $0<\alpha<\infty$	$1-e^{-\left(\frac{H}{\theta}\right)^{-\alpha}}=1-e^{-(e^{-b}\cdot H^{-a})}$
F.T.typeIII Weibull's threeparam. low range	L $\epsilon<H<\infty$ $0<\theta<\infty$ $0<\alpha<\infty$	$e^{-\left(\frac{H-\epsilon}{\theta}\right)^\alpha}=e^{-e^b(H-\epsilon)^a}$
F.T.tip III Weibull's threeparam. high range	U $-\infty<H<\epsilon$ $0<\theta<\infty$ $0<\alpha<\infty$	$1-e^{-\left(\frac{\epsilon-H}{\theta}\right)^\alpha}=1-e^{-e^b(\epsilon-H)^a}$
F.T.tip III Weibull's twoparamet. $\epsilon=0$	$0<H<\infty$ $0<\theta<\infty$ $0<\alpha<\infty$	$e^{-\left(\frac{H}{\theta}\right)^\alpha}=e^{-e^b\cdot H^a}$

Tab 5.1.2::I Long-term and extreme value probability distr. functions (p.d.f.) used to present the wave height random variable $\check{H}$ and other wave profile parameters $P(\check{H})=P(\check{H}\geq H)$ - probability of the random variable $\check{H}$ being higher or equal to the value of H; α - distribution shape parameter; θ -scale parameter for variable (x-axis); ϵ - location parameter which positions p.d.f. along x-axis, and in case of Weibull's three-parameter distribution, positions one end of the distribution beyond which another expression for p.d.f. stands; $a=y(1)-y(0)$ slope coefficient of p.d.f. in linear form $(y=ax+b)$ obtained by transformation of nonlinear scales for variable $\check{H}$ and probability P into linear scales x and y; $b=y(0)$ - intercept of the linear p.d.f. on y-axis [8,12,27,33].

(or α) has to be assumed for the transformation of scales. According to [26], ϵ is assumed as one of the lowest random variable values, and α, according to [26,46], within the 0.9-1.6 range. Thus transformed linear scales form the probability papers on which the discrete values of the empirical distribution from the available sample of the considered random variable (e.g. significant wave heights in Tables 4.1::I and 4.2::I) are governed by the regression line $y=ax+b$. "a" is the line slope tangent, and "b" is the y-intercept. Parameters "a" and "b" are determined by the least squares method from the empirical distribution. The accuracy of assuming the parameter ϵ (or α) is checked by subjective evaluation of the goodness of fit of the distribution function to the empirical distribution or by objective analysis of fit in an iterative testing procedure (Section 7).

The subjective graphical method of distribution function fitting, in the form of a straight line, to the empirical distribution is carried out on completed probability papers for the above 4 laws of distribution. They present the transformed scales for random variable values and probabilities, according to the expressions in Table 5.1.2::II, by means of which discrete values of the empirical distribution are plotted. A best fitting line is then drawn through thus obtained "cloud" of points.

distribution type	abscissa scale $x=$	ordinate scale $y=$	shape param. $\alpha=$	scale param. $\theta=$	locat. param. $\epsilon=$
Log-normal	ln H	from Gauss integral table by equalling $$P(\hat{H})=0.5-\int_0^y e^{-\frac{y^2}{2}}\,dy$$	$1/a$	$-b/a$	-
FT type I Gumbel's	H	$-\ln\{-\ln[1-P(H)]\}$	-	$1/a$	$-b/a$
FT type II Frechet's	H	$-\ln\{-\ln[1-P(H)]\}$	a	$e^{-(b/a)}$	-
FT typeIII$_L$ Weibull's 3 parameter low range	ln(H-ϵ) H	$\ln\{-\ln[P(H)]\}$ $\{-\ln[P(H)]\}^{1/\alpha}$	a assumed	$e^{-(b/a)}$ $1/a$	select.small value $-b/a$
FT tip III$_U$ Weibull's 3 parameter high range	-ln(ϵ-H) H	$-\ln\{-\ln[1-P(H)]\}$ $-\{-\ln[1-P(H)]\}^{1/\alpha}$	a assumed	$e^{(b/a)}$ $1/a$	select.small value $-b/a$
FT tip III Weibull's 2parameter	ln H	$\ln\{-\ln[P(H)]\}$	a	$e^{-(b/a)}$	0

Table 5.1.2::II Relationships for linear x and y scales plotting on probability papers for values and probabilities of exceedence of random variable $\hat{H}$ or some other random variable [8,12,26,27,33]

Fitting the Law of Distribution to the Empirical Distribution Applying the Maximum Likelihood Method

This method is used to evaluate distribution parameters which with the highest probability enable the sample to be presented in its real form. The procedure is carried out by maximising the function of likelihood $L(H,\alpha,\theta,\epsilon)$ [8,44] where the random variable of any wave profile parameter may substitute the wave height random variable $\hat{H}$.

5.1.3 Return Period and Risk

In ocean engineering, the design criteria are more often characterised by the return period RP[yrs] or risk R[%] than by probability. Both these magnitudes, according to Borgman [45], may be derived in two ways: -in the analysis of the random variable of annual extremes, and by statistical modelling of extremes of the considered physical process with Poisson distribution. The return period is defined as the time interval in which the considered wave parameter $(H_s, H_{max}, \overline{T}_0 ..)$ is reached or exceeded once. It is obtained from the exceedence probability of the random variable of the considered wave parameter. According to the first principle (e.g. for the random variable of annual wave height extremes), the return period is:

$$RP = \frac{1}{P(\hat{H} \geq H^{RP})} \quad [yrs] \dots\dots\dots\dots\dots\dots\dots\dots\dots\dots\dots\dots\dots\dots\dots\dots\dots\dots\dots\dots\dots(5.1.3::1)$$

According to the second principle it is:

$$RP = \frac{\tilde{T}}{N_{LT}} \cdot \frac{1}{P(\hat{H} \geq H^{RP})} \quad [yrs] \dots\dots\dots\dots\dots\dots\dots\dots\dots\dots\dots\dots\dots\dots\dots(5.1.3::2)$$

where $\tilde{T}$ -the number of years of continuous observation from which the long-term distribution sample is formed; N_{LT}-range of sample (the number of empirical values of the considered random variable) in period $\tilde{T}$; $P(\hat{H} \geq H^{RP})$-probability of the random variable $\hat{H}$ reaching or exceeding the value H^{RP}, and H^{RP}-wave height reached or exceeded once within the return period of RP years. The same stands for any other random variable. Expressions (5.1.3::1 and 2) are also used for the calibration of the auxiliary RP[yrs] axis on the probability distribution diagram, as in Figures 5.1.4::1 and 5.1.5::1.

The risk is defined as the probability of the structure being subjected to design conditions defined with the return period RP, or more severe conditions, during the life of the structure. As shown in [45], it is derived from the exceedence probability of the extreme value random variable of wave parameters as:

$$R = 1 - \left(1 - \frac{1}{RP}\right)^{lt} \dots\dots\dots\dots\dots\dots\dots\dots\dots\dots\dots\dots\dots\dots\dots\dots\dots\dots\dots(5.1.3::3)$$

In accordance with the second principle, it is derived by means of the Poisson distribution from the probability that a low value probability occurrence takes place once within several possibilities:

$$R = 1 - e^{-lt/RP} \qquad \forall \quad lt \gg 1 \quad \dots\dots\dots\dots\dots\dots\dots\dots\dots\dots\dots\dots\dots\dots\dots(5.1.3::4)$$

where R-risk; RP[yrs]-return period of sea state to which the structure has been designed, and lt[yrs]-life time of the structure.

5.1.4   Example of a Long-term Distribution of Representative Wave Height H_s

On the example of the long-term distribution, the determination of the extreme significant wave heights in return periods of several years will be considered. The random variable sample is given in the form of the marginal distribution of significant wave heights in Table 4.1::I for the central part of the North Atlantic (Grid Point 149; 55.9°N, 26.6°W) from all directions based on 9.5-year (1959-1969) continuous instrumental wave observation [42]. To this sample, applying the least squares method, the fitting of 4 laws of distribution was executed, the log-normal, Gumbel's, Weibull's two-parameter, and Weibull's three-parameter distribution (Fig. 5.1.4::1). Frechet's distribution does not fit this sample. Weibull's plotting formula has been used for the L-N distribution, and formulae (5.1.1::3) for Gumbel's and the two Weibull's distributions.

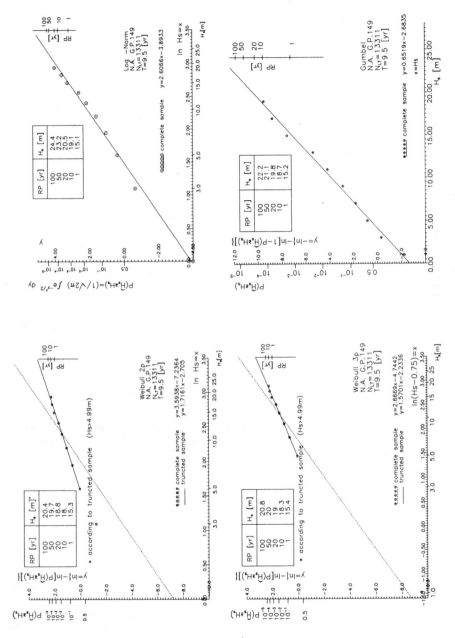

Fig 5.1.4:1 Long-term distributions of the sample from Tab 4.1::I

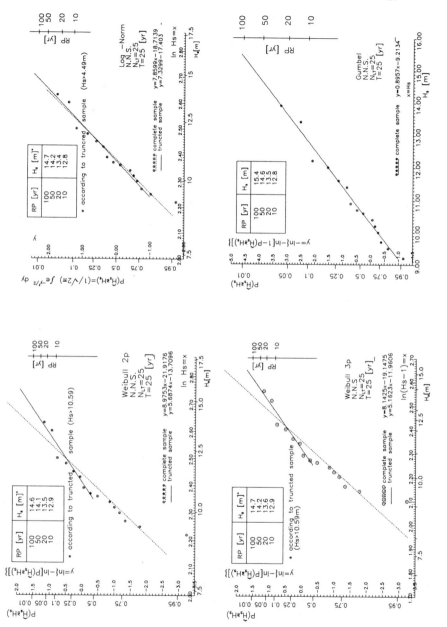

5.1.5:1 Extreme value distribution of the sample from Tab.4.2::1

5.1.5   Example of Annual Extreme Value Distrib. of Representative Wave Height H_s

On the example of the extreme values distribution, the determination of significant wave heights in return periods of several years, in accordance with the random variable sample given in Table 4.1::II for the northern North Sea from all directions based on 25-year (1949-1974) continuous wind observation, will be considered [12]. According to the methodology in Section 5.1.4, te result is shown in Fig. 5.1.5::1.

5.1.6   Example of Other Representative Wave Heights Forecast upon Forecasting Significant Wave Height H_s Extremes

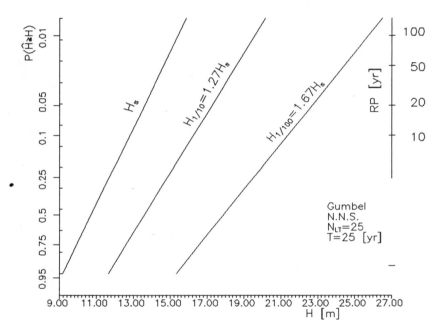

Fig. 5.1.6::1 Principle of representative wave heights prediction if the long-term significant wave height distribution is known (according to sample in Table 4.2::I)

The extreme significant wave heights on a certain location may be predicted by means of extrapolation procedures on long-term or annual extremes significant wave height distributions given above. If the predicted significant wave height of an extended return period is thought of as a representative of a certain sea state, then, by using relationships derived from Rayleigh's short-term wave heights distribution, other representative wave heights may also be obtained, as in Fig. 5.1.6::1.

5.2   Fitting the Law of Probability Distribution to Samples of Long-term Random Variables of Individual Wave Heights

5.2.1   Long-term Individual Wave Heights Distributions

The wave process within a long period of $\widetilde{T}$ years can be considered as the sum of all short-term sea states within this period of time. In it, the random variable of long-term individual wave heights Ĥ can be analysed. If the short-term sea state is defined by the significant wave height H_s, then $P_{LT}(\hat{H} \geq H)$, the probability of the random variable of long-term individual wave heights Ĥ reaching or exceeding value H, may be expressed as the sum of products of $p_{LT}(\hat{H}_s)$, which is the probability that a certain sea state defined by the representative wave height H_s occurs within an extended period of time, and $P_{ST}[(\hat{H} \geq H) | H_s]$, which is the probability of the individual wave height H being exceeded within that short period of time:

$$P_{LT}(\hat{H} \geq H) = \int_0^{\bullet} P_{ST}(\hat{H} \geq H | H_s) \cdot p_{LT}(\hat{H}_s) \; dH_s \quad \dots\dots(5.2.1::1)$$

This concept for the prediction of the expected maximum wave height has been introduced by Yasper [21]. Here, $P_{ST}[(\hat{H} \geq H) | H_s]$ is the short-term conditional Rayleigh's distribution of individual wave heights of a stationary sea state defined by the parameter H_s.Nordenstro̎m [20] has shown that, by introducing 2p Weibull's probability density function of long-term significant wave heights $p_{LT}(\hat{H}_s)$ in the Yasper's expression, a new 2p Weibull's distribution of long-term individual wave heights is obtained:

$$P_{LT}(\hat{H} \geq H) = e^{-\left(\frac{H}{C \cdot \theta^{1/d}}\right)^D} \quad \dots\dots(5.2.1::2)$$

It is a two-parameter distribution with parameters α and Θ. α does not apper in expression (5.2.1::2), but C and D are functions of α (Fig. 5.2.1::1). "d" is a const ant, where d=1.33 for visually observed wave heights and d=1 for significant wave heights. The parameter of shape α and the parameter of scale Θ are parameters of the

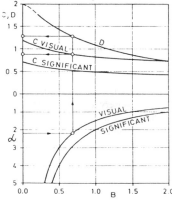

Fig. 5.2.1::1 Parameters of the distribution of individual wave heights according to Nordenstro̎m [20]

2p Weibull's distribution of the long-term random variable $\hat{H}_v$ or $\hat{H}_s$, meaning that the individual wave height prediction according to Nordenstro̎m is only possible when the distribution of one of the above representative longterm wave heights is known. The return period is determined according to the expression (5.1.3::2) but the sample volume N_{LT} here is substituted with the number of waves $n_{LT} = \overline{T} \cdot 365 \cdot 24 \cdot 3600 / T_{o,LT}$ in $\overline{T}$ years. $T_{o,LT}$ is the average zero crossing period of all long-term individual waves which, for a certain location, are obtained from successive instrumental measure-ments or by evaluation from the general knowledge of the sea [13]. If (5.1.3::2) is entered in

(5.2.1::2) and solved for H^{RP}, the explicit expression for the direct computation of individual wave height H with return period RP is obtained:

$$H^{RP} = C \cdot \theta^{1/d} \left[\ln \left(\frac{RP}{\bar{T}/n_{LT}} \right) \right]^{1/D} \quad \text{...(5.2.1::3)}$$

The symbols are the same as in equation (5.1.3::2). It has been shown that this method somewhat overestimates the reality, consequently further improvements were made. Pedersen [16] transformed the expression (5.2.1::1). Thus, according to [18], its numerical form for discrete observations H_s is:

$$P_{LT}(\hat{H} \geq H) = \sum_{i=1}^{N_{LT}} P_{ST}(\hat{H} \geq H | H_{si}) \cdot \frac{1}{N_{LT}} \quad \text{...(5.2.1::4)}$$

where N_{LT} is the volume of the discrete sample $\hat{H}_s$ taken in fixed time intervals. Battjes [17] modified the basic approach in the equation (5.2.1::1) by introducing the probability density functions $p_{LT}(\hat{H}_s, \hat{T}_o)$ of the bivariate distribution $\hat{H}_s$ and $\hat{T}_o$:

$$P_{LT}(\hat{H} \geq H) = \int_0^\infty \int_0^\infty P_{ST}(\hat{H} \geq H | H_s) \cdot p_{LT}(\hat{H}_s, \hat{T}_o) \, dH_s \, d\bar{T}_o \quad \text{...(5.2.1::5)}$$

Based on the above expression, by convolution of the contingency table of the long-term random vector $(\hat{H}_s, \hat{T}_o)$ and the short-term Rayleigh's distribution for every sea state defined by the magnitude $H_{s,i}$ from the pair $(H_{s,i}, T_{o,j})$, the discrete long-term distribution of the individual wave height is obtained:

$$P_{LT}(\hat{H} \geq H) = \sum_j^{N_j} \sum_i^{N_i} P_{ST}(\hat{H} \geq H | H_{si}) \cdot p_{LT}(H_{s,i}, \bar{T}_{o,j}) \quad \text{...(5.2.1::6)}$$

If the probability of occurrence of all sea states of "i,j" type defined by parameters $(H_{s,i}, \bar{T}_{o,j})$ in $\bar{T}$ years of continuous observation is $p_{LT}(H_{s,i}, \bar{T}_{o,j}) = n_{i,j}/n_{LT}$ if the number of waves of all sea states of i,j type is $n_{i,j} = (\bar{T} \cdot 365 \cdot 24 \cdot 3600 \cdot z_{i,j}/N_{LT})/T_{o,j}$ and if the number of all waves is $n_{LT} = \Sigma\Sigma n_{i,j}$ then:

$$P_{LT}(\hat{H} \geq H) = \frac{\displaystyle\sum_{j=1}^{N_j} \sum_{i=1}^{N_i} P_{ST}(\hat{H} \geq H | H_{si}) \cdot \frac{z_{i,j}}{\bar{T}_{o,j}}}{\displaystyle\sum_{j=1}^{N_j} \sum_{i=1}^{N_i} \frac{z_{i,j}}{\bar{T}_{o,j}}}$$

$$\text{...(5.2.1::7)}$$

$$P_{LT}(\hat{H} \geq H) = \frac{\displaystyle\sum_{j=1}^{N_j} \sum_{i=1}^{N_i} e^{-2(H/H_{si})^2} \cdot \frac{z_{i,j}}{\bar{T}_{o,j}}}{\displaystyle\sum_{j=1}^{N_j} \sum_{i=1}^{N_i} \frac{z_{i,j}}{\bar{T}_{o,j}}}$$

where H-individual wave height; $z_{i,j}$-frequency of sea state of i,j type in the contingency table; $H_{s,i}$-significant wave height of i-th class; $\overline{T}_{o,j}$-average wave period of j-th class; N_i and N_j-number of classes of the significant wave height and average period. For a series of values of H, probabilities of exceedence P_{LT} are calculated according to (5.2.1::7) and numerical values of functions of the probability distribution of individual wave heights are obtained. If these are fitted to Weibull's 2p law, the extrapolation and prediction of individual (i.e. expected maximum) wave heights with desired return periods is possible. The return period is determined according to expression (5.1.3::2), but the sample volume N_{LT} is substituted with the number of waves n_{LT} in $\overline{T}$ years according to equation (5.2.1::7). Kokkinovrachos [19] has shown the expression which also includes the wave arrival direction angle. Numerous commentaries are given in [12 and 18].

5.2.2   Examples of Long-term Individual Wave Parameter Distributions

In the first example, Weibull's 2p distribution of individual wave heights $\hat{H}$ applying the Nordenstro"m's method will be produced in accordance with the 1.-term random variable $\hat{H}_s$ sample given in Table 4.1::I using the parameters of Weibull's 2p distribution of $\hat{H}_s$ (Fig. 5.1.4::1) for the same sample. The heights of individual waves; i.e. the expected maximum wave heights for the desired return periods are given in the Table 5.2.2::I. The second example also pertains to the sample in Table 4.1::I, but the expected maximum heights of large return periods are predicted using the whole contingency table of the long-term random vector $(\hat{H}_s, \hat{T}_o)$. Applying expression (5.2.1::7), the

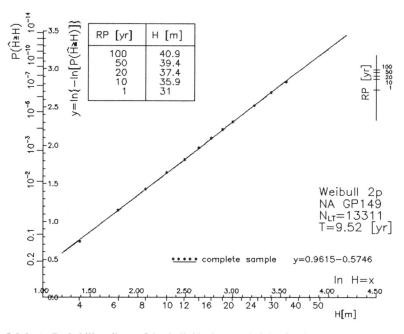

Fig. 5.2.2::1 Probability distr. of the individual wave height for the sample in tab. 4.1::I

empirical distribution of the individual wave height is defined and is fitted to Weibull's 2p law in the form of a straight line (Fig.5.2.2::1). Thus the extrapolation and prediction of individual (i.e. expected maximum) wave heights with the desired return periods becomes possible. The return period is determined according to expression (5.1.3::2).

RP [yr]	5	10	20	50	100
H [m]	35	40	42	44	45

Table 5.2.2::I Expected maximum wave heights applying Nordenstrom's method
according to the long-term random variable H_s sample given in Table 4.1::I

6   BIVARIATE LONG-TERM PROBABILITY DISTRIBUTIONS
     OF RANDOM VECTOR

In practice, the most frequently considered random vectors are $(\hat{H}_s, \overline{T}_o)$ and $(\hat{H}_{max}, \hat{T}_{Hmax})$. The empirical bivariate probability distribution of a long-term random vector is presented numerically in the form of a contingency table (Tab. 4.1::I). Laws of bivariate distribution expressed by mathematical formulae could be made to fit it. The bivariate probability distribution of a random variable of the significant wave height and average zero crossing period $(\hat{H}_s, \overline{T}_o)$ has been developed by Ochi in the form of a log-normal bivariate distribution [6,48]. Kimura developed it in the form of a bivariate two-parameter Weibull's distribution [30]. Houmb and Overvik [15] have established that a bivariate distribution may be presented as a combination of the marginal distribution of significant wave height H_s (two-parameter Weibull's) and the conditional distribution of the average zero crossing wave period caused by significant wave height $(\overline{T}_o | H_s)$ (also two-parameter Weibull's). Haver [7] has presented the bivariate distribution of the random vector $(\hat{H}_s, \hat{T}_p)$ as a combination of the marginal distribution of significant wave height H_s (two-parameter Weibull's) and the conditional distribution of spectre peak period caused by significant wave height $(T_p | H_s)$(also two-parameter Weibull's or log-normal). Other researchers have analysed different combinations of Weibulls' and log-normal distribution [14,24,25,31]. The bivariate probability distribution of the random vector of maximum wave height and simultaneous wave period $(\hat{H}_{max}, \hat{T}_{Hmax})$ was developed by Krogstad [32] in the form of the marginal distribution of maximum wave height and the conditional distribution of wave periods as above. The problem was also investigated in [14].

7   SELECTION OF THE BEST PROBABILITY DISTRIBUTION TYPE

Regardless of all attention and achievements of computational techniques, different types of distribution render different wave parameters forecasts on the same sample for the same return periods. This means that for each sample several types of probability distributions should be fitted and the best selected. There is no definite answer to this problem! However, certain procedures are normal. One of the objective possibilities is to test the goodness of fit of a mathematical distribution function to the empirical distribution of the available sample of the considered random variable by means of the χ^2 test or the Kolmogorov test. The testing procedure should be preceded by a personal evaluation of the adequacy of volume, i.e. sample variability, as well as an unbiased evaluation of the parameters. Indeed, inaccuracies which may occur due to sample variability are usually greater than the inaccuracies due to probability distribution type selection (see [18,26,47,49]). Tests show which of the distribution functions do not fit the sample, but they do not show which is the right one if several of them fit well. Even more dubious is the application of these two tests to the problems of defining extremes by extrapolation beyond the tail of the empirical distribution, since tests bear primarily the information on the central part of the distribution. There are also other ways to test the reliability of fit of the tail of the distribution, such as the method of the weighted distances, described in [52]. Further contribution to increasing the reliability is the definition of the confidence level of predicted extremes by means of the extrapolation method [26,52].

The selection of distribution according to the conservative criterion (predicted values are to be on the safe side) implies the selection of the distribution which yields the most unfavourable wave parameters for the design.

LITERATURE

[1] Longuet-Higgins,M.S.: On the Statistical Distribution of the Hights of Sea Waves: Journal of Marine Research (UK) 11 (1952.): p. 245-266

[2] Brebbia,C.A.;Walker,S.: Dynamics of Analysis of Offshore Structures. London: Newnes-Butterworths, 1979.

[3] Cartwright,R.E.;Longuet-Higgins,M.S.: The Statistical Distribution of the Maxima of a Random Function: Proceedings of Royal Society (UK) Series A (1956.) 273: p.212-232

[4] Bretschneider,C.L.: Sea Motion and Wave forecasting: Handbook of Ocean and Underwater Engineering, Myers,J.S.; Holm,C.H.; Mc Allister,R.F.Editors, New York: McGraw-Hill, 1969. -P. 1.61-1.63 & 12.14-12.16

[5] Longuet-Higgins,M.S.: On the Jiont Distribution of the Periods and Amplitudes of Sea Waves: Journal of Geophysical Research (UK) 80 (1975.): p.2688-2694

[6] Ochi,M.K.: On Long-Term Statistics for Ocean and Coastal Waves: Proceedings of the 16-th Coastal Engineering Conference Vol 1. New York: ASCE,1978.p.59-75

[7] Haver,S.: Wave Climate off Northern Norway: Applied Ocean Research Vol.7 (1985.) No.2: p.85-92

[8] Isacson,M. de St. Q.;Mac Kenzie,N.G.: Long Term Distributions of Ocean Waves: Journal of the Waterway, Port, Coastal and Ocean Division, A.S.C.EProceedings Vol.107 (May 1986.) No.WW2: p. 93-109

[9] Davenport, A.G.:Note to Distribution of the Largest Value of a Random Function with Application to Gust Loading: Proceedings Institution of Civil Engineering Vol 28 (1964.)

[10] Borgman,L.E.: Probability of Highest Wave in Hurricane: Journal of the Waterways, Harbours and Coastal Engineering Division, ASCEProceedings Vol II (1973.)

[11] Pršić,M.: Optimizacija lukobrana u uvjetima jadranskog valnog spektra. Zagreb: Fakultet gradevinskih znanosti Sveučilišta u Zagrebu. Disertacija obranjena 1987. -189 p.

[12] Houmb,O.G.: Latest Developments in Wave Statistics: Design and Construction of Mounds for Breakwaters and Coastal Protection. Bruun,P. Amsterdam :Elsevir,1985. - p. 220-237

[13] Houmb,O.G.: Wave Statistics for Design: Port Engineering. Bruun,P. Houston:Gulf Publishing Company,1976. p.131-138

[14] Guedes Soares,C; Lopes,L.C.:Computer Modelling in Ocean Engineering, Edited by Schrefler,B.A. & Zienkiewicz,O.E., Roterdam:Balkema 1988. -p. 169-175

[15] Houmb,O.G.; Overvik,T.: Parametrization of Wave Spectra and Long-Term Joint Distribution of Wave height and Period: Proceedings of the 1st Int. Conference on Behavior of Offshore Structures Vol 1, August 1976. -p. 144-169

[16] Pedersen,B.: Prediction of Long-Term Wave Conditions with Special Emphasis to the North Sea: Proceedings of the 1st Internat.Conf.on Port and Ocean Engineering Under Arctic Conditions Vol.II, Trondheim August 1971. -p.979-992

[17] Battjes,J.A.:Long-Term Wave Height Distribution at Seven Stations Around British Isles: N.I.O. Internal Report No. A.44. Wormley. National Institute of Oceanography 1970.

[18] Nolte,K.G.:Statistical Methods for Determining Extreme Sea States: Proc. of the Second Int. Conference on Port and Ocean Engineering Under Arctic Conditions, Reykjavik 1973. Reykjavik: University of Island 1973. -p.705-742

[19] Kokkinovrachos,K.: Offshore Bauwerke. Hamburg: Institut für Schiffbau der Universität Hamburg, 1980. -p.161

[20] Nordenstro"m,N.: A Method to Predict Long-Term Distributions of Wave and Wave-Induced Motions and Loads on Ships and Other Floating Strucrures: Publication No.81 (April 1973.): Det Norske Veritas, Oslo - 40 p.

[21] Jasper,N.H.: Statistical Distribution Patterns of Ocean Waves and Wave Induced Ship Stresses and Motions with Engineering Applications: Proceedings of the Society of Naval Architects and Marine Engineers

Vol.6 (1956.): p.41

[22] Carter,D.J.T.;Collenor,P.J.: Methods of Fitting Fischer-Tippett type I Extreme Value: Proceedings of the Society of Naval Architecture and Marine Engineering Vol.6 (1956.): p. 41

[23] Ochi,M.K.;Whalen,J,E.: Prediction of the Severest Significant Wave Hight: Proceedings of the 17-th Coastal Engineering Conference Vol.1, Sidney, March 1980. New York: ASCE,1980. -p. 587-599

[24] Burrows,R.;Barham,A.S.: Statistical Modelling of Long-Term Wave Climates: Proceedings of the 20-th Coastal Engineering Conference Vol.1, 1986. New York: ASCE,1986.-p. 42-56

[25] Mathisen,J.;Bitner-Gregersen,E.: Joint distributions for significant wave height and wave zero-up-crossing period: Applied Ocean Research Vol.12 (1990.) N0.2: -p. 93-103

[26] Petrauskas,C.;Aagaard,P.: Extrapolation of the Historical Storm Data for Estimating Design Wave Heights: Journal of the Society of Petroleum Engineers Vol.11 (1971.): -p. 23-37

[27] Borgman,L.E.: Extremal Statistics in Ocean Engineering: Civil Engineering in the Oceans III 1975. New York: ASCE,1975. - p. 117-131

[28] St. Denis,M.: Determination of Extreme Waves: Topics in Ocean Engineering Vol.1, Bretschneider, C.L., editor. Texas. Gulf Publishing Co., 1969. -p. 37-41

[29] Draper,L.;Driver,J.S.: Winter Waves in the Northern Sea at $57^{\circ}30'$N $3^{\circ}0''$E Recorded by M.V.Famita:Proceedings, The First Int.Conf. on Port and Ocean Engineering Under Arctic Conditions, Vol.II, Trondheim, August 1971.: -p.966-978

[30] Kimura,A.: Joint Distribution of Wave Heights and Periods of Random Sea Waves: Coastal Eng.in Japan Vol.24 (1989)

[31] Zhongsheng,F.;Shunsun,D.;Chenhyi,J.:On the long-term distribution of characteristic wave hight and period and its application: Acta Oceanologica Sinica, China Ocean Press, Vol.8 (1989.) No.3: -p. 315-325

[32] Krogstad,H.E.:Hight and Period Distribution of Extreme Waves: Applied Ocean Research 7 (1985.): -p. 158-165

[33] Chakrabarti,S.K.: Hydrodynamics of Offshore Structures. Southampton:Computational Mechanics Pubs.,1987: -p.155-157

[34] Pauše,Ž.: Probabilistic'ki aspekti problema sigurnosti i pouzdanosti: SEITH. Cavtat (1988.)

[35] Thom, H.C.S.:Extreme Wave Hight Distribution Over Oceans: Journal of the Waterways, Harbours and Coastal Engineering Division, A.S.C.E.Proceedings Vol.99 (1973.) No. WW3: -p. 355-374

[36] Hogben,N.;Dacunha,N.M.C.;Olliver,G.F.:Global Wave Statistics. Feltham UK: British Maritime Technology Limited, 1986. -p. 661

[37] Hogben,N.;Lamb,F.E.:Ocean Wave Statistics. London: HMSO, 1967. - p. 263

[38] Gorskov,S.V.: World Ocean Atlas, Vol.1-5.:Pergamon Press 1988

[39] US Navy Marine Climatic Atlas of the World, Vol I-IX. Washington: US Government Printing Office, 1969.-1981.

[40] Yamanouchi,Y.;Ogawa,A: Statistical Diagrams on the Winds and Waves on the North Pacific Ocean. Tokyo: Japanese Ship Research Institute, 1970.

[41] MIAS (Marine Information and Divisory Service) Catalogue of Wave Data. Wormly UK: Institute of Oceanographic Siences, 1982.

[42] Cummins,W.E. & all :Hindcasting for Engineering Applications: Proceedings of Internatonal Symposium on Hydrodynamics in Ocean Engineering, Vol.1, Trondheim, August 24-28 1981. : Norwegian Inst. of Technology Trondheim 1980. p. 72-89

[43] Chow, Ven Te: Handbook of Applied Hydrology. New York: Mc Graw - Hill Book Company, 1964.

[44] Muir, L.R.;El-Sharawi, A.H.:On the calculation of extreme wave heights:Ocean Engineering 13(1986.) No.2:-p.93-118

[45] Borgman,L.E.: Risk Criteria: Journal of the Waterways, Harbours and Coastal Engineering Division, A.S.C.E.Proceedings Vol.89 (1963.) No. WW3: -p. 1-35

[46] Le Mehaute,B;Wang,S.: Wave Statistical Uncertainties and Design of Breakwaters: Journal of the Waterway, Port, Coastal and Ocean Division, A.S.C.E.Proceedings Vol.111 (1985.) No.5: -p. 921-938

[47] Wang,S.;Le Mehaute,B: Duration of Measurements and Long-Term Wave Statistics: Journal of the Waterway, Port, Coastal and Ocean Division , A.S.C.E.Proceedings Vol.109 (1983.) No.2: -p. 236-249

[48] Ochi,M.: Wave statistics for the Design of Ships and Ocean Structures: Transactions, Society of Naval Architects and Marine Engineers, Vol.86(1978.), -p. 47-76

[49] Davidan,I.N.; et al.:The Results of Experimental Studies on the Probabilistic Characteristics of Wind Waves: Proceedings, Symposium on the Dynamics of Marine Vehicles and Structures in Waves, (1974.)

[50] Ward,E.G.;et al.:Extreme Wave Heights Along the Atlantic Coast of the United States: Proceedings, Offshore Technology Conference, OTC 2846, Houston, Texas (1977.)

[51] Yra,Y.:On the methodology of selecting design wave height: Proceedings of the 21-st Coastal Engineering Conference, 1988. New York: ASCE,1988. -p. 899-913

[52] Labeyrie,J.;Huther,M.;Parmentier,G.: Confidence Interval in Reliability Analysis of Marine Structures: 4-th Conference on the Integrity of Offshore Structures, Glasgow, July 1990. -p. 493-509

Wave Propagation in a Small Harbour: A Numerical Model Study

O.A. Kuye, J.W. Kamphuis

Dept. of Civil Engineering, Queen's University at Kingston, Ontario, K7L 3N6 Canada

ABSTRACT

An efficient numerical model to compute wave propagation in a small harbour is presented. The model is based on the parabolic approximation of the refraction-diffraction equation. Since the parabolic equation is valid only when the direction of wave propagation nearly coincides with a coordinate, the parabolic equation was written for curvilinear coordinates. The coordinate system was based on simple bathymetric refraction. Wave reflection was included by using a mirror imaging technique.

Two types of breakwater configurations were modelled: a straight, semi-infinite breakwater and a gap between collinear, straight breakwaters.

Laboratory experiments using unidirectional, irregular waves over an irregular harbour bathymetry were performed for reflecting and non-reflecting harbour perimeters. The numerical model results were compared with these hydraulic model results. Constant depth harbours were also calculated numerically and compared with available analytical solutions.

INTRODUCTION

The wave height distribution in a harbour depends on the bathymetry and orientation of the breakwaters. It can be obtained using hydraulic model tests, but may also be derived numerically by solving a refraction-diffraction equation. Berkhoff[1] derives such an equation known as the "mild slope" equation and has applied it in a variety of situations. Tsay and Liu[2], and Houston[3] used the mild slope equation to study long wave propagation in the vicinity of islands; Berkhoff, Booy and Radder[4] have used it to study short wave propagation over arbitrary variations in bottom topography. Berkhoff[5] has used it to investigate wave propagation in harbours.

The present numerical model is based on the parabolic approximation of the mild slope equation. The equation was solved for a curvilinear coordinate system. Wave reflection is introduced by using the mirror imaging technique after Carr[6], since linear theory, on which the mild slope Equation is based, allows superposition. The numerical model was used to simulate the transformation of waves in a small harbour of varying bathymetry. The accuracy of the present model is verified by comparing numerical results with analytical and hydraulic model results.

DESCRIPTION OF THE MODEL

For assumed irrotational flow, the motion of small amplitude surface gravity waves may be described by the complex velocity potential Φ :

$$\Phi(x,y,z,t) = \phi(x,y) f(z,h) e^{-i\omega t} \tag{1}$$

where

$$f(z,h) = \frac{Cosh\ k(h+z)}{Cosh\,kh} \tag{2}$$

The X-axis is in the wave propagation direction, the Y-axis is orthogonal to the X-axis in the horizontal plane and the Z-axis points upward from the still water surface. The angular frequency, ω, is related to the wave number, k, and the local water depth, h, by the dispersion relationship

$$\omega^2 = gk\tanh kh \tag{3}$$

Berkhoff[1] derives a differential equation which describes refraction-diffraction. The equation governs the horizontal variation of the velocity potential Φ:

$$\nabla.\ (CC_g\nabla\phi(x,y)) + k^2 CC_g\phi(x,y) = 0 \tag{4}$$

where ∇ is the horizontal gradient operator. The wave celerity, C, and group velocity, C_g, are defined as ω/k and $d\omega/dk$ respectively. The solution of Equation (4) is assumed to be periodic in time and weakly dependent on z since the variation in depth can be accounted for in the function f(z,h), defined by Equation (2).

Equation (4) is elliptic with complex variables. Its numerical solution for short waves in a large area makes large demands on computing time and storage.

For most applications, approximate methods of solving the mild slope equation may be used. In the parabolic equation method, the elliptic equation is approximated by a parabolic equation. The problem may be stated as an initial value problem and solution by a marching method is possible. The parabolic equation uses less computing time and storage than the elliptic equation. For a square model grid of N nodes along each side, the number of computations required for the solution of the elliptic equation is about N^4 (Booij[7]). The number of computations required for the solution of the parabolic approximation is about N^2 to obtain a solution without a reflected wave.

In deriving the parabolic equation, the elliptic equation is split into forward and reflected wave motion. The reflected wave is neglected and the resulting equation is (Radder[8]):

$$\frac{\partial \phi}{\partial x} - \left(ik - \frac{1}{2kCC_g} \frac{\partial}{\partial x}(kCC_g)\right)\phi + \frac{i}{2kCC_g}\frac{\partial}{\partial y}\left(CC_g\frac{\partial}{\partial y}\phi\right) \qquad (5)$$

For the parabolic method the direction of wave propagation must be close to the X-axis. If the governing equation is written for curvilinear coordinates generated by simple bathymetric refraction, the wave propagation direction will correspond closely to the wave ray direction.

Equation (4) may be expressed in the curvilinear coordinates (σ, ρ) as (Isobe[9])

$$\frac{1}{h_\sigma h_\rho}\left[\frac{\partial}{\partial \sigma}\left(CC_g h_\rho \frac{\partial \phi}{h_\sigma \partial \sigma}\right) + \frac{\partial}{\partial \rho}\left(CC_g h_\sigma \frac{\partial \phi}{h_\rho \partial \rho}\right)\right] + k^2 CC_g \phi - 0 \qquad (6)$$

and using Radder's simplification the parabolic approximation in curvilinear form is

$$\frac{1}{CC_g h_\sigma}\frac{\partial}{h_\rho \partial \rho}\left(CC_g h_\sigma \frac{\partial \phi}{h_\rho \partial \rho}\right) + 2ik\frac{\partial \phi}{h_\sigma \partial \sigma}$$

$$+ \left(\frac{i}{CC_g h_\rho}\frac{\partial}{h_\sigma \partial \sigma}(kCC_g h_\rho) + 2k^2\right)\phi - 0 \qquad (7)$$

where h_σ and h_ρ are the scale factors applicable when using a non-cartesian coordinate system.

The curvilinear coordinate system over which the parabolic equation is solved is based on simple refraction

$$\frac{\partial \theta}{\partial s} = \frac{1}{k} \frac{\partial k}{\partial n} \tag{8}$$

where s is the wave direction, n is orthogonal to s, and θ is the angle between the local s and x axes. The wave ray, which forms the σ-axis coordinate, is projected forward a single step in the direction determined by θ. The step size is a phase distance determined from the wave number at the mid-point of each ray. The orthogonal to the σ-axis which is the line of equal phase forms the ρ-axis. The information required for refraction calculation include the wave period, wave direction and depth inside the harbour.

To permit the use of a coarser finite difference grid, the solution to Equation (7) may be slowed by the transformation

$$\phi = \psi \exp(i k_* \sigma) \tag{9}$$

where σ is the main propagation direction, k_* is the average wave number in the region under consideration and ψ is the complex two-dimensional potential function. Substitution of Equation (9) into Equation (7) gives

$$\frac{1}{CC_g h_\sigma} \frac{\partial}{h_\rho \partial \rho} \left(CC_g h_\sigma \frac{\partial \psi}{h_\rho \partial \rho} \right) + 2 i k \frac{\partial \psi}{h_\sigma \partial \sigma}$$

$$+ \left(\frac{i}{CC_g h_\rho} \frac{\partial}{h_\sigma \partial \sigma} (k CC_g h_\rho) + 2 (k - k_*) k \right) \psi = 0 \tag{10}$$

This parabolic equation is solved in the numerical model.

For a computational grid consisting of J wave rays as defined by Equation (8), lateral boundary conditions need to be specified for rays 1 and J which were assumed to propagate along the harbour boundaries adjacent to the harbour entrance. Assuming complete reflection yields

$$\frac{\partial \psi}{h_p \partial \rho} - 0 \qquad (11)$$

as the curvilinear boundary condition.

METHOD OF SOLUTION

The incident wave conditions were transformed outside the harbour by simple linear refraction theory. The incident wave height to the harbour was chosen at a location on the grid closest to the mid-point of the harbour entrance.

A Crank-Nicholson scheme was used to solve the parabolic equation. The curved distances between the nodes along the phase lines (i=constant) are approximated by integrated cubic spline functions passed through the nodes j—1, j and j+1 for both rows i and i+1, Figure 1. The distances along the wave rays (j=constant) are known from the simple refraction procedure.

The numerical scheme determines the curvilinear grid one phase line ahead and solves the parabolic equation over this new phase line to obtain the amplitude and phase values. The mirror imaging technique unfolds the harbour as calculation progresses, presenting essentially an infinite harbour. A second reflection combination model combines the amplitude values obtained in the different image harbours with the original harbour to produce a final output.

Wave Amplitude and Phase

The parameters of interest in the numerical scheme are the amplitude and the phase which are related to the velocity potential, ϕ as

$$\phi(\sigma,\rho) - A(\sigma,\rho).\exp[iS(\sigma,\rho)] \qquad (12)$$

where $A(\sigma,\rho)$ is the modulus of $\phi(\sigma,\rho)$, which after the transformation of Equation (9) can be written as

$$A(\sigma,\rho) - \text{mod}[\psi(\sigma,\rho)\exp(ik_*\sigma)] - |\psi(\sigma,\rho)\exp(ik_*\sigma)| \quad (13)$$

and $S(\sigma,\rho)$ is the argument of $\phi(\sigma,\rho)$, denoted by

$$S(\sigma,\rho) - \text{arg}[\psi(\sigma,\rho)\exp(ik_*\sigma)] - \tan^{-1}\left[\frac{\text{Im}\{\psi(\sigma,\rho)\exp(ik_*\sigma)\}}{\text{Re}\{\psi(\sigma,\rho)\exp(ik_*\sigma)\}}\right.$$

$$(14)$$

The modulus of $\phi(\sigma,\rho)$ determines the amplitude of the waves, and the argument determines the wave pattern due to refraction-diffraction.

Wave Reflection

Wave reflection is determined by multiplying the wave in an incident harbour by the reflection coefficient of the boundary through which the reflection occurred

$$\psi_j{}' - K_r \cdot \psi_j \quad (15)$$

where K_r is the reflection coefficient which was determined from other tests. The reflected amplitude A_r is then calculated from this modified $\psi_j{}'$.

EXPERIMENTAL WORK

Experiments were conducted in the wave basin at the Coastal Engineering Laboratory of Queen's University. The portion of the basin used was 10.2m wide and 15.0m long. Irregular waves with a JONSWAP spectrum were used. An irregular bathymetry with an average slope of 1:60 was installed, Figure 2. The slope rises from a region of constant depth near the wave generator to the harbour end of the basin.

Two types of breakwater layout were investigated, a semi-infinite breakwater and a breakwater gap. Both were subjected to the same wave conditions.

The semi-infinite breakwater arrangement was built as in Figure 3a. The breakwater which consisted of 100 gram stones with an impermeable plywood core was rotated about Point Q to produce 75, 90, and 105 degree angles of incidence for the different tests.

Energy absorbing material was placed at the back of the breakwater (wall a in Figure 3a) and along one side of the harbour (wall b), to reduce wave reflection in the harbour. Additional tests showed that the reflection coefficient, K_r, of the energy absorber is 10%.

Two interchangeable boundaries were used at the back of the harbour : a gravel beach for limited reflection, and a vertical wall for full reflection. These generated two basic sets of data.

The breakwater gap configuration was similar to the semi-infinite arrangement described above, with another breakwater added in line with the first. A metre gap was maintained between the two breakwaters. The breakwaters were placed at an angle of 65, 75 and 90 with the incident direction of the wave; the breakwater-gap configuration at 90 degrees is shown in Figure 3b. The two interchangeable boundaries were again used at the back of the harbour and energy absorbing material was placed along walls a, b, d, and e of the harbour.

RESULTS

Comparison with Analytical Results

The numerical model was compared with analytical solution of the Fresnel's integral similar to the Shore Protection Manual[10] solution. A constant harbour depth for breakwater-gap configuration, B/L (gap to wavelength) ratios of 1.0 and 1.2 were used.
A 4.8m square harbour with 53mm incident wave height and 0.72 and 0.83 second period were used for this pure diffraction analysis. The water depth for all configurations was 125mm and the gap width was 800 mm.

When comparing the analytical results and the numerical results in Figures 4a and 4b, higher values were observed in the centre of the harbour in the propagation direction for the parabolic model. This observation is consistent with the findings of MacDonald[11]. It is attributed to the governing parabolic equation's neglect of diffraction in the wave propagation direction.

Pos and Kilner[12] compared the results of their finite

element program of the complete mild slope equation to similar analytical solutions, for a 4.9m long by 4.5m wide harbour. The finite element program predicted smaller values than the analytical model. This observation indicates that diffraction in the wave propagation direction is indeed important.

Comparison with Hydraulic Model Results

The numerical model was run for each laboratory configuration. First, wave reflections were not considered in the numerical model for these configurations. The results obtained from the numerical model were made dimensionless and compared with the hydraulic model results.

Figures 5a, 5b and 5c show the semi-infinite breakwater results, while Figures 5d and 5e show the results with the breakwater-gap configuration. Wave periods of 0.8 second, 1.0 second and 1.3 seconds were used. The numerical model results generally match the hydraulic model results when the angle between the wave propagation direction and the curvilinear grid is limited to 30 degrees. Since the direction of wave propagation must be close to the main axis, less accurate results should be expected when deviation from the grid increases.

Computed surface elevations, using the amplitude and phase, for the 90 degree breakwater-gap are shown in a three-dimensional format in Figure 8.

Reflection

Figures 6a, 6b and 6c show the results for one complete reflection, that is, the wave ray computation is stopped when the last wave ray on the computation grid crosses a second reflecting surface.

Input to the numerical model includes the reflection coefficients of the boundaries of the harbour. Wave absorbers were placed on two sides of the harbour; K_r was measured as 0.1. Vertical walls were used on the other two sides for the semi-infinite configuration in the hydraulic model (refer to Figure 3a). When K_r for the vertical walls was assumed to be 1.0, the numerical model gave too high wave amplitudes in the harbour.

The waves were reflected by the vertical walls in the hydraulic model but these waves dissipate some energy as they travel around the harbour. The mirror image technique does not include any wave energy dissipation. To match the hydraulic model results and to account for energy dissipation in the numerical model, K_r values were multiplied uniformly by 0.4 on all the boundaries of the harbour in order to obtain the results shown in Figures 6a to 6c.

The numerical model predicted considerably smaller wave heights than measured when K_r was multiplied by 0.3, conversely considerably higher wave heights were predicted when K_r was multiplied by 0.5. For the breakwater-gap configuration, K_r values were also multiplied by 0.4 to obtain the results shown in Figures 6d and 6e.

The waves in the hydraulic model with the vertical back wall, undergo several reflections before the end of data acquisition. However, specifying more than one reflection in the numerical model gave only slightly better results. This was because wave absorbers were placed along the other harbour boundaries of the hydraulic model. Wave height contribution from a second-image harbour to the original harbour was small.

Increasing the number of wave reflections did however result in appreciable differences in the numerical model results when vertical walls were assumed to exist on all sides of the harbour. Figures 7a and 7b show the numerical model results for 0, 1, 2 and 3 reflections for a semi-infinite breakwater at 90 degrees. Reflection coefficients of 0.4 and 1.0 were used for Figures 7a and 7b respectively. The results show little increase after the first reflection when reflection coefficients of 0.4 were used. While the results generally continued to increase with the number of wave reflections when reflection coefficients of 1.0 were used on the harbour boundaries. The latter case is unrealistic.

These numerical results could not be verified since hydraulic model tests with all vertical walls lead to unrealistic "hot spots" throughout the harbour, even with irregular waves.

OTHER APPLICATIONS OF THE MODEL

The model has been applied for refraction-diffraction around large coastal protection structures.

SUMMARY AND CONCLUSIONS

The parabolic equation in the numerical model assumes that diffraction is more rapid tangential to wave fronts than along the wave rays.   This led to a curvilinear formulation which was expected to increase the allowable range of wave angles. The study shows that, for high precision, the allowable angle between the wave propagation direction and the curvilinear coordinate grid is limited to about 30 degrees.   It also shows that wave diffraction in the wave propagation direction cannot be neglected entirely.

Wave reflection was introduced into the numerical model through the mirror image technique. Comparison between the hydraulic model results and the results from the numerical model, using the measured boundary reflection coefficients of 0.1 for absorbing boundaries, and assumed 1.0 for vertical walls show that wave agitation in the harbour is over predicted.   More realistic results taking into account wave energy dissipation were obtained by multiplying all boundary reflection coefficients by 0.4.

REFERENCES

1. Berkhoff, J.C.W., (1972), **"Computation of Combined Refraction-Diffraction"**,   Proc.   of   the   13th International Conference on Coastal Engineering, ASCE, Vancouver, Canada, pp. 471-490.
2.   Tsay, T.K. and Liu, P.L.-F., (1983), **"A Finite Element Model for Wave Refraction and Diffraction"**, Applied Ocean Research, Vol. 5, No. 1, pp. 30-37.
3.   Houston, J.R., (1981), **"Combined Refraction and Diffraction of Short Waves using the Finite Element Method"**, Applied Ocean Research, Vol. 3, No. 4, pp.163-170.
4.   Berkhoff, J.C.W., Booy, N. and Radder, A.C., (1982), **"Verification of Numerical Wave Propagation Models for Simple Linear Harmonic Waves"**, Coastal Engineering, Vol. 6, pp. 255-279.
5. Berkhoff, J.C.W., (1976), **"Mathematical Models of Simple Harmonic Linear Water Waves. Wave Diffraction**

and Refraction", Publication No.163, Delft Hydraulics Laboratory, Holland.

6.   Carr, J.H., (1952), **"Wave Protection Aspects of Harbour Design"**, California Institute of Technology, Hydrodynamics Laboratory Report No. E-11.

7.   Booij, N., (1981), **"Gravity Waves in Water with Non-uniform Depth and Current"**, Report No.81-1, Dept. of Civil Engineering, Delft University of Technology, The Netherlands.

8.   Radder, A.C., (1979), **"On the Parabolic Equation Method for Water-Wave Propagation"**, Journal of Fluid Mechanics, Vol. 95, Part 1, pp. 159-176.

9.   Isobe, M., (1986), **"A Parabolic Refraction-Diffraction Equation in the Ray Front Coordinate System"**, Proc. of the 20th International Conference on Coastal Engineering, ASCE, Taipei, Taiwan, pp.306-317.

10. U.S. Army Corps of Engineers, (1984), **"Shore Protection Manual"**, 4th Edition, Coastal Engineering Research Centre, Vicksburgh, Mississippi.

11. MacDonald, N.J., (1989), **"A Numerical Model for the Calculation of Combined Refraction-Diffraction in a Small Craft Harbour"**, M.Sc. Thesis, Department of Civil Engineering, Queen's University at Kingston, Ontario, Canada.

12. Pos, J.D. and Kilner, F.A., (1987), **"Breakwater Gap Wave Diffraction: An Experimental and Numerical Study"**, Journal of Waterway, Port, Coastal and Ocean Engineering, Vol. 113, No. 1, pp. 1-21.

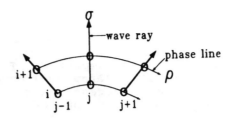

Fig. 1.   Curvilinear grid.

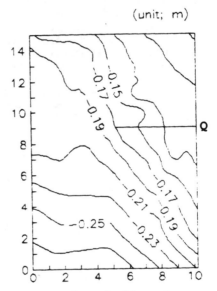

Fig. 2.   Basin bathymetry

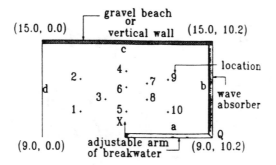

Fig. 3a. Semi-infinite breakwater configuration.

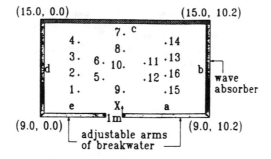

Fig. 3b. Breakwater-gap configuration.

J

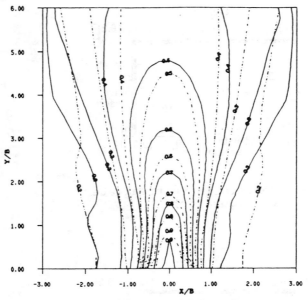

Present Model——— Analytical Model ﹍﹍﹍

Fig. 4a. Normalized wave height contours for breakwater-gap. (B/L = 1.0).

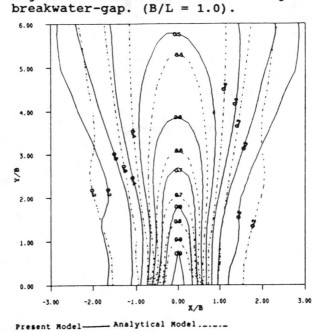

Present Model——— Analytical Model ﹍﹍﹍

Fig. 4b. Normalized wave height contours for breakwater-gap. (B/L = 1.2).

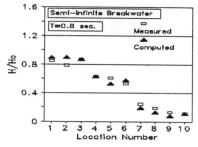

(a) 90° incident angle

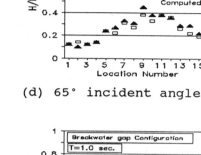

(d) 65° incident angle

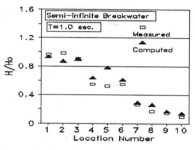

(b) 105° incident angle

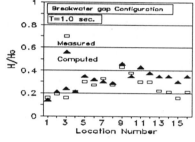

(e) 75° incident angle

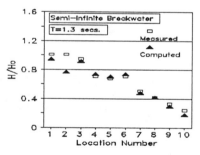

(c) 75° incident angle

Fig. 5. Comparison between measured and computed wave heights (No reflection)

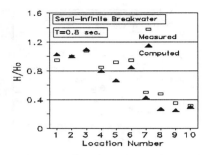

(a) 90° incident angle

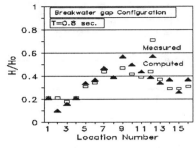

(d) 65° incident angle

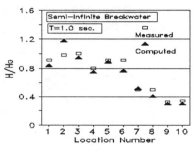

(b) 105° incident angle

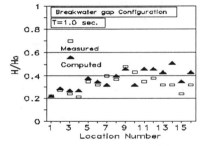

(e) 75° incident angle

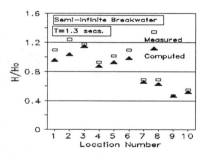

(c) 75° incident angle

Fig. 6. Comparison between measured and computed wave heights (With reflection)

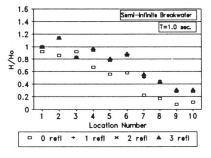 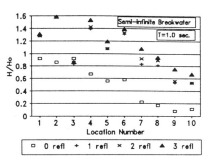

(a) $K_r = 0.4$ (b) $K_r = 1.0$

Fig. 7. Numerical model results for 0, 1, 2, and 3 reflections

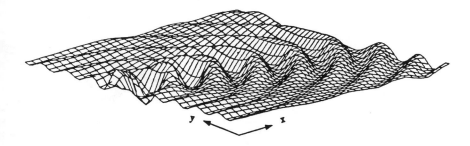

Fig. 8. Instantaneous surface elevation. Breakwater-gap configuration.

SECTION 6: FLUSHING CHARACTERISTICS AND WATER CIRCULATION

Numerical Model Study of Tidal Currents for a Port Development Using Weakly Reflective Open Boundary Conditions

R.A. Falconer (*), N.E. Denman (**)

(*) University of Bradford, Department of Civil Engineering, Bradford, West Yorkshire, BD7 1DP U.K.

(**) ABP Research & Consultancy Ltd.

ABSTRACT

The paper describes the application of a numerical model to determine the impact of a proposed reclamation and channel dredging on the flow regime of the River Test at Southampton. In modelling the proposed reclamation at Royal Pier, Southampton, the diametrically opposing open boundary conditions led to severe numerical reflection problems using the original field measured data. To overcome this problem various absorbing numerical boundary representations were considered. The final scheme adopted resulted in good agreement between the measured and predicted velocities at four measuring sites within the domain, with the model then being used to predict the impact of the geometric and bathymetric changes on the flow structure.

INTRODUCTION

In the application of numerical models to relatively small coastal basins, such as ports and marina developments, problems frequently arise in the treatment of open boundary conditions. Since numerical model open boundaries are not physical boundaries, short period wave type disturbances generated within the model domain are commonly reflected back into the domain at the open boundary, rather than being allowed to propagate out through the boundary.

In this study, undertaken by ABP Research and Consultancy Ltd on behalf of ABP Southampton, severe wave reflection problems were initially encountered in the application of a two-dimensional depth integrated numerical model, used to predict the impact of a reclamation and channel dredging proposal on the tidal flow regime along the River Test, UK. Various methods of absorbing the reflected waves at the open boundaries of the numerical model were considered, with a modified form of the Sommerfield radiation boundary condition appearing to give the most accurate results.

The region included in the numerical model to determine the hydraulic effects of proposed developments at Royal Pier, Southampton, is shown in Figure 1, including the extent of the intended reclamation at Royal Pier. Also shown is an area of channel deepening, which was included in the design to restore the cross-sectional area of the waterway and thus possibly compensate for any adverse effects that may arise due to the reclamation alone.   The channel deepening would also provide a wider and straighter navigation channel to the Western Dock complex. The diametrically opposite open boundaries of the model are indicated at the south eastern and north western extremities of the domain, together with the boundary schematization using a 40m grid.

Field measurements of current speed and direction and concentration of suspended silt loads were made at four sites within the domain to calibrate the numerical model. Following calibration and verification, the model was used to provide comparisons of tidal flows before and after the construction of the reclamation and also to predict the effect of widening the channel.

MODEL DETAILS

The mathematical model used in this study was of the depth integrated type and included the equations of momentum in the horizontal plane and the mass conservation, or continuity, equation.   Full details of the derivation of these equations are given in Falconer[1], with the corresponding equations of continuity and momentum respectively being given here for completeness:-

$$\frac{\partial \zeta}{\partial t} + \frac{\partial U H}{\partial x} + \frac{\partial V H}{\partial y} = 0 \tag{1}$$

$$\frac{\partial U H}{\partial t} + \beta \left[\frac{\partial U^2 H}{\partial x} + \frac{\partial UVH}{\partial y} \right] - \Omega V H + gH \frac{\partial \zeta}{\partial x}$$
$$- \rho_a \frac{C_* W_x W_s}{\rho} + \frac{g f U V_s}{2} - \epsilon H \left[\frac{\partial^2 U}{\partial x^2} + \frac{\partial^2 U}{\partial y^2} \right] = 0 \tag{2}$$

$$\frac{\partial V H}{\partial t} + \beta \left[\frac{\partial UVH}{\partial x} + \frac{\partial V^2 H}{\partial y} \right] + \Omega U H + gH \frac{\partial \zeta}{\partial y}$$
$$- \rho_a \frac{C_* W_y W_s}{\rho} + \frac{g f V V_s}{2} - \epsilon H \left[\frac{\partial^2 V}{\partial x^2} + \frac{\partial^2 V}{\partial y^2} \right] = 0 \tag{3}$$

where ζ = water surface elevation above datum, t = time, x,y = co-ordinates in horizontal plane, U,V = depth average velocities in x,y directions, H = total depth, β = momentum correction factor (= 1.017 for Seventh Power Law or 1.2 for wind induced second order parabolic profile, see Falconer and Chen[2]), Ω = Coriolis parameter, g = gravity, ρ_a = air density, C_* = air-water resistance coefficient, W_x, W_y = wind velocity components, W_s = wind speed, ρ = water density, f = Darcy-Weisbach friction factor, V_s = depth average current speed and ϵ = depth average viscosity (= $0.15 U_* H$ after Fischer[3], where U_* = shear velocity). For the Darcy-Weisbach friction factor the Colebrook-White equation for open channel flow was used, giving:-

$$f = - 4 \, log \left[\frac{k_s}{12H} + \frac{2.5 \, f}{2 \, R_e} \right] \tag{4}$$

where k_s = equivalent sand grain roughness and R_e = Reynolds number (= $4U$

H/ν, where ν = kinematic viscosity).

The finite difference equations corresponding to the differential equations (1) to (4) were expressed in an alternating direction implicit form, with all terms being centred in space and time centred by iteration in a similar manner to the predictor-corrector method. The advective accelerations were represented using the third order upwind difference scheme, with full details of the basic scheme being given in Falconer[4].

The model was set up to simulate the area shown in Figure 1 using a 40m grid schematisation and with a timestep of 24s. The model was driven using a measured tidal curve at the downstream (or north west) boundary and measured velocities at the upstream (or south east) boundary. The bathymetry in the model was derived from the Admiralty Chart (2041) and more recent detailed surveys provided by ABP, Southampton.

The model was operated using a repeating spring tide of range 4m, with this range being typical of the mean spring tide conditions at Southampton. For each test the model was run for two consecutive tides starting from rest at high water, with results being taken from the second tide. A test run showed that the third and fourth tides repeated the results of the second tide, indicating that the model had stabilised.

TREATMENT OF REFLECTING BOUNDARIES

In early test runs of the model it was noted that the original model simulations gave rise to numerical wave reflections at the open boundary, which resulted in short period wave-type oscillations in the velocity and water elevation fields. Although these short period wave-type oscillations could be smoothed out in interpreting the numerical results, it was felt desirable to eliminate these oscillations numerically by introducing absorbing boundary conditions.

For this purpose several methods were considered and coded up in the model. The first method considered involved evaluating the reflected wave train from the receding characteristics and the incident wave train, with the open boundary water elevation then being modified to include both the incident and reflected waves, see Kobayashi et al[5]. The second method involved using the first approximation of the widely used absorbing boundary conditions for the numerical simulation of waves given by Engquist and Majda[6], with full details of the application of this scheme to practical model studies being given by Copeland[7]. However, for the current study neither of these schemes eliminated the oscillations generated at the boundaries and a modified form of the Sommerfield radiation boundary condition was adopted, similar to that described by Blumberg and Kantha[8].

The modified boundary condition proposed by Blumberg and Kantha involves using a radiation boundary condition similar to that given by Engquist and Majda[6], but with the inclusion of a damping term which tends to force the value of the open boundary water elevation (or velocity) to a known, or measured, value within a time scale of order T_r. Thus, the modified boundary condition used in the current study for a water elevation boundary can be written as:-

$$\frac{\partial \zeta}{\partial t} + C \frac{\partial \zeta}{\partial n} = -\left[\frac{\zeta - \zeta_m}{T_r}\right] \tag{5}$$

where C = mean phase speed of the wave, given as $\sqrt{(g \ H_m)}$ where H_m = mean depth along the wave path, n = direction normal to the planar boundary, ζ_m = measured water elevation at the boundary and T_r = time of travel of the incident wave to the opposite boundary and back, i.e. twice the wave path from one boundary to the other divided by the mean wave speed C. If the time scale $T_r = 0$ then this corresponds to a clamped boundary in which no disturbances are allowed to pass out through the boundary and if $T_r \to \infty$ then this corresponds to the pure radiation condition[6] which renders a boundary transparent to waves travelling in the positive n direction with a phase speed C. The equilibrium condition occurs when $\zeta = \zeta_m$, with a similar relationship being used for a velocity boundary.

The finite difference scheme used to represent Eq.5 was of the fully centred explicit type giving, for an upper boundary:-

$$\frac{\zeta_I^{n+1} - \zeta_I^n}{\Delta t} + \frac{C}{2\Delta x}\left[\zeta_I^{n+1} + \zeta_I^n - 2\zeta_{I-1}^{n+\frac{1}{2}}\right]$$

$$= -\left[\frac{\zeta_I^{n+1} + \zeta_I^n - 2\zeta_m}{2 T_r}\right] \tag{6}$$

where n = time step number, Δt = time step and Δx = grid size. For the current study the wave celerity C was set to a constant of 8.4ms^{-1}, with the mean depth being set to 7.2m. For the time scale T_r, this was set to 0.1hr, based on estimating the time taken for the incident wave to travel from one boundary to the other and then back again at the constant wave celerity C. This method of overcoming the problems associated with wave reflection at the open boundaries was found to give good agreement with the field data and eliminate the oscillations.

MODEL APPLICATION

Site Description

The area of the River Test reproduced in the model is dominated by a 10m deep navigation channel flanked to the south west by a broad shallow area, much of which dries at low water springs. Along the entire northeast side of the dredged channel lie port facilities. Generally depths at the berths are 11.7m below Chart Datum, but in the vicinity of Royal Pier and Town Quay they are more shallow and vary between + 1m to -5.5m CD.

The tidal range typically is 4m on a spring tide and 1.9m on a neap tide. The flood tide is characterised by two periods of rising water level separated by a mid-tide stand. Current speeds are generally low with peak surface ebb currents reaching about 0.75ms^{-1}. Weak tidal currents, together with low silt concentrations, give rise to an environment dominated by deposition of fine sediment, albeit at a slow rate. Dredging is therefore infrequent with the exception of some berths and at the margins of the navigation channel where maintenance dredging is required twice each year.

Field Measurements
        In order to calibrate and verify the model, ABP Research and Consultancy's
Hydrographic Unit and Staff from the Harbour Master's department at Southampton
Docks undertook field measurements for 2 days in October 1989.   Speed and
direction of tidal currents and suspended load silt measurements were made at
four sites throughout a spring tide.   Likewise, silt concentrations were recorded at
two of the four velocity sites.   Full details of the measured data are given by ABP
Research and Consultancy[9].

Model Calibration
        Following modifications to the water elevations and velocities at the open
boundaries, using the reflecting boundary scheme outlined previously, the model
accurately reproduced tide level variations across the domain.   Furthermore,
current speeds were compared at the four field measurement sites and were
found to agree closely.   Current directions were also found to agree closely with
the full scale measurements when the current speed exceeded $0.15ms^{-1}$.   At
lower current speeds the agreement is less satisfactory, although this was thought
to be mainly due to the difficulty in aligning the current meter with the flow
direction at such low flows.

        It has therefore been concluded that with the modified boundary
representation the model accurately reproduced the full scale tidal current speeds
and directions at all four measuring sites, with the result that the model has been
used as a predictive design tool with some degree of confidence.   Full details of
the comparisons between the field measured and predicted data are given by
ABP Research and Consultancy[9], with a typical comparison being given here for
completeness at site 2.

MODEL RESULTS

        The model was first run for the existing geometry and bathymetry with a
typical velocity distribution being as indicated for peak ebb flow conditions in
Figure 3.   The main findings from a time series of velocity distribution plots
throughout the tidal cycle showed that the flood tide peak currents were lower
than the equivalent currents for the ebb tide, with the highest currents not being
exclusively associated with the deep water channel.

        The model was then run with the reclamation being included by removing
the corresponding reclaimed cells.   The main findings obtained in comparing the
velocities with the original geometry results were that the ebb flows immediately
opposite Royal Pier were redirected around the reclamation, giving rise to a
maximum increase in the current speed from $0.12ms^{-1}$ to $0.55ms^{-1}$ in the main
navigation channel.   A slow moving eddy was also formed in the lee of the
reclamation.   Flows were similarly redirected around the reclamation during flood
tides, but as current speeds were generally lower the impact of the reclamation
was less marked.   Figure 4 shows the modified velocity distribution with the
reclamation, at the same tidal phase as Figure 3.

        A detailed analysis and comparison of a series of time history velocity
plots, both pre and post development, indicated that the changes to the flow
regime as a result of the proposed reclamation were not significant.   On the other
hand, calculations of sedimentation at a point based upon a method described by
Delo[10] were undertaken for 32 velocity sites led to the conclusion that changes to

the present pattern of deposition could be expected in the regions indicated in Figure 5.

Further simulations were then undertaken with the region proposed for channel dredging, indicated in Figure 1, being dredged to 10.2m below CD and the inshore edge suitably graded into the natural river bank. The resulting velocity increases associated with the reclamation alone were now predicted to be restored nearer to their former values. In the dredged region the velocity changes were variable, with an increase occurring during ebb tides and a decrease for flood tides. The extent of the velocity changes during the second flood tide was surprisingly large compared with other states of the tide, but the magnitude of the velocity change was generally small everywhere except close to the reclamation and in the dredged area.

CONCLUSIONS

In the application of a two-dimensional depth integrated numerical model to study the effects of reclamation and channel dredging on the tidal flow regime and rates of sediment deposition in the River Test, at the Port of Southampton, severe numerical problems were initially encountered at the open boundaries. These problems arose as a result of wave type disturbances generated at the open boundaries being reflected back at the other open boundary and giving rise to short period oscillatory water elevations and velocities within the domain and centred about a time varying mean. Several numerical schemes were studied to eliminate these oscillations, with a modified form of the Sommerfield radiation boundary condition proving to be most effective. The modified Sommerfield condition included a damping term, which was related to the time that the boundary generated wave took to travel across the domain and back to the open boundary following reflection.

This enhanced model was then applied to the River Test, where good agreement was obtained between the predicted and measured velocities at four sites within the domain. The main findings from the model application to this site specific study can be summarised as follows:-

a) The significant changes in the current speeds arising from construction of the reclamation alone would be confined to 500m upstream and 700m downstream of the development.

b) Immediately south of the development and in the main channel, the peak velocities increased by up to $0.12 ms^{-1}$, leading to maximum flood and ebb spring tide currents of $0.30 ms^{-1}$ and $0.55 ms^{-1}$ respectively.

c) An eddy formed in the vicinity of Town Quay on the ebb tide, together with a reduction in the velocity in the lee of the reclamation for both the flood and ebb tide.

d) Widening the navigation channel along the south side, opposite Royal Pier, extended the region of significant change in the current speed to 600m upstream and 900m downstream of the development.

e) The wider navigation channel also restored peak current speeds to the pre development levels.

f)   Finally, siltation was predicted to be increased to the east side of the reclamation by about 10% and to the west side by 5%. In the vicinity of channel widening, siltation was predicted to occur at an estimated level of about 20,000m^3 yr^{-1}.

ACKNOWLEDGMENTS

The study reported herein was undertaken by ABP Research and Consultancy Ltd for ABP Southampton. The authors are grateful to Mr D H Cooper, Managing Director of ABP Research and Consultancy Ltd and Captain M Ridge, Harbour Master, ABP Southampton, for permission to publish the result of this study.

REFERENCES

1.   Falconer, R A, "Review of Modelling Flow and Pollutant Transport Processes in Hydrualic Basins", Proceedings of the First International Conference on Water Pollution: Modelling, Measuring and Prediction, Southampton, UK, Computational Mechanics Publications, September 1991, pp.3-23.

2.   Falconer, R A and Chen, Y, "An Improved Representation of Flooding and Drying and Wind Stress Effects in a 2-D Tidal Numerical Model", Proceedings of the Institution of Civil Engineers, Part 2, Research and Theory, Vol.91, December 1991.

3.   Fischer, H B, "Longitudinal Dispersion and Turbulent Mixing in Open Channel Flow", Annual Review of Fluid Mechanics, Vol.5, 1973, pp.59-78.

4.   Falconer, R A, "A Two-Dimensional Mathematical Model Study of the Nitrate Levels in an Inland Natural Basin", Proceedings of the International Conference on Water Quality Modelling in the Inland Natural Environment, Bournemouth, June 1986, paper Jl, pp.325-344.

5.   Kobayashi, N, Otta, A K and Roy, I, "Wave Reflection and Run-Up on Rough Slopes", Journal of Waterway, Port, Coastal and Ocean Engineering, ASCE, Vol.113, No.3, May 1987, pp.282-297.

6.   Engquist, B and Majda, A, "Absorbing Boundary Conditions for the Numerical Simulation of Waves", Mathematics of Computation, Vol.31, Number 139, July 1977, pp.629-651.

7.   Copeland, G J M, "A Practical Alternative to the Mild Slope Wave Equation", Coastal Engineering, Vol.9, 1985, pp.125-149.

8.   Blumberg, A F and Kantha, L H, "Open Boundary Condition for Circulation Models", Journal of Hydraulic Engineering, ASCE, Vol.111, No.2, February 1985, pp.237-255.

9.   ABP Research and Consultancy Ltd, "Effects of Proposed Reclamation at Royal Pier", Research Report No.R.390, November 1989, pp.1-8.

10.   Delo, E A, "Estuarine Muds Manual", Hydraulics Research Ltd, Report No.SR164, February 1988.

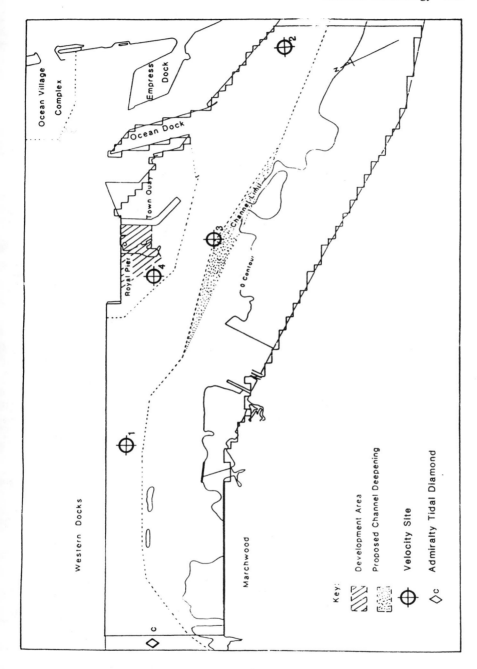

Figure 1.   Illustration of the Study Area, Showing the Area of
            Reclamation and Dredging, the Model Grid Boundary and
            the Measuring Sites.

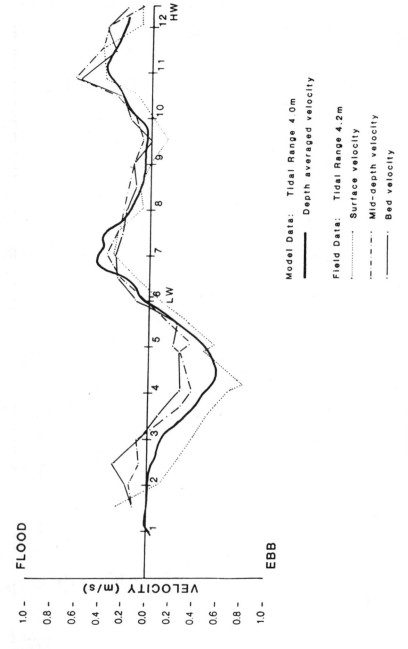

TIME (hours from HW – Southampton)

Model Data: Tidal Range 4.0m
              Depth averaged velocity

Field Data: Tidal Range 4.2m
              Surface velocity
              Mid-depth velocity
              Bed velocity

Figure 2.   Comparison of Measured and Predicted Velocities at
            Site 2 for a Spring Tide Simulation Using the
            Modified Boundary Representation.

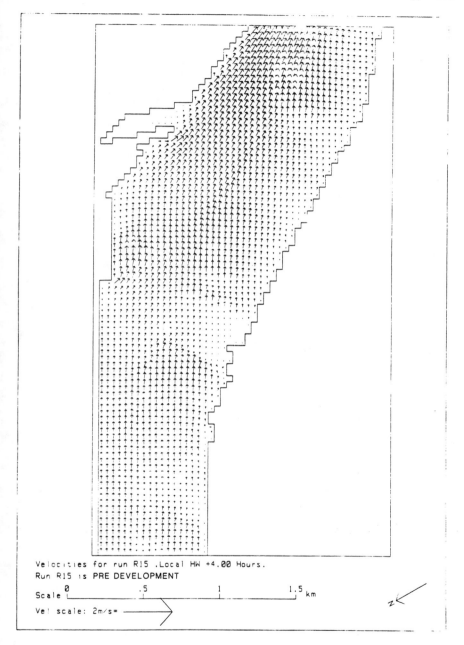

Velocities for run R15 .Local HW +4.00 Hours.
Run R15 is PRE DEVELOPMENT
Scale 0 .5 1 1.5 km
Vel scale: 2m/s=

Figure 3. Predicted Peak Ebb Tide Velocities for Pre Development
Conditions.

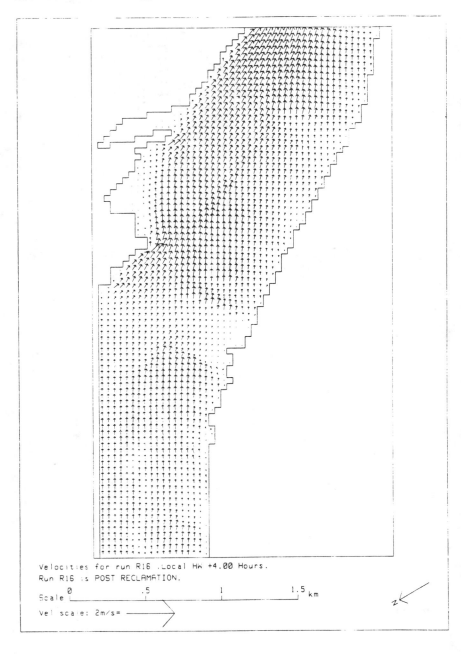

Velocities for run R16 .Local HW +4.00 Hours.
Run R16 is POST RECLAMATION.
Scale 0 .5 1 1.5 km
Vel scale: 2m/s=

Figure 4.   Predicted Peak Ebb Tide Velocities with Reclamation
at Royal Pier.

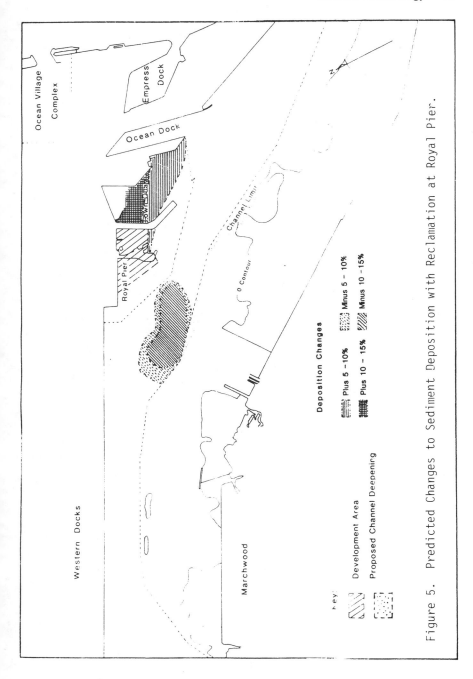

Figure 5.   Predicted Changes to Sediment Deposition with Reclamation at Royal Pier.

Studies for Improving Flushing Ability of Marmaris Marina

E. Ozhan (*), E. Tore (**)
(*) Coastal & Harbor Engineering Research Center, Middle East Technical University, Ankara 06531 Turkey
(**) Netsel Tourism Investments Inc., Nispetiye Cad., Ayyildiz Ishani, No. 4, kat 2, Levent 80620 Istanbul, Turkey

ABSTRACT

Various engineering aspects of Marmaris Marina were investigated at the Coastal and Harbor Engineering Research Center (KLARE) of METU by using physical and mathematical models in two consecutive studies conducted in 1985 and 1990. Among these, particular attention was given to water exchange and improvement of the flushing characteristics. Natural flushing due to tidal motion is far from being sufficient due to very small tidal ranges.

In physical model tests, water circulation patterns caused by waves in and around the marina were determined by observing the paths followed by drogues. The improvement of flushing due to installation of a wave pump or to provision of a gap at the main breakwater-quay for maintaining water exchange were studied. Forced flushing which is achieved by pumping out water from the surface layer at a location inside the marina was investigated also through model tests.

The paper discusses the results of mathematical and physical model studies and presents a case study in which a number of alternatives are investigated for improving the flushing ability of a marina which suffers from insufficient natural flushing due to weak tidal motion.

INTRODUCTION

Marmaris Marina, with a capacity of 676 berths, is one of the large and modern marinas in Turkey. The marina is located at a corner of Marmaris Bay which is connected to the southern Aegean Sea by two narrow entrances (Fig. 1). Marmaris Marina has been in operation since the summer of 1989.

Turkish coastline along the southern Aegean Sea and the eastern Mediterranean is increasingly popular among yachtmen. Marmaris Marina, which is strategically located along the sailing route, is busy all year around. The importance of this marina is also due to the reason that the town of Marmaris is one of the most developed coastal resorts along the Turkish coastline.

Various engineering aspects of Marmaris Marina were investigated by Coastal and Harbor Engineering Research Center (KLARE) of M.E.T.U. in two consecutive studies. (Ozhan and Ergin 1985 and 1990). Among these, particular attention was given to water exchange and improvement of flushing characteristics. This was important since natural flushing due to tidal motion was estimated to be far from being sufficient for maintaining acceptable water quality in the marina due to very small tidal ranges.

Pollution of water enclosed in marinas has received considerable attention in Turkey. Basic and applied research has been carried out by KLARE for over 10 years on marina pollution. (Ozhan, 1979, 1983, 1989, 1990; Ozhan and Ergin, 1985, 1987,1990; Mashroutechi, 1991). Remedies are available at three levels for combating marina pollution. Two of these, being basically managerial measures, are to minimize pollutant input and to carry out regular cleaning operations. The third is to increase flushing ability of the marina both by choosing advantageous plan shapes and/or by employing special design features. When the tidal ranges are small as they are in the Mediterranean, improvement on plan shape alone is usually not sufficient. In such cases, forced flushing to be caused by a water discharge entering or leaving the marina may be necessary.

In addition to tidal motion, summer breeze may play a role in the flushing of marinas. However, this process has not been sufficiently investigated.

This paper discusses the results of two successive engineering research projects carried out for assessing and improving flushing ability of Marmaris Marina as a case study.

PHYSICAL CHARACTERISTICS AND TIDAL FLUSHING OF MARMARIS MARINA

Marmaris Marina, located next to the old town of Marmaris, has a single basin bounded by two composite breakwaters (vertical concrete block walls placed on rubble fill), and a single entrance opening northwest (Fig.1). The mean water depths inside the marina range from 3.0 m to 17.0 m, with a mean value of 10.5 m at mean low tide level. The surface area is about 100 000 m^3, resulting at a mean low tide volume of 1 043 000 m^3. The mean length to width ratio (aspect ratio) of the marina is 1.4. The marina has floating concrete finger piers for berthing yachts up to 40 m length.

Tidal motion at the site of Marmaris Marina is typically semi-diurnal. The mean tidal range is about 0.15 m and the tidal period is 12 hours 25 minutes (Ozhan and Ergin, 1985). These figures indicate that the mean volume of water entering (or leaving) Marmaris Marina during half a tidal period (i.e. the tidal prism) is 15 000 m^3, and the mean tidal discharge is about 0.65 m^3/sec.

The wind rose constructed by using hourly averaged wind speeds and directions recorded at Marmaris Meteorological Station is shown in Fig.2. The ESE-SSE quadrant is seen to be the directions from which strong winds occur frequently. The winds from W to NNW are also frequent, though they are significantly milder.

Surface currents outside Marmaris Marina are generally weak with speeds typically less than 10 cm/sec in the summer season. (Ozhan and Ergin, 1985). They are due to combined effects of tidal motion and wind. Current patterns inside Marmaris Bay due to combined actions of winds and tide were investigated by using a depth averaged mathematical model in 1990 study (Ozhan and Ergin, 1990). An example is provided in Fig.3 where the formation of a large counterclockwise circulation cell by southeasterly winds is observed. Secondary cells also exist at some locations close to the coastline.

The tidal flushing ability of Marmaris Marina may be examined by employing the one-dimensional flushing model of Ozhan (1983,1989). According to this mathematical model which assumes complete mixing in the marina basin at all times, the number of tidal cycles 'n' necessary for exchanging 'p' percent of water volume inside the marina with water from outside is given as:

$$n = \frac{\ln (1-p/100)}{\ln a} \quad \ldots\ldots\ldots\ldots (1)$$

where ;
$$a = \frac{M-1}{M+1} \quad \ldots\ldots\ldots\ldots\ldots (2)$$

M being the flushing parameter (Ozhan, 1989) which is equal to the ratio of mean water depth to tidal amplitude.   For Marmaris Marina, the flushing parameter has the value of 140. Using this value in Eq.(1) and (2), one computes that it requires 48 tidal cycles (approximately 25 days) for 50% water exchange and 161 tidal cycles (63 days) for 90% water exchange. These figures indicate that flushing of Marmaris Marina by tidal motion is far from sufficient for self cleansing. Forced flushing by special design features would be very valuable.

HYDRAULIC MODEL TESTS ON CIRCULATION AND FORCED FLUSHING

Laboratory models of Marmaris Marina were constructed in both studies of 1985 and 1990. In the earlier study, the major purpose was to optimize the marina layout by paying attention to wave agitation. In addition, short term field measurements were carried out to obtain data on tidal motion and surface currents for use in the assessment of flushing ability of the marina. As shown in the previous section, the tidal flushing was judged to be far from being sufficient. Consequently, the undistorted laboratory model which had a scale of 1:50, was used to investigate special design features for forced flushing.

The special design feature studied was the so called "wave pump" arrangement of Bruun and Viggosson (1977). This wave pump   arrangement aims to introduce a water discharge into a marina by utilizing wave energy outside. The details of the wave pump designed for Marmaris Marina and the location   are shown   respectively in Fig. (4)

and (5). This structural arrangement amplifies the wave height in the region bounded by two converging walls. The waves with increased heights break on the ramp section, resulting in a volume flux towards the marina in the channel.

In the wave pump arrangement designed for Marmaris Marina, the ramp section had a slope of 5:8. The crest of the ramp section was 30 cm below mean water level. The channel was 10 m. wide with a bottom elevation of -1.5 m. The channel was continued parallel to the quay for a short length inside the marina for directing the current initially parallel to the quay in order to have a circulation cell as large as possible.

An example circulation pattern inside the marina generated by the wave pump is shown in Fig.5. This is the steady state circulation in surface layer caused by waves having 0.75 m height and 3.5 m period. Such waves are typical at Marmaris Bay for all seasons due to sever limitations on fetch lengths.

The current speeds caused by the wave pump are initially in the order of 7-8 cm/sec. They decrease to less than 5.0 cm/sec near the harbor entrance. A weak current arm is established directing outwards the marina from the entrance for balancing the mass flux through the wave pump. As the result of 1985 study, it was concluded that the wave pump would contribute to the flushing of Marmaris Marina at least in the same level as the tidal motion. It was recommended that the construction of the wave pump be carried out, and the water quality inside the marina be managed further by periodical cleaning operations as appeared necessary.

During the construction of the marina however, it was noticed that the wave heights in front of the wave pump walls were smaller than anticipated. The doubts which aroused on the level of the usefulness of the wave pump with so small wave heights led to the second investigation of 1990 (Ozhan and Ergin, 1990). It was readily found out that the wave heights at the location of the wave pump were indeed small due to the sheltering effect of a reclaimed sea area and a small jetty which was constructed as an independent project during 1988. This development is seen southeast of Marmaris Marina in Fig.6.

The 1990 study concentrated solely on the circulations caused by tidal motion, wind and waves and flushing of the marina. As it was mentioned in the previous section, the current patterns in the whole of Marmaris Bay as generated by tide alone or in combination with winds from various directions were computed by using a depth averaged mathematical model. Circulation patterns caused by waves were measured in laboratory tests carried out on the physical model which was built at a scale of 1:80 (Fig.6).

At the initial phase of the study, the wave pump was built on the model as recommended by the 1985 study. It was demonstrated in laboratory tests that the reclamation project carried out southeast of the marina did affect the wave conditions in front of the wave pump and the subsequent currents inside the marina (Fig.7). It was observed that deep water wave heights of 0.75 m could derive only very weak currents in the vicinity of the discharge channel and could not even form a noticeable circulation cell in the whole of the marina (Fig.7). Thus it was decided not to execute the construction of the wave pump which was about to start according to the construction schedule.

Further tests were conducted for establishing circulation patterns caused by waves in the vicinity of the marina. Current paths were determined by following the positions of surface drogues. At the start of test, wave driven surface currents were observed to be directed towards the marina (Fig.8), indicating that pollutants which may be present outside will be transported into the marina during the initial phases of storms. However, as the steady state was established, the current pattern outside the marina was seen to change significantly. Surface drogues which were deployed at the same point as in Fig.8, moved in opposite direction to form a large clockwise circulation cell (Fig.9). A concrete keel was placed at the location of the wave pump to divert this current from the shore in order to protect the recreational area just next to the marina from pollution.

Alternative schemes for establishing forced flushing of a marina are periodical pumping of clean water from outside into the marina, or of polluted water from the marina to an appropriate location outside. It has been argued that the later of these is preferable both from the point of hydraulic efficiency and for easier management procedures (Ozhan, 1983, 1989). It has also been argued that it is better to remove water from the surface layer as this layer is likely to be more polluted due to buoyant pollutants such as oil and debris. This may be achieved by a "morning glory" shaped intake structure. Details of this structure designed for Marmaris Marina is shown in Fig.10. The best location of such an intake for periodical removal of polluted marina waters from the surface layer was searched through laboratory experiments.

The inlet structure described in Fig.10 consists of a vertically placed conical shaft connected to a horizontal discharge pipe placed on the sea bed. A rectangular entrance is provided on the side of the shaft. This side entrance serves two purposes: a) To feed water into the inlet at the times of extreme low water, b) By disturbing the symmetry, to discourage the formation of a stable circulation cell inside the shaft. The top of the shaft and the side entrance is covered with mesh screens to prevent the clogging of the system by large size floating substances which may be sucked in.

In the model, the water surface inside the marina was covered by confetti before the tests to simulate a polluted marina. Various locations of the intake structure was tried and the removal of confetti during a fixed pumping duration was observed. In these tests, two narrow channels were provided across main and secondary breakwaters to allow water exchange with outside. It was found that a single intake would not be sufficiently effective over the whole marina area partly due to the size of the marina and to the blocking effect of floating pontoon piers. Thus, the use of two intakes simultaneously was investigated. The removal of confetti at the end of the tests with the most favorable intake location is described in Fig.11. In this figure, the narrow channels across breakwaters and the paths travelled by drogues from various basins of the marina are also indicated. There existed 2.0 m long gaps between the pontoons of the finger piers and the quay, and also narrower gaps between successive pontoon elements. Despite of these gaps, the presence of the floating pontoons was seen to adversely affect water transfer between neighboring basins so that complete removal of confetti could not be succeeded.

Two important questions which need to be answered for forced flushing alternative with pumping polluted water away from the marina are firstly the pump capacity required, and secondly the location of discharge. For the size of Marmaris Marina, the total pump capacity should be high enough to actuate a discharge rate in the range of 0.5 to 1.0 m^3/sec. In order not to carry the polluted water back to the harbor or to the adjacent recreational area, it was recommended that the discharge point to be close to the center of the circulation cell shown in Fig.9, some 300 m away from the corner which the main breakwater makes.

The function of two covered channels located across the breakwaters for increasing water exchange was tested with waves. It was found that water exchange through the channels caused by waves was local and insignificant. The effect of wind may be more significant for in and out water transfer through these openings. This phenomenon is included in the program of field observations.

As the result of studies carried out for Marmaris Marina, it was recommended that all measures should be taken to minimize pollutant input into the marina basins. Furthermore, regular cleaning of floating debris and oil patches should be exercised. Circulation patterns inside the marina by wind and the water exchange through the narrow channels across the breakwaters should be observed, and water quality within the marina is monitored. The decision for the construction of two intakes for pumping out polluted water from surface layer should be delayed until the results of field observations are available.

CONCLUDING REMARKS

Maintaining good water quality standards within marinas is often not an easy task. The problem is more severe when the tidal motion which is usually the main flushing agent, is weak. The best strategy to combat marina pollution is to exercise all possible three modes of action simultaneously. These are; minimize pollutant input, maximize flushing ability, and carry out periodical cleaning operations.

As it is demonstrated by this paper, alternatives which exist for inducing forced flushing in marinas are not simple solutions. They involve expensive arrangements. They should be fully investigated by model studies before implementation.

Quantification of the contribution of wind to circulation and flushing of harbors and marinas may allow simple modifications on plan shapes for improving flushing ability. Research along this line will be highly valuable.

REFERENCES

1.Bruun, P. and Viggosson, G.(1977), The wave pump: Conversion of wave energy to current energy, Jour. Waterway Port, Coastal and Ocean Div., ASCE, v.103, no WW4, Nov, pp:449-470.

2.Mashroutechi, F. (1991) , Mathematical Modeling of three-Dimensional Modeling of Coastal Circulation and Application to Marmaris Bay, M.S. Thesis, Dept. of Civil Engineering, M.E.T.U., February 1991, 95p.

3.Ozhan, E. (1979), "Pollution of Kemer Marina ", M.E.T.U., Coastal and Harbor Engineering Research Center (KLARE), Middle East Technical University, Ankara, Turkey, Tech. Rept. No. 21.

4.Ozhan, E. (1983), " Pollution Study of Kemer Marina, Antalya-Turkey", Proc. 2nd International Symposium on Environmental Pollution and Its Impact on Life in the Mediterranean Region, Iraklion (Crete), September 1983.

5.Ozhan, E. (1989), " Flushing of Marinas with Weak Tidal Motion", In: Marinas: Planning and Feasibility (Eds: W. R. Blain and N. B. Weber), Computational Mechanics Publications, Southampton, Boston, pp:485-498.

6.Ozhan, E.(1991), "Water quality improvement measures for marinas subjected to weak tidal motion", Proc. Third Int'l Conf. on Coastal and Port Engineering in Developing Countries, 16-20 Sep. 1991, Mombasa, Kenya, pp: 1337-1350.

7.Ozhan, E. and Ergin, A.(1985), "Marmaris marina project", Coastal and Harbor Engineering Research Center (KLARE), Middle East Technical University, Ankara, Rept.no. 33. (In Turkish).

8.Ozhan, E. and Ergin, A.(1987), "Datca Marina Project", Coastal and Harbor Engineering Research Center (KLARE), Middle East Technical University, Ankara, Rept. No. 37. (In Turkish).

9.Ozhan, E. and Ergin, A.(1990),"Circulation patterns of Marmaris Bay and Pollution of Marmaris Marina", Coastal and Harbor Engineering Research Center (KLARE), M.E.T.U., Ankara, Rept.No.44 (In Turkish).

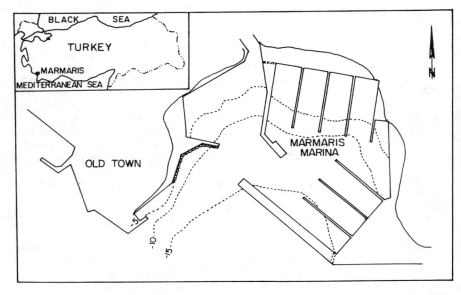

Fig. 1 - The location and layout of Marmaris Marina

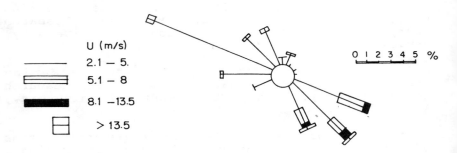

Fig. 2 - Wind rose for Marmaris Bay

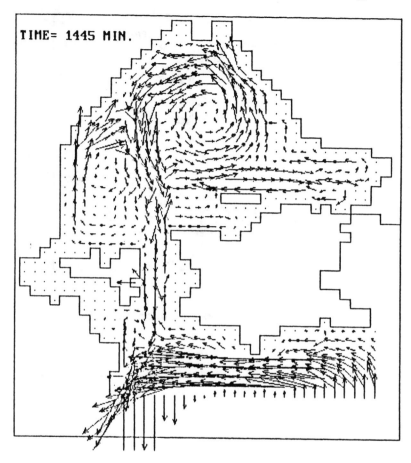

Fig. 3 - Current pattern in Marmaris Bay caused
by tidal motion and southeasterly wind

K

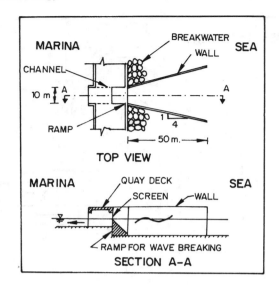

Fig. 4 - Details of wave pump

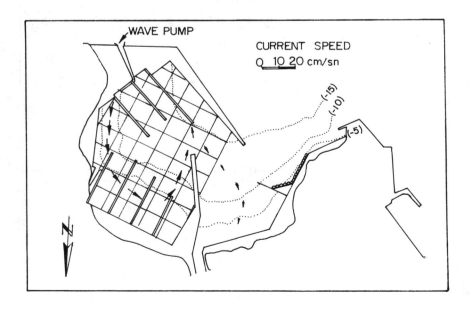

Fig. 5 - Current pattern generated by wave pump

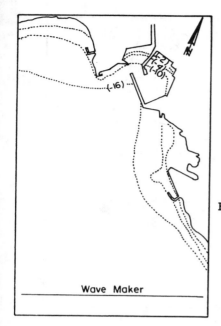

Fig. 6 - Layout of laboratory
         model in   the   study
         of 1990

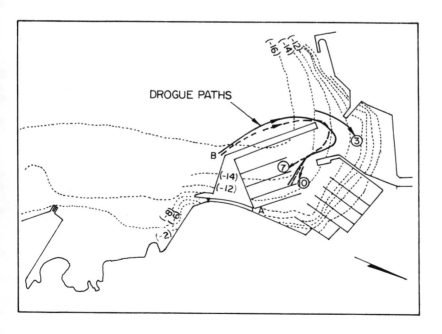

Fig. 7 - Paths of weak currents generated by wave
         pump

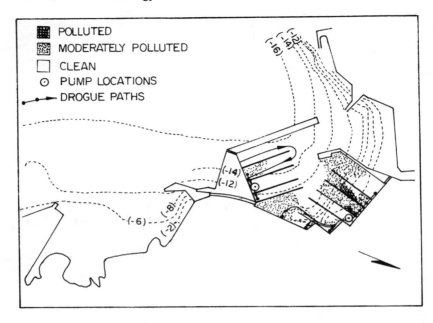

Fig. 8 - Currents entering the marina at the start of a storm

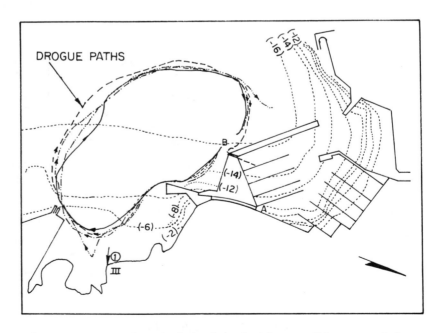

Fig. 9 - Steady state circulation cell caused by waves

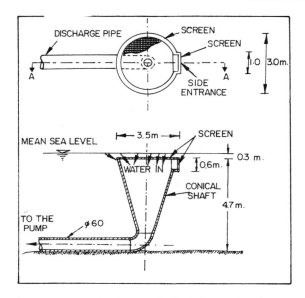

Fig.10 - Details of intake structure

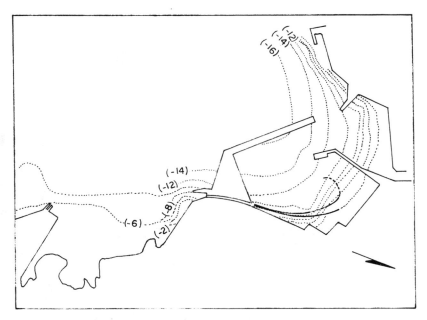

Fig.11 - The level of cleaning at the end of test
with the most favorable intake locations
(2 intakes)

Hydraulic Mini-Model Studies

T.K.H. Beckett

Beckett Rankine Partnership, 270 Vauxhall Bridge Road, London, SW1V 1BB U.K.

ABSTRACT

After discussing the effect of fluid shear force on sediment movement the difference in shear force conditions for uni-directional and tidal flow are examined. It is concluded that the fluid shear force and consequent bed lift can be as high or higher at the turn of the tide as at mid-tide when the net velocity is greatest. Examples of prototype measurements supporting this theory are given.

The difficulty of modelling tide turn shear forces is examined and the advantages and disadvantages of different types of model are explained. An economical type of mobile bed model is described together with the method used for proving its capacity for prediction of siltation movements.

INTRODUCTION

Before any hydraulic model is constructed the problem of similarity has to be addressed. The difficulty is to ensure that the model behaves similarly to the prototype in the respect which it is required to study. No hydraulic model, except a full scale one, can be similar in all respects to the prototype.

In a model intended to study sediment movements we must recognise the mechanism of sediment lift, transportation and deposition before determining the critical factors for similarity in a model. While the behaviour of a sediment is complex the critical factors determining whether a particular particle moves or not are the shear force exerted upon it and its mass. The shear force arises from a greater fluid velocity above the particle than below it which gives rise to hydrodynamic lift. If the lift is great enough to overcome the gravitational attraction of the particle then the particle will rise into the moving fluid and if it finds conditions of suspension will be transported at the mean velocity of the surrounding fluid.

Hydrodynamic lift or drag on a particle will always tend to attract the particle away from areas of high shear into an adjoining zone of lower shear. Thus even under laminar flow conditions particles may rise from the bed up into the body of the flow or move laterally across the flow into areas of little or no shear (e.g. into a dock entrance adjacent to a flowing river).

Of course turbulent flow will greatly increase the mixing of the sediment within the flow but even in a model without turbulent flow shear force mixing is capable of distributing and depositing sediment.

Figure 1 shows the velocity distribution with depth in a channel with laminar flow, the slope of the curve gives the shear force at any depth. Under these conditions the velocity curve is parabolic and the shear force extends throughout the flow although it is at a maximum near the bed. In Figure 2 is the corresponding curve for turbulent flow. In this case the shear is again greatest at the base although it effectively reduces to zero below mid-height. In turbulent flow conditions eddy currents are also present which not only result in sediment mixing above bed level but also produce their own small scale flow system.

TIDAL FLOW

So far I have only considered uni-directional flow which is relatively well researched and understood. To my knowledge there are over 20 accepted formulae for uni-directional sediment transportation which give reasonable results. What then happens in reversing tidal flow and why do all the uni-directional sediment transportation equations become so unreliable for tidal flow? (inaccuracy factors of 100 are common).

It is accepted that increasing shear force results in increased bed lift and from Darcy Weisbach the bed shear stress is proportional to the square of the velocity:-

$$t_o = \frac{feu^2}{8} \qquad \text{Where} \qquad \begin{aligned} t_o &= \text{bed shear stress} \\ f &= \text{friction factor} \\ e &= \text{fluid density} \\ u &= \text{mean velocity} \end{aligned}$$

Many modellers assume from this relationship that the maximum bed lift occurs when the flow velocity is greatest i.e. at mid-flood or mid-ebb. This however is not necessarily the case.

Figure 3 shows the changing velocity profile in a channel at the turn of the tide, the curves run from A to E. Under the influence of the reversed surface slope the slower moving water near the boundaries of the channel changes direction first so that at high or low tide the bottom and edges of the water mass are travelling in opposite directions to the top centre flow. While the net velocity of the river flow is close to zero the shear forces are clearly high and, for certain river locations, can exceed the mid-tide shear values. In particular it should be noted that the curve's slope, i.e. the shear force, at D and E is greater near the bed than in the steady state condition at A or F.

It is therefore to be expected that while the greatest transportation of sediment will occur at mid-tide, sediment will also be lifted into suspension at the turn of the tide. Do measurements in the field support this theory?

One of the most comprehensive tidal river studies of sediment transportation was carried out by Sir Claude Inglis and Fergus Allen in their work on the Thames in the 1950s. Figure 4 shows graphically the result of their measurements of current velocities and suspended solids concentrations throughout the tidal cycle. The measurements were taken in Long Reach at a level one foot above the bed. The most noticeable feature of these curves is the massive rise in sediment concentration just after low water. There is a similar but smaller rise just after high water. The figures show that 1 foot above the bed the solids concentration just after low water was over 2,000 ppm whereas the maximum mid-tide concentration was no more than 1,000 ppm. The effect at high water shows the same bed lift occurring at slack water although the greater depth of the river and slower tide turn makes it less pronounced.

Manz and Wakeling found a similar effect during their studies of sedimentation in Bridgwater Bay during 1980. At two of their six sampling stations sediment concentrations 1.5 metres above the bed showed a double peak (Figure 5) whereas if the concentrations were in-phase with the sinusoidal velocity profiles then only one peak would occur per tidal cycle. At the other 4 sampling stations which were located further offshore in deeper, slower moving water the sediment concentration curves were closer to a sinusoidal profile although minor secondary peaks were still observed.

MODELLING SEDIMENTATION

If we are to correctly model sedimentation effects it follows from the above that we must achieve similarity of fluid shear force throughout the tidal cycle. Measuring shear force is not easy particularly around the turn of the tide where the situation is highly dynamic. The flow direction and velocity are not only very different across the river's section but they also vary every minute.

The computer models available today are hampered by the poor results of uni-directional sedimentation equations when applied to reversing flow. They make little or no attempt to model the critical turn of tide shear forces owing to the enormous computer power required to model 3-D flows in the degree of detail required for normal features found in tidal rivers.

Large scale fixed bed Froudian models encounter similar difficulties with the turn of the tide. Measurement of shear forces over a range of depths and times is not a practical proposition especially as these models frequently rely on artificial roughening which not only adversely affects the geometrical similarity of the model but also interferes with the shear force similarity between model and prototype.

So, given the difficulty of measuring shear force in a model, how can we ensure shear force similarity with the prototype. The answer is to follow nature and use a sediment in the model to detect and react to the shear force in the same way as the prototype. If the model sediment can be made to lift and to settle in the same pattern as in nature we can be confident that we have got not only shear force similarity but also sedimentation similarity.

Live bed sedimentation models used to be common but have tended to fall from popularity on account of the length of time needed to tune and run them with the consequent high cost. They are now most frequently used for very localised studies such as for scour around bridge piers. There is however a method of significantly reducing the cost of live bed models by reducing their scale. By being smaller 'mini-models' are not only less expensive to build but the tide cycle times are reduced so that several runs of the model can be carried out in a single day.

Reducing model scale brings with it its own problems which have to be addressed. Firstly turbulent flow similarity is clearly impractical even if artificial roughening were to be used. In fact this does not present a serious problem providing shear forces are sufficient to provide shear force mixing of the sediment into the flow. For geometric similarity the models are in any case more accurate without artificial roughening.

The second difficulty is that at Froudian velocity scales no movement of sediment will occur. In the prototype mid-tide velocities of around 0.5m/sec will be necessary before bed lift will be initiated. On a model with a vertical scale of 1:250 the corresponding Froudian scale velocity will only be $0.5/\sqrt{(250)} = 0.03$m/sec. Which is clearly inadequate to initiate any sediment movement.

To overcome this problem two measures are employed, use of a less dense sediment and shortening the tidal period. In the model a sediment with a specific gravity of 1.3 is used as opposed to a specific gravity of around 2.6 in the prototype. While bed lift initiation formulae are numerous it is generally accepted that the critical shear stress for bed movement τ_c is proportional to $(e_s - e)/e$ where e_s = sediment density. By reducing e_s in the model τ_c is reduced by a factor of $1.6 \div 0.3 = 5.3$ for the model.

Thus from the Darcy Weisbach equation above the velocity scale for the model should be $\sqrt{(V_p \div 5.3)}$ where V_p = fluid velocity in the prototype, which for the 1:250 vertical scale example above gives a model velocity of 0.3m/sec for initiation of bed lift, a factor of 10 greater than the Froudian velocity scale.

This velocity scale calculation is only truly applicable to the mid-tide uni-directional flow condition. In practice, both in the prototype and the model, we know that much of the bed-lift takes place at the turn of the tide. Turn of the tide shear forces are less dependent on maximum flow velocities than on tidal period. Thus by progressively shortening the tidal period from the Froudian scale until bed lift initiation is observed the correct model period can be empirically established. In practice a tide period of around half the Froudian period is generally found to be effective although it is model specific and needs to be established for each individual model.

SCALES AND MODEL CONSTRUCTION

Before building a model the first decision is the scale to be used. In a mini-model we want the physical size of the model to be manageable and this usually restricts the physical size of the modelled area to a maximum of 1.5 x 1.5 metres. The area modelled will depend on the complexity of the project under study. With a proposed marina in a river it would generally be necessary to model the whole of the reach in which the marina lies together with the bends at either end of the reach. If flows in the region are split by an island or other complex feature then the whole of that feature will need to be modelled. In some cases it is found that the model scale necessary to encompass all the significant features adjacent to the site is too small to enable a detailed study of the site. In this case two models are made, the first to study the general location and the second to examine the site in detail to a larger scale. The first model is then used as a control to tune the flows and general siltation of the detailed model.

In order to prevent surface tension and other viscous effects from over influencing the model it is desirable to have a model tidal height which is not too small.   A height of 20mm is often used which for a 5 metre prototype tidal height gives a vertical scale of 1:250.

All small siltation models need to be distorted by having a larger vertical scale than horizontal in order to maintain significant velocities at the edges of the channel.   Nature distorts small channels herself; in a sediment bottomed river large channels always have flatter side slopes than small channels and low tide channels are always steeper than at high tide providing no hard boundary is present.

The degree of necessary distortion depends on the scale used but for a live bed model experience shows that a distortion of 6 works well for a vertical scale of 1:250 giving a horizontal scale of 1:1500.

Before building the model it must be decided whether it is to be a truly live bed model or a fixed bed silt injection model.   A true live bed model has all the prototype's hard boundaries and features reproduced in rigid material but the bed of the model is left flat and at a level below the deepest part of the river.   Before running the model silt is screeded in to a flat mean level throughout.   The model is then tuned so that when run it develops the flat bed into the present day contours found in the prototype.

A true live bed model gives a very good control as once it is tuned to reproduce the prototype contours it has also proved itself for prediction. However, tuning a true live bed model is time consuming as each run will take several hundred tide cycles to develop the bed.   For this reason fixed bed silt injection models have been developed.   These models are rigidly formed to represent either today's prototype contours or recent historic contours.   Areas where it is proposed to make alterations (such as proposed dredge areas) are modelled low and built up with 'plasticene'.   Silt injection models are run by introducing model silt into the flow either upstream, downstream or on both sides of the area of interest.   Proving a silt injection model requires historic records of the bathymetry to be available so that the model may be tuned to reproduce the prototype's changes.

The construction of either type of model is similar.   Individual contours are cut from plywood then glued and pinned together to form a 3-dimensional form.   Steps between contours are filled with waterproof filler, sanded and then painted.   The model is placed in a tilting tray with upstream reaches of rivers reproduced by labyrinths while downstream areas are reproduced by a reservoir area (Figure 6).   Significant upstream flows are reproduced by pumps circulating water from the reservoir.   It is unusual for the size of the total model including labyrinths and reservoir to exceed 3.5 x 1.5 m.

To simulate tidal action the tray in which the model sits is tilted about the central axis by a motor driven cam. The cam is adjustable to reproduce different tidal curves while the electro-hydraulic motor is designed not only to be easily variable in speed but also to deliver a fixed speed irrespective of torque.

MODEL TUNING AND PROVING

The 'proving of models' is a term which has different meaning for different modellers and is thus a cause of confusion. 'Proving' is often used for both physical and mathematical models to describe their ability to reproduce tidal and velocity phenomena as observed in the prototype. Such 'proving' does not give any assurance as to the ability of the model to forecast consequential changes in the prototype that will result from anticipated construction works.

Strictly there is only one absolute proof of effectiveness for any model which is to compare its forecasts with the prototype after the works have been constructed and enough time has passed for the full effects to be evident. This is a long term feed-back process that we have been carrying out for mini-models since 1950.

While this feed-back is essential for confidence in a system how can we ensure that our newly constructed model is tuned correctly and how can we prove that it is capable of predicting siltation changes correctly?

The method we use for mini-models is to find some historic documented alteration in the vicinity of the proposed works and tune the model so that it accurately reproduces this change. The alteration may be the natural growth of a sand spit, alteration to sedimentation as the result of past civil engineering works or characteristic patterns of sedimentation occurring as the result of dredging. The reproduction of dredging effects is particularly useful as it permits ready calibration of the model for the forecast of maintenance dredging volumes.

Once the mini-model is tuned to reproduce such consequential changes in the sediment regime it has demonstrated its ability to forecast and the proposed works are examined. I believe that the mini-modelling technique is the only siltation modelling technique currently in use which proves every model for its ability to predict changes before using them for testing proposed works.

INSTRUMENTATION AND CONTROLS

Working at smaller scales does require more accurate and less intrusive measuring devices than are used on larger scale models. We have developed a PC supported laser distance measurer for measuring water levels and/or wave heights in models which is illustrated in Figure 7. The laser measures the distance to a float which may be as small as 2mm in diameter. The instrument takes readings six times a second to an accuracy of 0.01mm. Which on a 1:250 vertical scale model is equivalent to 2.5mm on the prototype. To avoid giving an unrealistic expectation of accuracy we round the instruments reading off to the nearest 0.1mm in the software. Figure 8 shows a screen shot of the output from the instrument. Successive tide profiles can be superimposed on each other to examine the effect of tuning alterations such as modifications to the labyrinth or downstream reaches. The model tidal curves can be compared with the prototype on screen or via a print-out.

Measurements of siltation are currently carried out by photographic recording and by physical removal of the silt from measured areas of the model, drying and weighing. We are however developing a device, based on the laser equipment, for scanning across the model while measuring the bed profile. The resulting sections taken before and after a run will be able to be plotted out and superimposed to reveal changes.

Velocities and flow paths in the model are measured by using multiple floats and time-lapse photography. Velocities at depth are not usually measured when working to such small scales owing to the difficulty of distorting readings by the intrusion of the measuring equipment.

MINI-MODEL COSTS

The cost of any model study is not only site specific but also depends on the amount and detail of hydrographic data that has already been gathered. Data collection is costly and can double the cost of a study. However many Port Authority or Admiralty base surveys are sufficiently accurate for the construction of a mini-model so in many cases the data gathering exercise is limited to float tracking, sediment analysis and concentration measurements. If this is the case then the typical cost of a mini-model study is likely to be in the range of £20,000 to £60,000 depending on the number of models which need to be constructed and the number of construction options to be analysed. If no reliable control information is available for the site and a true live bed model is required then the cost of the study may exceed this range.

CONCLUSIONS

In a tidal river sediment bed lift can occur both during the mid-tide run and during the turn of the tide. Field measurements indicate that in some rivers the turn of tide bed lift is appreciably greater than the lift occurring during periods of greatest net velocity. If a model study is to accurately reproduce tidal sedimentation then the turn of the tide shear force effects need to be represented as well as the mid-tide effects. Physical models using a mobile sediment are currently the only types of model available that can reproduce tide turn sediment effects.

While large scale mobile bed models are time consuming and costly to run small scale mini-model studies can confidently predict sedimentation effects at a cost which is similar to a mathematical study.

REFERENCES

1.   Beckett A.H. and Marshall S., Design and Construction of Muara Deep Water Port, Brunei. ICE Proceedings Part 1 1983 No. 74.

2.   Beckett A.H., Conditions Initiating Bed Lift and the Effect of Tide Cycle Time. Proceedings International Association for Hydraulic Structures Research, Grenoble 1949.

3.   Inglis, Sir Claude and Allen F.H., The Regimen of the Thames Estuary as Affected by Currents, Salinities and River Flow. ICE Proceedings 30 May 1957.

4.   Mantz P.A. and Wakeling H.L., Aspects of Sediment movement near to Bridgwater Bar, Bristol Channel. ICE Proceedings Part 2 1981 No.71.

5.   Delo E.A., The Behaviour of Esturine Muds during Tidal Cycles.   Hydraulics Research, Wallingford, Report No. SR138 February 1988.

6.   Weare T.J., Mathematical Models. pp15-31 Proceedings of ICE Conference on Hydraulic Modelling in Maritime Engineering, London 1981.

7.   Allen J., Scale Models in Hydraulic Engineering. Longmans 1947.

8.   Yalin M.S., Theory of Hydraulic Models. Macmillan 1971.

9.   Sharp J.J, Hydraulic Modelling. Butterworths 1981.

10.   Sleath J.F.A., Sea Bed Mechanics. John Wiley & Sons 1984.

11.   Lamb H., Hydrodynamics, 5th edition. Cambridge University Press 1924.

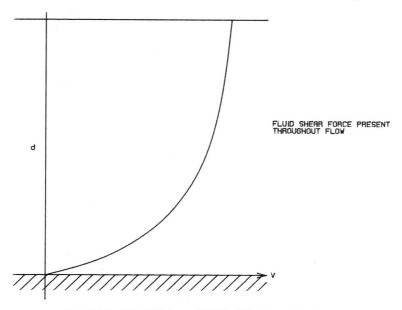

FLUID SHEAR FORCE PRESENT
THROUGHOUT FLOW

FIGURE 1. VELOCITY vs DEPTH FOR LAMINAR FLOW

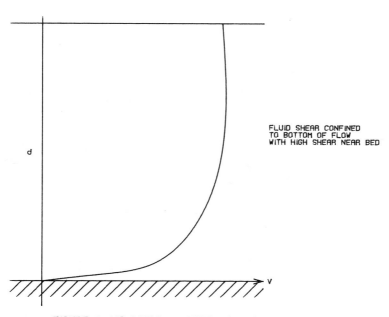

FLUID SHEAR CONFINED
TO BOTTOM OF FLOW
WITH HIGH SHEAR NEAR BED

FIGURE 2. VELOCITY vs DEPTH FOR TURBULENT FLOW

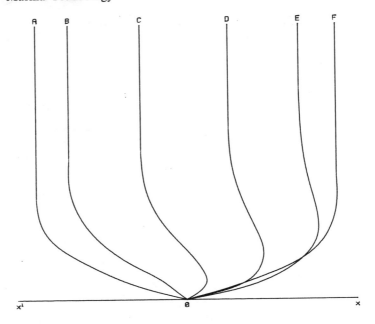

FIGURE 3. VELOCITY vs DEPTH AT TURN OF TIDE   AFTER LAMB

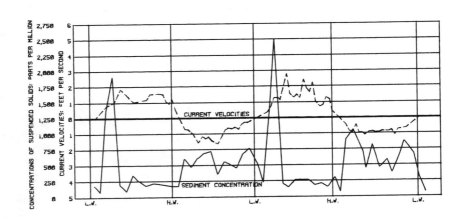

FIGURE 4. RIVER THAMES – LONG REACH   SEDIMENT CONCENTRATION AND
         VELOCITY 1 FOOT ABOVE BED LEVEL   AFTER INGLIS & ALLEN

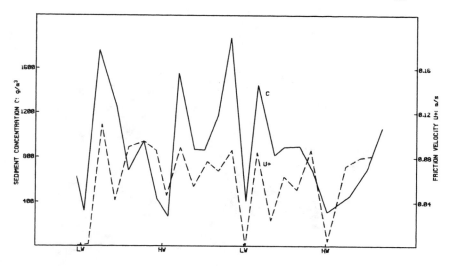

FIGURE 5. BRISTOL CHANNEL – BRIDGEWATER BAR   SEDIMENT CONCENTRATION
AND VELOCITY 1.5m ABOVE BED LEVEL   AFTER MANTZ & WAKELING

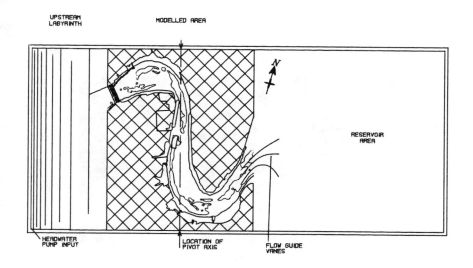

FIGURE 6. GENERAL ARRANGEMENT OF A MODEL USED TO STUDY DREDGING
PROPOSALS FOR THE RIVER MEDWAY
MODEL VERTICAL SCALE 1:250 HORIZONTAL SCALE 1:1500

FIGURE 7. THE LASER DISTANCE MEASURER RECORDS THE LEVEL
OF A LIGHT WEIGHT FLOAT

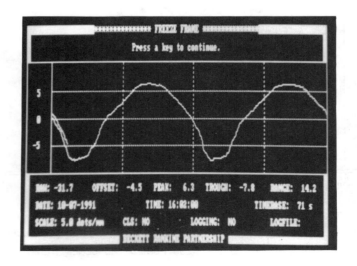

FIGURE 8. READINGS TAKEN SIX TIMES PER SECOND ARE DISPLAYED IN
REAL TIME TO GIVE THE MODEL TIDAL CURVE

Gulf of La Spezia: An Analysis of the Mechanisms of Water Exchange between the Inner Gulf and the Open Sea

A. Borella (*), M. Cambiaghi (*), P. Marri (**),
R. Meloni (***), M. Miserocchi (****), R. Nair (**),
M. Tomasino (*****)

(*) CISE - Tecnologie innovative, Milano,
(**) ENEA-CREA, La Spezia,
(***) C.N.R.-ISDGM-S.O., La Spezia,
(****) ENEL-ULP, Piacenza,
(*****) ENEL-DSR, Venezia

ABSTRACT

The experimental results of a wide study aimed at characterizing the hydrodynamics and the water quality of the Gulf of La Spezia show a low contaminant concentration in spite of the high population density and industrial activity. A 3-dimensional hydrodynamical mathematical model, calibrated with the experimental data, has been applied to investigate the flow field. The most important mechanisms of the water exchange between the internal Gulf and the open sea are the internal anticlockwise residual circulation, the tide and the 'pumping' effect of a local seiche. This seiche (with a period of 1.2 h) causes wide level oscillations and high current velocity variations at the entrances to the harbour.

INTRODUCTION

The use of large quantities of water (60 m^3/s) from the Gulf of La Spezia as a coolant for the local electrical power station and it's subsequent impact on the marine environment necessitated a thorough understanding of the environmental processes governing this area. A study was initiated to characterise the circulation within the Gulf. A mathematical model describing the hydrodynamical processes in the waters of the Ligurian sea adjacent to the Gulf helped to define the external forcing terms while the internal ones were evaluated on the basis of data obtained from measurements of some of the major meteorological and oceanographical parameters over long periods of time. This paper describes the results of the analysis of the data acquired from physical measurements and those from the application of the model.

METHODS AND METHODOLOGY

In-situ measurements were carried out during three periods (30 may-1 june 1989; 29 august-1september 1989; 30 january-2 february 1990). The main goal of these measurements were to provide information on the currents regulating the movements of the waters both within and outside the harbour and also some knowledge of the spatial distribution of various physico-chemical parameters in the waters of the Gulf. Temperature and salinity were measured with a Beckman RS-11 thermosalinometer while the velocities and directions of the different currents at various points in the Gulf were obtained with Martek SD-4 currentmeters moored at the surface and at the bottom. Current drogues were placed at the entrances of the harbour at varying depths and their paths were followed and noted. Simultaneously, the trajectories of the different masses of water were measured with a number of Lagrangian currentmeters at different depths.

Time series measurements - Time series data were obtained from moored currentmeters (Aanderaa currentmeters of the RCM4 or RCM7 types, positioned 4m from the surface and 1m from the bottom and preset for data aquisition every 10'), sea-level recorders (Aanderaa WLR preset for data aquisition every 5') and transmitting buoys (Idronaut 501 probes giving a vertical profile of the water column in terms of temperature, conductivity, dissolved oxygen and pH every 4 h); the necessary meteorological data were collected at three stations that remained unchanged for the entire duration of the study. The positions of the measuring instruments are shown in fig. 1.

RESULTS

The analysis of the data obtained from the time-series and the in situ measurements allow a tentative description of the circulation within the Gulf and its determining factors. The factors discerned comprise: the oscillation of the sea level, the wind, a density gradient and the discharge from the power station.

Tide and seich - The analysis of the sea-level measurements show a frequency oscillation of a period of 70' and a mean amplitude of 10 cm (similar to the amplitude of tide) overlapping the tidal oscillation (Fig. 2). This is the frequency of a stationary wave 8 km long, that in its length corresponds to the length of the Gulf of La Spezia in the longitudinal dimension. The node of the wave is located along an imaginary line uniting Tellaro with the island of Tino, while the maximum amplitude is seen in a zone within the confines of the port. The maximum velocity of the water mass was measured at the entrance of the harbour at the extremities of the breakwater. This velocity (at a depth of 5m) having its origin in the seiche, exceeded many times, 70 cm/s (Fig. 3) while the velocity of tidal origin measured only a few centimeters. The large amplitude of the seiche may be the result of particular weather and sea conditions as can be seen from the fact that strong winds from the south, with the subsequent rough seas, produced oscillations greater than those produced at other times (Fig. 4). For example, it was seen that strong north winds, which do not greatly affect sea conditions, have no great effect on the seiche. It was also possible to estimate, from current velocities measured at different points in section at the entrances to the harbour and the area of the vertical section involved, the turnover of the water mass, for every seiche cycle, at 10 million cubic meters. On the basis of this value, it may be hypothesized that, in the event of complete mixing, about 10-20 days are necessary for a total exchange between the water within the harbour and the sea. However, this mixing occurs only in the first 500-600m from the entrances, after which the water mass is primarily affected by current velocities derived from density variations and the wind.

Density gradients and wind - The wind and the density gradients existing between the harbour and the open sea are the primary factors determining the residual circulation within the Gulf. The analysis of the consecuetive vectors obtained from successive singular current measurements at C1, C2 and C3 (fig. 1) permitted the calculation of the average vector velocities over a long period of time both at the surface and at the bottom. Within the Gulf, the circulation tended to be anticlockwise with water entering at the bottom and exiting at the surface at the entrances to the harbour. It was noted that the waters at the western entrance exhibited velocities about five times greater than those at the eastern one. A representation of the residual currents at the surface and at the bottom of the Gulf is shown in fig. 7. The letter "S" denotes a zone displaying conditions of stagnancy where the waters exhibit an anomalous horizontal distribution in the various physicochemical parameters measured (for example, the salinity in fig. 8). This situation defines the conditions of stratification and maintains an "island" of less saline water that enters the Gulf as urban discharges. The bottom currents at the western entrance were found to follow a channel extending, diagonally, all the way across the port. The velocities of the currents in the Gulf are governed, to a great extent, by the distribution of the density. Fig. 9, for example, shows a density gradient parallel to the direction of the surface currents (fig. 5). The wind forcing on the surface of the water, causing a movement similar to the circulation of the barocline current, amplifies the total circulation. The diurnal sea breeze, from the south, driving the water along the eastern coast of the Gulf and the nocturnal sea breeze, from the north, driving the water along the western coast, amplifies the anticlockwise nature of the current generated by the density gradient and the effluent discharge of the electrical power station.

The coolant discharge of the electrical power station - The cooling discharge of the electrical power station, shown with an arrow in fig. 5-6, reinforces the baroclinic general circulation by its jet velocity (1 m/s at the outlet) and by increasing the density gradient due to the immission of hot superficial coolant water. The real role of the coolant discharge will be evaluated with the simulations below.

MATHEMATICAL MODELIZATION

A hydrodynamic mathematical model TRIMDI has been applied to the gulf of La Spezia to define the terms forcing the local circulation already described and to estimate the flushing time of the water of the harbour . It is worth noting that, in our analysis, we have summed up the singular effects of the forcing terms even if the non linearities are often quite substantial.

Main characteristics of the TRIMDI model - The calculus code is intended for numerically simulating the heat dissipation in free-surface
bodies of water. It accounts for spatial variations in water density
resulting for variations in temperature (or concentration) so that
density convection currents and stratification phenomena may be reproduced. The mathematical base consists of general fluid-dynamic
and motion equations in a three-dimensional form modified on the basis of the following assumptions: a) all discarges and intakes respectively add or remove water by mainly horizontal motion, so that the vertical component of the acceleration of the water particles is negligible
with respect to the acceleration due to gravity at every single point.
Thus, the vertical pressure distribution follows the law of hydrostatics
(taking into account the vertical variation in density); b) density variations due to temperature (or concentration) are only introduced into the vertical equilibrium equation (Boussinesq theory); this allows
the pressure gradients to be correctly evaluated; c) the typical values
for the depth are assumed to be small compared to those for the horizontal distances. Consequently, the vertical components of water velocity should be equally small compared to the horizontal components.
The TRIMDI code, thus derived, is based on the following equations:
- the mass balance equations for incompressible fluids;

$$\frac{\partial u}{\partial x} + \frac{\partial v}{\partial y} + \frac{\partial w}{\partial z} = 0$$

- the balance equations for the momentum components in the orthogonal horizontal directions x and y, in terms of the velocity components u, v, w , averaged out with respect to the turbulent fluctuations;

$$\rho_0 \frac{\partial u}{\partial t} + \rho_0 \frac{\partial u^2}{\partial x} + \rho_0 \frac{\partial uv}{\partial y} + \rho_0 \frac{\partial uw}{\partial z} = -\frac{\partial p}{\partial x} + (\frac{\partial}{\partial x}\mu_t \frac{\partial u}{\partial x} +$$
$$+ \frac{\partial}{\partial y}\mu_t \frac{\partial u}{\partial y} + \frac{\partial}{\partial z}\mu_t \frac{\partial u}{\partial z}) + (\frac{\partial}{\partial x}\mu_t \frac{\partial u}{\partial x} + \frac{\partial}{\partial y}\mu_t \frac{\partial v}{\partial x} + \frac{\partial}{\partial z}\mu_t \frac{\partial w}{\partial x})$$

$$\rho_0 \frac{\partial v}{\partial t} + \rho_0 \frac{\partial uv}{\partial x} + \rho_0 \frac{\partial v^2}{\partial y} + \rho_0 \frac{\partial vw}{\partial z} = -\frac{\partial p}{\partial y} + (\frac{\partial}{\partial x}\mu_t \frac{\partial v}{\partial x} +$$
$$+ \frac{\partial}{\partial y}\mu_t \frac{\partial v}{\partial y} + \frac{\partial}{\partial z}\mu_t \frac{\partial v}{\partial z}) + (\frac{\partial}{\partial x}\mu_t \frac{\partial u}{\partial y} + \frac{\partial}{\partial y}\mu_t \frac{\partial v}{\partial y} + \frac{\partial}{\partial z}\mu_t \frac{\partial w}{\partial y})$$

- the balance equation for the momentum components in the vertical
direction z, which due to assumption "a" is reduced to the hydrostatic
equilibrium   equation;

$$\frac{\partial p}{\partial z} = -\rho g$$

- the heat balance   equation;

$$\rho_0 c_p \frac{\partial T}{\partial t} + \rho_0 c_p \frac{\partial uT}{\partial x} + \rho_0 c_p \frac{\partial vT}{\partial y} + \rho_0 c_p \frac{\partial wT}{\partial z} = \frac{\partial}{\partial x}K_T \frac{\partial T}{\partial x} + \frac{\partial}{\partial y}K_T \frac{\partial T}{\partial y} + \frac{\partial}{\partial z}K_T \frac{\partial T}{\partial z}$$

- the state equation which defines the law governing the dependence
of the water density on the water temperature ;

$$\rho = \rho \quad (T,......)$$
$$\mu_t = \mu_t \quad (u,v,...)$$
$$K_T = K_T \quad (u,v,...)$$

- the kinematic condition on the free surface;

$$w(x,y,t) = \frac{\partial \eta}{\partial t} + u(x,y,\eta,t)\frac{\partial}{\partial x}\,\eta + v(x,y,\eta,t)\frac{\partial \eta}{\partial y}$$

- the appropriate conditions for the solid and the open boundaries. The listed equations have been treated as follows. An initial discretising process was caried out by dividing the domain occupied by the water into 'L' number of superimposed horizontal layers, all of constant thickness (except for the surface layer whose thickness depends on the space-time configuration of the free surface). In this way, the following variables were obtained for each index layer k (k = 1 ...,L): velocity components u, v, w,; temperature T, pressure P, density ρ (as average values along the thickness of each layer) and, just for the surface layer (k = L), the level of the free surface. All the variables were functions of the horizontal coordinates x, y and the time t. Each equation resulting from the vertical integration was then rendered discrete along the horizontal plane x, y, by means of a method of finite differences supported by a system of three offset rectangular grids. This produced a system of ordinary differential equations which was integrated with respect to time by means of the Crank-Nicholson implicit method. An extension of the well known SIMPLE method was used to solve the equations thus obtained . The Reynolds forces, which appear in the balance equations for the quantities of horizontal motion, were treated as viscous turbulent forces, while equivalent terms in the heat balance equation were treated as diffusive turbulent flows. A simple turbulence model, based on the mixing length and subgrid scale turbulent viscosity concepts make it possible to fix the values of local viscosity v_t and turbulent diffusivity α_t for each layer as functions of the position.

$$\sigma_x = \rho_0 v_t\ (\frac{\partial u}{\partial x} + \frac{\partial u}{\partial x})$$

$$\dot{\tau}_{xy} = \rho_a v_t\ (\frac{\partial u}{\partial y} + \frac{\partial v}{\partial x})$$

$$\alpha_{th} = v_{th} = l_h^2\ \sum_{i=l}^{2}\ \sum_{j=l}^{2}\ \left[\left(\frac{\partial u_i}{\partial x_j}i + \frac{\partial u_j}{\partial x_i}j\right)\frac{\partial u_i}{\partial x_j}i\right]^{1/2}$$

These values help to define the forces and turbulent diffusive fluxes which are trasmitted horizontally (across the vertical planes). The local values for v_t and α_t from the turbulence model were replaced by the corresponding v_{tm} and α_{tm} values for environmental turbulence, when the former were less than the latter. Semi-emperical relations that reduced the v_t and the α_t values with the increasing Richardson number, were used to define the momentum and diffusive turbulent fluxes responsible for stratification.

$$\alpha_t' = \alpha_t\left(1 + a'R_i\right)^{b'}$$

$$v_t' = v_t\left(1 + a''R_i\right)^{b''}$$

$$R_i = - \frac{g}{\rho} \frac{\frac{\partial \rho}{\partial z}}{(\frac{\partial u}{\partial z})^2 + (\frac{\partial v}{\partial z})^2}$$

The code also takes into account the heat exchange with the atmosphere through the free surface by means of a suitable term, representing the heat flow per unit of surface area, appearing in the equation for the heat balance in the surface layer. The latter is assumed to be proportional to the difference between the local temperature in the surface layer and the temperature of the unperturbed water.

Preliminary simulations - In the first simulation, only the sea breeze or the semidiurnal tide was considered as the forcing element. To quantify the single forcing effects on the field of motion in the area, a simplified grid was used (Fig. 10) The water column was divided into 4 layers; the surface and bottom layers had a smaller thickness than the middle layers.

Sea breeze effects - The wind shear, evaluated by the Kondo relation (3), has been computed using the wind measurements in winter.

Component	Velocity (m/s)	Beginning time	Stop time	
North	+3.5	0	12	Values of the
North	-3.5	12	18	components of
North	-3.5	18	24	the wind used
East	-2.0	0	12	in the simu-
East	0.0	12	18	lation.
East	-2.0	18	24	

The boundary conditions on the line from Tino island to Tellaro (Fig. 1) are as follows: a) a homogenous and time constant temperature equal to 13 degrees centigrades; b) the sea level is constant with time; c) the derivative of the horizontal component of the current velocity in the direction normal to the boundary is 0. Fig. 11 shows the component of the current perpendicular to the entrance of the breakwater. The results of the simulation show a current oscillation of a period of 70' but with values of an order of magnitude less than the experimental data.

Effects of the tide - In this case the open boundary conditions are: a) a homogenous and time constant temperature equal to 13° C; b) a variable sea level with a period of 12 h and 25' and an amplitude of 15 cm; c) the derivative of the horizontal component of the current velocity in the direction normal to the boundary is 0. The results of the simulation show two different oscillations (Fig. 12) the first with the tidal frequency and the second with the period of the seiches. The first oscillation causes current values of about 5-10 cm/s, the second about 1-2 cm/s. These values which are less than the experimental data show that the tide is not the only forcing factor.

Effects of the external current - To study the effects of the external current in the Gulf circulation, we have used a simulation in the area described in Fig. 13. On the left upper side there is the outflow and on

the right upper side is located the La Spezia Gulf. The left side describes the position at the 100 meter bathymetry and its length is about 60
nautical miles. The inflow is from the bottom. For the boundary conditions in this side, we have used the data obtained over two weeks by a
chain of currentmeters moored 30 nautical miles south-east of La Spezia at a bathymetry of 100m with a sampling time of 10' (4). They showed a mean direction of the current towards the north-west and two
principal periodicities: the first caused by the inertial component (a
period of 17 h) and the second by the semidiurnal tide (a period of 12 h
and 25'). The velocity on this side was forced by the mean values of the
component of the current described previously and two periodic
components for the inertial oscillations and the semidiurnal tide
which had an amplitude equal to the standard deviation of the
measured data. The nodes at the bottom of the grid have similar
velocities that change linearly with the depth which was divided
vertically into 9 layers. The first two layers (each of a thickness of 5m)
from the surface comprise the La Spezia Gulf. On the left upper side,
the derivative of the components of the velocity perpendicular to the
boundary was taken to be 0. The maximum amplitude of the sea level,
assumed to be mainly due to the semidiurnal tide was calculated using
the following relation (5): $Q = c\,h$ where: Q is the flow per unit of length of boundary; $c = (g\,H)^{1/2}$ is the wave velocity ; h is the elevation of
the free surface respect to the undisturbed level; H is the sea depth.
The Figs. 14 and 15 show the field of motion in the surface layer during conditions of maximum current velocities. Fig. 16 shows the time
series of the component of the current perpendicular to the breakwater at the western entrance and at Tellaro while fig. 17 shows the time
series of the sea level in the same areas. The graphs show results
which resemble, to a good degree, the experimental data for the period
and amplitude of oscillation. The variability of the mesoscale currents
seems to be the most important "activator" of the free oscillation of the
gulf of La Spezia.

Effect of the external currents on residual circulation - To evaluate the
residual component of the circulation in the Gulf due to the mesoscale
dynamic, a simulation of the current in the stationary state was
carried out. Fig. 18 shows the residual field of motion in the surface
layer. This current is prevalently longshore and turns to the northwest near the Gulf. The value calculated at Tellaro was 1 mm/s whereas
the value measured was 2 mm/s. Thus, the mesoscale circulation has a
poor influence on the residual circulation of the innermost part of the
harbour.

*Effect of the sea breeze and the coolant discharge from the electric po-
wer plant on residual circulation* - To study the residual circulation
solely in the harbour, we have taken the sea breeze and the coolant
discharge from the electric power plant for the forcing factors. The
area considered is shown in fig. 19 where the water column is divided
into 9 layers of similar thicknesses. The derivative of the normal component of the velocity at the nodes at the open boundary was assumed
to be zero. The coolant discharge of the electric power station has a
flow rate of 60 m^3/s and a velocity of 1 m/s with temperature 8.5°C
higher than the basin temperature. As this model cannot be used for

stratified waters, the results obtained describe only a typical winter situation when the water column is mixed. The figg. 20 and 21 show that, when the wind blows from the North, the superficial current exiting the harbour is greater at the western mouth than at the eastern one; the same contrast in the velocities at the mouth is seen for the bottom current entering the harbour. When the wind blows from the South an anticlockwise circulation is observed in all the layers with the current moving towards the open sea at the western mouth and towards the harbour at the eastern mounth. The values of the velocities are lower than those computed for the nocturnal sea breeze. This indicates that the flushing of the harbour waters into the open sea takes place during the night and mainly at the western mouth. These results agree with the experimental data with the exception that the simulations underestimate the current flow at the bottom. This is probably because the forcing factors with high frequencies have not been considered. But these results demonstrate that the action of the sea breeze and the coolant discharge are the most important forcing factors affecting the residual circulation in the harbour.

CONCLUSION

From the results obtained, it is possible to identify two major forcing factors affecting the circulation within the harbour. The first is a local seiche affecting the whole of the Gulf that causes currents of about 50 cm/s at the entrances. These favour the flushing of the harbour waters into the open sea. The second is the coolant discharge from the power station which produces a circulation of an anticlockwise nature within the harbour. This circulation helps to carry the waters from the internal harbour to the entrances. The simulations carried out over a wide area have also demostrated that the seiche is produced by the currents in the external part of the Gulf. The effect of the sea breeze is particularly important during the night and it augments the observed anticlockwise circulation; in fact, the circulation in the absence of the wind and the coolant discharge is of an order lower than that found in the presence of the same.

REFERENCES

Bogani V., Di Monaco A., Dinelli G., Leoncini A.: Analisi termoidraulica della dispersione di inquinanti in corpi idrici complessi; applicazione alla foce del Tevere. ENEL, Rassegna Tecnica, n. 3, 1983.
Rodi W.: Turbulence Models and their applications in hydraulics - A State of the Art Review. IAHR Publications, Delft, 1980.
Kondo J.: Air-Sea Transfer Coefficient in Diabatic Conditions. Boundary Layer Meteorology, 9, 91-112, 1975.
CNR, ISDGM, S. O., S. Teresa: Current and Temperature Measurements in front of the Ligurian Coast near La Spezia. CNR, Progetto Finalizzato - Oceanografia e Fondi Marini, La Spezia 1983.
Roed L.P., Cooper C.K.: Open Boundary Conditions in Numerical Ocean Models. From: Advanced Physical Oceanographic Numerical Modelling. Edited by J.J. O'Brien. D. Reidel Publishing Company, 1986.

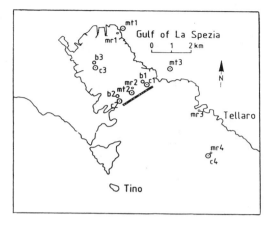

Fig. 1: Positions of fixed stations
b = buoys for physico-chemical
    parameters
c = currentmeters
mt = meterological stations
mr = sea level recorders

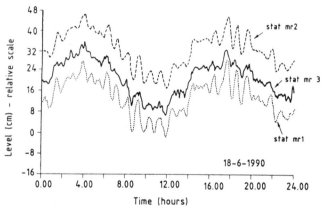

Fig. 2: Sea level registrations at 3 points in the Gulf.
The ocsillations of the seiche and of the tide are shown

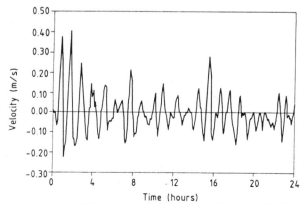

Fig. 3: The component of the water velocity perpendicular to the breakwater
at the western entrance at a depth of 5 meters on 30/05/1989

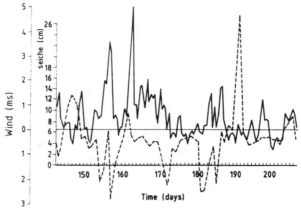

Fig. 4: Seiche (continuous line) and the N-S component of the wind (dotted line)

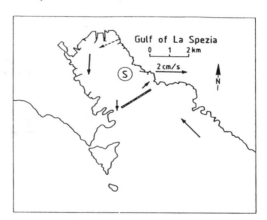

Fig. 5: Vectors of residual current velocity in the surface layer (-4m)

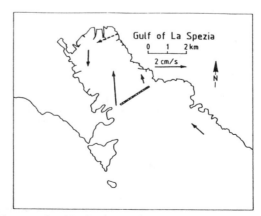

Fig. 6: Vectors of residual current velocity in the bottom layer (-9m)

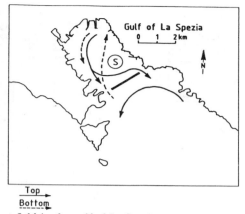

Fig. 7: Current field in the gulf of La Spezia.
The cyclonic vortex present both at the surface and at the bottom are shown along with the stationary zone (s)

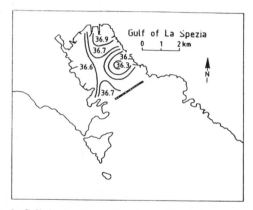

Fig. 8: Salinity in the surface layer (0-1.5m) on 22/02/1990

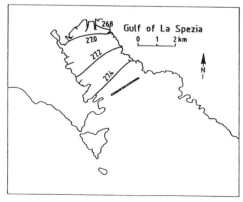

Fig. 9: Density in the surface layer (0-1.5m) on 22/02/1990

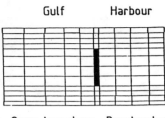

Gulf   Harbour

Open boundary   Breakwater

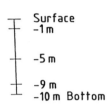

Surface
−1 m

−5 m

−9 m
−10 m Bottom

Fig. 10: Horizontal and vertical
schematizations of the gulf of La Spezia

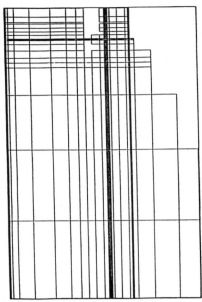

Fig. 13: Eastern ligurian basin:
A grid of the surface layer

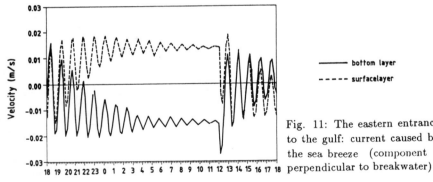

—— bottom layer

----- surfacelayer

Fig. 11: The eastern entrance
to the gulf: current caused by
the sea breeze (component
perpendicular to breakwater)

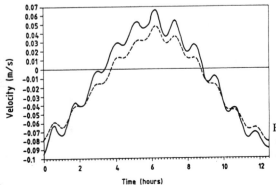

—— Eastern entrance

------Western entrance

Fig. 12: The eastern and western
entrances to the gulf: current
caused by the tide (component
perpendicular to the breakwater
in the superficial layer)

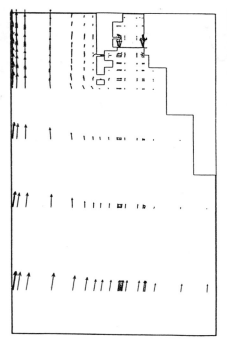

Time = 4583700 s

Fig. 14: Eastern ligurian basin: Field
of motion in the surface layer during
conditions of maximum current velocity
entering the gulf

Time = 4585500 s

Fig. 15: Eastern ligurian basin: Field
of motion in the surface layer during
conditions of maximum current velocity
exiting the gulf

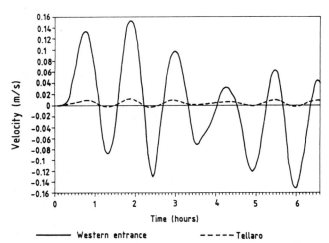

Fig. 16: Time series of the component of the current perpendicular to the
breakwater at the western entrance and at Tellaro

L

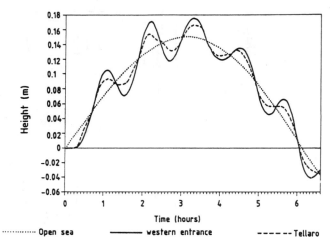

············· Open sea   ——— western entrance   ------- Tellaro

Fig. 17: Time series of the sea level at the western entrance to the gulf and at Tellaro

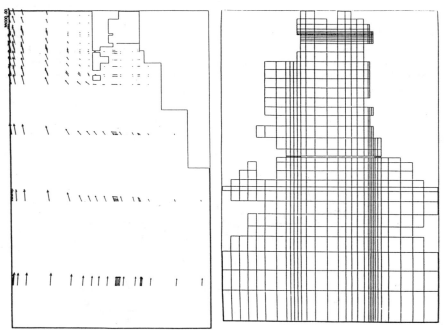

Fig. 18: Eastern ligurian basin:
The residual field of motion in
the surface layer

Fig. 19: Gulf of La Spezia:
A grid of surface layer

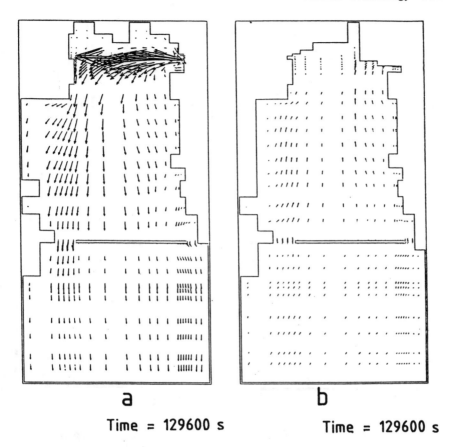

a

Time = 129600 s

b

Time = 129600 s

Fig. 20: Gulf of La Spezia: The residual current due to the sea breeze at night
(a) Surface layer
(b) Bottom layer

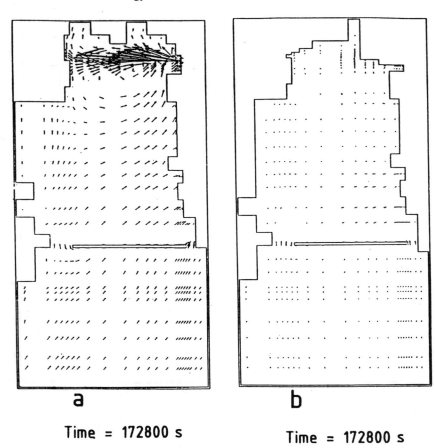

Time = 172800 s   **Time = 172800 s**

Fig. 21: Gulf of La Spezia: The residual current due to the sea breeze during the day
(a) Surface layer
(b) Bottom layer

SECTION 7: GENERAL MODELLING OF MARINAS AND MARINE STRUCTURES

Hydraulic Modelling Studies for the Hartlepool Marina Development

J.V. Smallman

Hydraulics Research Wallingford, Wallingford, Oxon, OX10 8BA U.K.

ABSTRACT

Hydraulic models are an invaluable tool for the designer of a marina. Different hydraulic modelling techniques can be used at all stages in the development of a marina through from feasibility to final design. An integrated approach to modelling will permit the marina layout and structures to be optimised so that safe conditions at the moorings are achieved, whilst still allowing cost savings to be made. An example of this type of approach to hydraulic model studies, is provided by the work carried out at Hydraulics Research to examine the proposed marina development at Hartlepool. This paper describes the studies carried out for the Hartlepool development, and indicates the benefits of using an integrated set of hydraulic models in the design of a marina at a coastal site.

INTRODUCTION

In 1989 it was proposed by the Teesside Development Corporation that a marina complex should be developed in the Union, Jackson and Coal Docks at Hartlepool. (See Fig 1) The marina complex was to include significant areas of new housing and associated residential and marina facilities. Vessel access to the main marina berths within the three docks was to be through the West Harbour. The West Harbour was formed from a series of vertical breakwater piers of varying ages, which were to be refurbished as part of the marina development.

The preliminary design for the marina at Hartlepool was undertaken for the Teesside Development Corporation by Mott MacDonald, Consulting Engineers. They identified a number of areas where hydraulic model investigations were required to assist in the design process. These included definition of the wave climate, optimisation of the layout of the breakwaters surrounding the outer harbour, positioning of

the approach channel to a locked entrance to the inner
harbour, and examination of the cross sectional design of the
outer harbour breakwaters and the seawalls adjacent to the
marina.

The modelling approach which was recommended to investigate
all of these aspects comprised two mathematical and three
physical models. The mathematical models used measured wind
and wave data at the site to predict the annual and extreme
wave climate. A three dimensional wave basin model was used
to examine the overall layout of the marina, and some aspects
of they hydraulic performance and stability of the armouring
of the outer harbour breakwaters. Two dimensional physical
models were used to optimise the selected breakwater cross
section, and examine the behaviour of the proposed sea wall
section adjacent to the marina.

MODEL SELECTION

For any study where hydraulic modelling is to be undertaken a
careful selection of the most appropriate technique must be
made. The studies for Hartlepool were directed towards
achieving a final design for the marina and outer harbour,
therefore the models which were selected needed to provide
an accurate representation of the hydraulic effects. With
this in mind it was decided, at an early stage in the planning
of the study, that a physical model should be used to
investigate wave disturbance. Whilst, a mathematical model
could have been used to compare a variety of designs, it was
recommended that a physical model would provide the absolute
accuracy required for the final design. In addition, the
physical model would also be used to explore some aspects of
the structural performance of the South Breakwater and
Southern seawall, which were to be refurbished as part of the
marina development.

In parallel to the wave disturbance studies two separate sets
of two dimensional tests were carried out. The first of these
was to optimise the crest height of the proposed refurbished
South Breakwater (Figure 1). The other was to examine
alternative designs for the Southern seawall, with the aim of
minimising overtopping and ensuring that their would be no
detrimental effect on the existing sand beach adjacent to the
wall. Both of these models were at a larger scale than that
in the wave disturbance model, allowing a more detailed
investigation of the hydraulic aspects to be made.

The two mathematical models were used to determine the wave
climate at the site. The first of these predicted the deep
water conditions, and the second modified the deep water wave
climate to allow for shallow water effects at the site. These
models were specifically selected to make the best possible

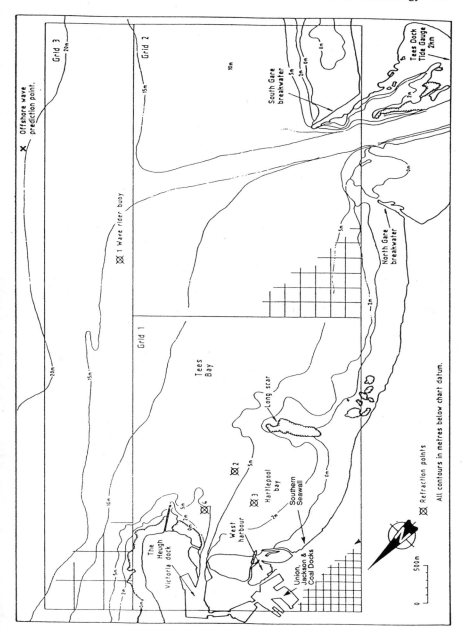

Figure 1   Location map and wave refraction model grids

use of the wind and wave data available at the site.

A more detailed description of the individual models, and a discussion of the way in which they interact with each other follows.

ANALYSIS OF WAVE CLIMATE

An accurate prediction of wave climate is an essential pre-requisite to any study of wave effects for a coastal development. For the studies for Hartlepool this analysis was required to provide incident wave conditions for the physical models, and to define wave climate for the preliminary design of the refurbishment of the outer harbour breakwaters and the Southern seawall.

The method of wave prediction which is chosen for a particular site should ensure that the major physical phenomena are represented, and that the best possible use is made of available data. Hartlepool Bay is situated on the east coast of the U.K., and is open to waves generated by the action of the wind across the North Sea (see Figure 1). There is no significant swell wave activity as, for example may be experienced on the Cornish Coast, and therefore a computational model representing wind generated waves is appropriate to the site. The shallow water region adjacent to Hartlepool extends only a short distance offshore, and the depth contours are fairly regularly spaced. It is therefore appropriate to use a computational wave refraction model to represent shallow water effects as waves travel inshore.

The models which were used for the Hartlepool study satisfied these requirements. The deep water wave climate was derived using the HINDWAVE model. This takes as input details of the geometry of the area in which waves are generated, and hourly measured wind records. Using this information, and a wave forecasting method based on the JONSWAP spectrum, a set of site specific offshore wave forecasting tables are produced. These are converted into a synthetic wave climate corresponding to the period of the wind data. This deep water wave climate was transformed to an equivalent climate in shallow water using the OUTRAY wave refraction model. In this model the bathymetry is represented by specifying depths at the intersection points of the grid lines. The OUTRAY model then follows a series of rays, lines everywhere perpendicular to wave crests, seawards from an inshore point to the offshore edge of the grid system. Each ray gives information on how energy is transformed between the seaward edge of the grid system and the nearshore point of interest. By considering a large number of ray paths the transformation of energy for a wide range of frequency and direction components can be determined. This allows a spectral representation of wave climate at an inshore point to be calculated.

Three sets of data were available for the wave climate
analysis. These were:

- Wave data from the Tees and Hartlepool Port Authority
  (THPA) wave-rider (see Figure 1) which had measured half
  hourly records for the period 4 February 1988 to 30
  September 1989.

- Wind data from the THPA anemometer at South Gare
  breakwater (see Figure 1) for the period 4 February 1988
  to 30 September 1989.

- U.K. Meteorlogical Office measured hourly wind speeds and
  directions from the aneomometer station at South Shields
  for the period from January 1976 to September 1989.

To make the best use of this data the modelling procedure
which was followed was to first select a wave prediction point
in deep water for the HINDWAVE model. (See Figure 1). This
was positioned on the -20m CD contour east of Hartlepool Bay.
Fetch lengths were then measured radially from this point to
provide a description of the wave generation area. In
parallel, a grid of depth points covering the whole area of
Tees and Hartlepool Bays, extending out to the -20m CD
contour, was set up. This provided an accurate description of
the bathymetry for use in the OUTRAY model.

Having set up both models, HINDWAVE was used to predict an
offshore wave climate using South Shields wind data for the
period of the wave measurements. This climate was then
transferred inshore to the precise location of the wave
recorder using the OUTRAY model. This provided an hour by
hour prediction of wave height, period and direction at a
single (high) water level. The hour by hour record of
predicted wave heights was then compared with the wave
measurements. Following this, the HINDWAVE/OUTRAY model
parameters were adjusted until good agreement was achieved at
higher water levels. This procedure effectively calibrates
the model against the measured data.

On completion of the calibration the HINDWAVE model was run
for the full 14 years of wind data from South Shields. This
provided an annual directional offshore wave climate for the
area. From this climate, directional extreme wave conditions
of prime interest for this site were then derived. This was
done for the 1/1 year, 1/10 year, 1/50 year, 1/100 year and
1/500 year events for 30° sectors between north and
south-east. The wave spectra corresponding to these events
were then run through the OUTRAY model to give wave spectra,
height, period and direction at the wave generator locations
in the physical model for all the required extreme offshore
wave conditions.

PHYSICAL MODEL OF WAVE DISTURBANCE

The main purpose of the physical model of wave disturbance was
to examine the layout of the breakwaters with regard to the
levels of wave activity in the West harbour. Vessels may need
to moor temporarily in this area whilst waiting to enter the
marina, and it is also possible it will be used for visiting
vessels that cannot be accommodated within the marina. At the
time of the studies it was intended that swing moorings would
be used in this area. Therefore the significant wave height
criterion adopted for acceptance levels of wave activity was
0.6m. This is the criterion normally adopted for a marina
with swing moorings. Various re-furbishment schemes were
tested in the physical model with the aim of achieving a
layout which matched this criterion. In addition to measuring
wave activity, the three-dimensional physical model also
provided incident conditions for the two dimensional models of
the Southern Breakwater and Southern seawall, and was used to
examine some aspects of the performance of the Southern
Breakwater.

The extent of the model constructed for the study was from the
Heugh Breakwater to the Southern Seawall. The seabed was
represented over an area equivalent to 1.75km by 1.6km. The
harbour layout was constructed basically as shown in Figure 2,
although this figure also shows modifications to the layout
which were introduced during the test programme. The model
scale was 1 to 57.18 in both the horizontal and vertical
planes. Froude scaling was employed so that the time and
velocity scales were $1:(57.18)^{\frac{1}{2}}$. This means that events
occurred approximately 7½ times faster in the model than in
nature. The scale chosen was somewhat unusual because of the
necessity to represent correctly the weight of Accropode
armour units chosen as one form of armouring on the proposed
new South Breakwater for the West Harbour.

A random sea was produced in the model by a mobile wave
generator consisting of a 15m paddle driven by an electro
hydraulic system. The wave conditions generated corresponded
to those calculated using the wave refraction model. In fact,
the spectra from the OUTRAY model are used directly by the
wave generation system to create the random sea in the
physical model.

At the outset of the project it was clear that the South
Breakwater would need to be substantially re-furbished in
order to afford the necessary shelter for vessels entering the
marina and those moored in the Outer Harbour. It was
initially proposed to test four main designs for the new South
Breakwater. The options were as follows:

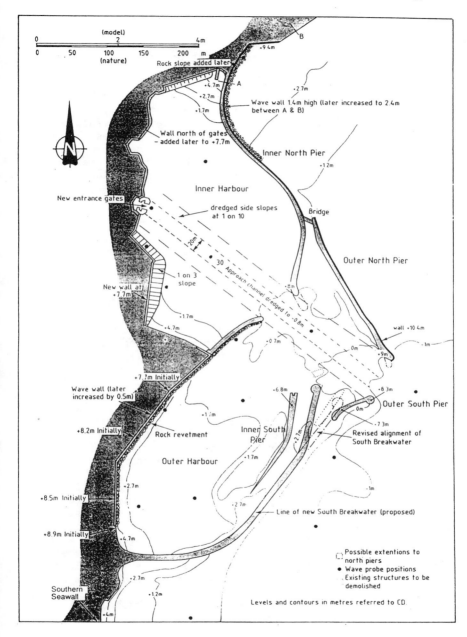

Figure 2   Proposed layout of Western Harbour

(1)   Rock armoured structures
(2)   Accropode armoured structure
(3)   Vertically faced structure
(4)   Vertical face/rock armoured structure

By a process of elimination, through observations and
measurements of wave heights in the harbour, this number was
to be reduced to a single option. The performance of the most
likely form for the South Breakwater was then to be further
explored in terms of changes to the layout of the West Harbour
in general. These changes to the layout were designed to
provide more acceptable conditions in the anchorage areas and
along the sea frontage. Ultimately, the best solution for the
new breakwater was to undergo testing in a wave flume, with
subsequent checking for three dimensional effects in the wave
disturbance model.

Before testing commenced, Option 4 above was discounted by the
Consulting Engineers due to cost consideration, leaving three
basic designs for the South Breakwater to be tested. The
breakwater alignment for all the proposals was essentially the
same. This alignment is shown in Figure 2 as the proposed
harbour layout.

It was found that Options 1 and 2 provided very similar
shelter in the West Harbour, with Option 3 giving larger wave
heights in this area. Option 2 gave a marginally better
performance than Option 1, and was the preferred method of
construction. During testing it was also found that the Inner
South Pier had very little effect on wave conditions in the
West Harbour, but that the Outer South Pier had a generally
detrimental effect in concentrating wave activity and causing
waves to plunge onto the seaward face of the South Breakwater.
On removal of the Outer South Pier there was a significant
decrease in wave activity within the West Harbour.

On completion of the wave disturbance studies a range of
possibilities for the West Harbour had been examined, and the
optimum layout determined. The final layout for the West
Harbour provided conditions which were acceptable during
storms with return periods up to 1 year. During storms with
return periods much in excess of this, vessels in the Outer
Harbour will have to seek more sheltered waters in the Inner
Harbour (or in the marina). Here they should experience safe
mooring conditions in storms with return periods close to 50
years. For the intended purpose of the outer harbour, these
conditions were considered to be acceptable.

BREAKWATER ARMOUR STABILITY AND HYDRAULIC PERFORMANCE

Two series of tests were carried out to examine the
performance of the South Breakwater. Firstly a series of two
dimensional tests were undertaken at a scale of 1:37.81 to

investigate the overtopping performance of the main trunk of
the breakwater. This scale was constrained by the size and
availability of the model Accropode units. This section was
armoured with 6 tonne Accropode units and was tested for three
crest heights of increasing elevation. Once an acceptable
crest level had been obtained (see Fig 3), two stability tests
were completed in the wave flume. These tests explored the
influence, on stability, of relaxing the placement
specification provided by Sogreah, the inventor and patent
holder for Accropode units.

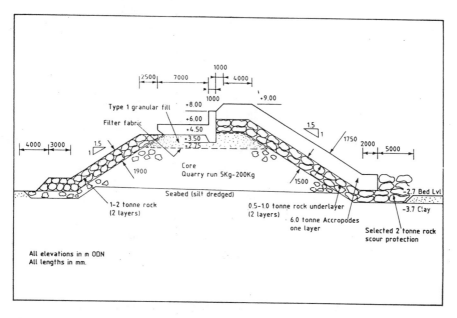

Figure 3   Cross section of the main breakwater trunk

Finally, two three dimensional stability tests were carried
out. The first test investigated the stability of the 9.6
tonne Accropode units at the breakwater roundhead. Also
determined during this test was the stability of the
intersections between these large units and the 6 tonne
Accropode armour on the breakwater elbow, see Figure 4.

The two dimensional model tests were carried out in the deep
random wave flume at HR. The bathymetry used in the tests was
identified from observations in the wave disturbance model.
It corresponded to the steepest bathymetry, in the predominant
wave direction, for waves that are likely to impinge on the
structure. Waves travelling along this bathymetry result in
an almost normal angle of attack at the structure. Wave
conditions for the two dimensional model were taken directly
from measurements made in the wave disturbance model.

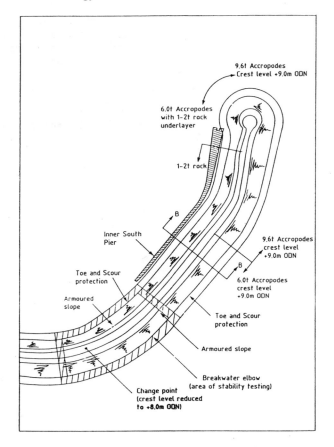

Figure 4   Distribution of armour along the breakwater

The overtopping discharge tests indicated that an acceptable breakwater crest level was achieved at 9m ODN. Extremely low levels of Accropode armour unit displacement were observed during the two dimensional test in which the 100 year wave height at the structure was increased by 16%. There was some slight movement of the toe rock and rear face armour but this mainly occurred for poorly placed rocks that settled down into a more stable position. This level of damage was not sufficient to threaten the stability of the breakwater toe or rear face armour layer.

The level of Accropode unit displacement observed in the three dimensional stability tests was also extremely low. A small number of toe and rear face armour rocks along the main breakwater trunk exhibited movement. There was however considerable movement of the toe rock at the end of the roundhead coincident with an increase in seabed level at the breakwater toe.

The hydraulic model tests described above confirmed that the proposed Accropode armour in 9.6 tonne and 6 tonne sizes would experience very low levels of armour movement under the design conditions considered. The tests also indicated that the proposed toe and rear face armour rock sizes along the main breakwater trunk were acceptable, and would not endanger the long term stability of the breakwater. It was strongly recommended, however, that the toe rock at the roundhead should be increased in size, or that other measures are taken to ensure the stability of the toe rock in this area.

HYDRAULIC MODEL TESTING OF THE SOUTHERN SEAWALL

The seawall south of Hartlepool West Harbour, see Figure 1, prior to the studies was in a poor state of repair. Beach levels were very low, allowing waves to impact directly on the wall during the higher stages of the tide. As a consequence the wall experienced severe overtopping during storms, whilst wave reflections, particularly from the vertical lengths of wall, cause a further lowering of beach levels. In this condition the seawall is incapable of providing the level of protection from wave activity necessary to allow the adjacent land to be developed. Hydraulics Research were therefore asked to assess a number of alternative designs for a seawall along this frontage.

The assessment was to concentrate on two main aspects, namely:

(a) To ensure that wave overtopping discharges would be within acceptable limits, and
(b) To ensure that the scheme chosen would not have a detrimental effect on existing sand beach levels adjacent to the wall.

The model tests were carried out under random waves in a flume at a scale of 1:25. Wave conditions for this model were taken directly from the results of the OUTRAY wave refraction model, supplemented by observations made in the physical model of wave disturbance. The bathymetry for the model was taken from the same sources as the three dimensional model. From the existing seawall to around 110m seaward, the profile was moulded approximately 1.0m below the correct beach level. This represented the sand clay interface at the toe of the proposed seawall as indicated by the Consulting Engineer's observations and borehole records. At the seaward end a smooth 1:10 transition brought the recess back to the existing seabed level. This shallow recess was subsequently filled with a carefully scaled mobile bed material moulded to the correct contour values.

In order that a mobile bed physical model may accurately simulate natural beach processes, it is necessary to ensure that the sediment used in the model is representative of that occurring in nature. At Hartlepool the beach to the south of the West Harbour is predominantly sand with an occasional localised covering of fine coal particles. Analysis of surface and sub-surface sediment samples, revealed a relatively poorly graded sand with a median size (d_{50}) of 200 µm. This grading is compatible with the beach being located in a disturbed and relatively unstable environment in front of a highly reflective sea wall. It was noticeable that there was slight fining of the material in the onshore direction, as well as with vertical depth through the beach, and that the sub-surface samples are generally less well graded than the surface samples. Coal particles were apparent in only two of the samples supplied and were not considered further in the study.

Previous work at HR has demonstrated that mobile beach models are capable of realistically reproducing micro-scale variations in beach sediment patterns. It therefore follows that the beach grading curve adopted for the model sediment should be representative of the complete sand beach at Hartlepool rather than just the surface samples. To this end, all the available sand beach grading curves were combined to produce a single "typical" grading for Hartlepool. For a sand beach, sediment response in the model can be correctly scaled if two main requirements are satisfied, that is: the accurate reproduction of the threshold of motion and the sediment settling velocity. For this particular study, calculations suggested that in order to reproduce the required sand grading at a model scale of 1:25, the ideal model sediment would have a specific gravity of about 1.20, with a median size of 200 µm. These requirements can be satisfied by the use of granulated perspex which is readily available in the required size range. To ensure the correct grading, the perspex was first sieved into a number of factions and then re-combined in the proportions necessary to give the model grading curve.

A total of four seawall sections were tested as detailed in Figure 5. These were made up of two original rock designs (sections 8c and 9c), a smooth impermeable slope (section 9a) and an optimized rock design (section 8d). For the design conditions it was found that greenwater overtopping of the main sea wall could only be reduced to acceptable limits with a crest raised to at least 11m OD. However, an acceptable overtopping performance could be achieved if the promenade and garden wall were utilised as an integral part of the sea defences. Such a design, with a crest of 9.0m OD, was tested (section 8d) and was found to have negligible greenwater overtopping for the 100 year wave condition with a water level of 3.55m OD. Raising the water level to 3.85m OD, to allow

for possible 'greenhouse' sea level rise, resulted in a
maximum mean overtopping discharge of 1.89 x 10^{-5} m³/s/m run
of wall. According to published guidelines this would result
in conditions immediately behind the garden wall being rated
as uncomfortable but not dangerous for pedestrians, and likely
to result in only minor damage to the fittings of buildings.

It was advised that drainage through the recurve should be
incorporated into the final scheme if the promenade area is
utilised as a reservoir area. In the model 0.15m diameter
drains located at 2m intervals were found to be satisfactory.

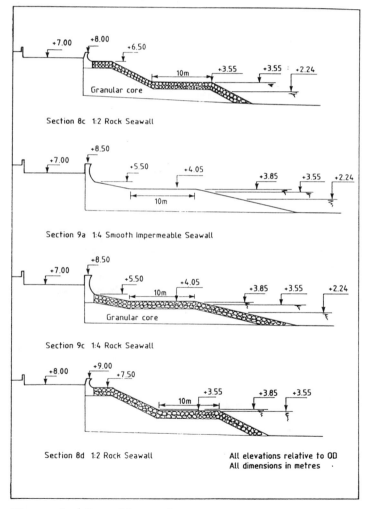

Figure 5   Seawall sections

Although sea wall section 8d will reduce greenwater overtopping to acceptable levels, spray overtopping will still be considerable, particularly under the conditions tested. Counteracting this effect would be very difficult;requiring either a much wider wall, or the addition of a wide beach in front of the wall, to increase the distance between the breaking waves and the sea front properties.

All sea wall sections tested were found to be beneficial to the existing sand beach when compared with a vertical wall. Rock armoured walls are however likely to promote a more rapid build-up of beach material than smooth sloping walls, provided a suitable supply of beach sand to the frontage exists.

CONCLUSIONS

The studies carried out for Hartlepool resulted in an optimum design for the West Harbour and the adjacent Southern seawall. Hydraulic modelling allowed possible changes to the layout and design of the structures to be examined, with a high degree of confidence being associated with the outcome. The integrated approach to modelling allowed the most appropriate tools to be used for each part of the study. At the time of writing the marina development is proceeding well, with a significant amount of the refurbishment of the West Harbour breakwaters being complete.

ACKNOWLEDGEMENTS

The studies for Hartlepool Marina were commissioned by Mott MacDonald on behalf of the Teesside Development Corporation. Their permission to publish this paper is gratefully acknowledged. The contribution of collegues in the Ports and Harbours, and Coastal Group who carried out the studies is also gratefully acknowledged.

REFERENCE

1.   Smallman J.V.   The use of physical and computational models for the hydrualic design of coastal harbours. PIANC Bulletin, 1986, No 53, p 16-36.

Numerical Modelling of Bathymetric Variability

M. Anwar, A.A. Khafagy, A.M. Fanos
Coastal Research Institute, 15 El-Pharanna Street, El-Shalalat, Alexandria, Egypt

ABSTRACT

This article presents a numerical model for predicting bathymetric changes under the influence of wave action. The model takes into consideration the changes in N-Contour lines due to both longshore and onshore-offshore sediment transport. The model consists of three main submodels namely; a backrefraction model to establish the deep water conditions of the wave field, the wave transformation model due to varying topography and presence of coastal structures, and the sediment transport and changes of bottom topography model. The model has been applied for a specific situation and the main conclusions are that it produces the general features of erosion and deposition along the coast.

INTRODUCTION

It is evident that shoreline models for predicting the changes in beach profiles at various locations along the coast are of extensive practical use. In recent years, such models have been widely used for practical application in many areas world wide. But, there are many limitations such that they neglect onshore-offshore sediment transport, assume parallel movement of the beach profile and most of such models are either one line or two lines ones. Thus to overcome such limitations several models have been developed. For example the numerical model by Fleming and Hunt [1], and the N-lines model by Perlin and Dean [2].

The present article documents the development and application of an N-line numerical model to predict the bathymetric variability along the Nile-Delta coast due to time varying wave conditions and non parallel contours. The model takes into account both longshore and onshore-offshore sediment transport. The model is developed along the general outlines of that presented by Perlin and Dean [2] but is not restricted to almost parallel contours.

MODEL DESCRIPTION

The model consists of three main submodels namely; the backrefraction model to predict the deep water wave conditions; the wave transformation model to establish the wave climate along the coast, and the sediment transport and bathymetric response model to estimate the variability in the bottom topography and the shoreline. In the present investigation a fixed longshore space step but with a variable cross-shore distance is allowed to vary according to the existing contour lines. Figure (1) gives a schematic representation of such a computational grid. The present system allows the modeler to think of bathymetric changes in terms of the effects exerted on the contour lines. Thus, the bathymetry is represented by a set of N-contour lines, each of a specified depth and the variability of such lines in the offshore direction is mainly governed by the Equation of continuity that takes into account both the longshore are and onshore- offshore sediment transport. The implementation of this Equation requires the determination of the wave field within the study area and the knowledge of corresponding equilibrium profile.

For the present application, wave information is available in shallow water only, thus it is necessary to predict the characteristics of deep water waves from the available measurements. Subsequently, from the deep water wave parameters, the wave climate within the study area can be estimated using the wave transformation model.

The following sections give brief accounts for main submodels that consists of the overall model.

(a) Backrefraction Model

The available wave data are collected at two locations within the shallow water region at Abu Quir station and at Ras El-Bar station. It is well known that several practical techniques for producing refraction diagrams are available in the literature. A modified technique based on the work of Dobson [3] to account for the special features of the present computational grid is adopted. The two basic governing equations for the propagation of a particular wave ray are:

- The ray curvature equation

$$R = \sin\theta \frac{\partial c}{\partial x} - \cos\theta \frac{\partial c}{\partial y} \tag{1}$$

where

R	ray curvature
c	wave speed
x,y	Cartesian coordinates of the point on the ray

- The ray separation equation

$$\frac{d^2B}{dt^2} + p\frac{dB}{dt} + qB = 0 \tag{2}$$

where

B   ray separation factor

$$p = 2\left[\cos\theta \frac{\partial c}{\partial x} + \sin\theta \frac{\partial c}{\partial y}\right]$$

$$q = c\left[\sin^2\theta \frac{\partial^2}{\partial x^2} + \sin 2\theta \frac{\partial^2\theta}{\partial x \partial y} + \cos^2\theta \frac{\partial^2 c}{\partial y^2}\right]$$

The backrefraction procedure is initiated from the known particular point where the wave data is measured. This initial point is to be denoted by (x_i, y_i) and the location of the next point along a ray is determined iteratively by computing its coordinates from the expressions

$$x_{i+1} = x_i + \Delta s \cos\theta$$
$$y_{i+1} = y_i + \Delta s \sin\theta \tag{3}$$

where Δs is a small projection distance along the ray.

It should be noticed that c is a function of x and y and Snell's law could be used here since the depths are known at all computational nodes. The small distance Δs must be controlled so that the incremental backward projection along the ray does not leave the computational cell bounded by two adjacent cross lines. Hence the ray curvature R_{i+1} at this new point can be calculated from Equation (1) using finite difference approximation. The process is repeated using the following parameters;

$$\theta_{i+1} = \theta_i + \Delta\theta$$
$$\Delta\theta = R\Delta s$$
$$\theta = \frac{1}{2}(\theta_i + \theta_{i+1}) \tag{4}$$
$$R = \frac{1}{2}(R_i + R_{i+1})$$

and Equations (3) are used to recompute the modified location of the $(i+1)$th. point. The process is carried on until two consecutive answers for the ray curvature differ by a small acceptable tolerance. Then, we project the ray backward to the next offshore point until the deep water limit is reached.

As for the ray separation equation it is solved using finite difference approximation subject to the initial conditions that in deep water $B = 1$ and $dB/dt = 0$.
Finally, based on linear wave theory the height at any point i is given by

$$H_i = H_o K_s K_r \tag{5}$$

where
$\qquad H_o \qquad$ deep water wave height

K_s shoaling coefficient
K_r refraction coefficient

and

$$K_r^2 = \frac{1}{|B|}$$

$$K_s^2 = \frac{c_o}{c_{gi}}$$

where

c_o wave speed in deep water
c_{gi} group velocity at point i

This completes the determination of the deep water wave conditions. Figures (2) - (5) display these conditions due to measurements at both Abu Quir and Ras El-Bar stations.

(b) Wave Transformation Model

In this model the deep water wave characteristics are transformed to the shallow water region taking into account the irregular bottom bathymetry. A numerical model of wave refraction and shoaling compatible with the variable offshore grid spacings was developed. The model, like most existing techniques neglected the nonlinear effects, bottom friction, turbulent dissipation and bottom percolation. The procedure involves solving a pair of simultaneous equations for the wave direction and the wave height. The first equation is derived from the fact that the wave number vector field k is irrotational, thus in cartesian coordinates we have;

$$\cos\theta \frac{\partial\theta}{\partial x} + \sin\theta \frac{\partial\theta}{\partial y} = \frac{1}{k}\left(\cos\theta \frac{\partial k}{\partial y} - \sin\theta \frac{\partial k}{\partial x}\right) \tag{6}$$

The second equation is obtained from the conditions of the steady-state conservation of energy, namely

$$\left(\frac{2}{H}c_g\right)\left[\cos\theta \frac{\partial H}{\partial x} + \sin\theta \frac{\partial H}{\partial y}\right] + \cos\theta \frac{\partial c_g}{\partial x}$$

$$+ \sin\theta \frac{\partial c_g}{\partial y} - c_g \sin\theta \frac{\partial \theta}{\partial x} + c_g \cos\theta \frac{\partial \theta}{\partial y} = 0 \tag{7}$$

The numerical solution of these equations is achieved using finite difference approximations of the above terms over the chosen computational grid with fixed longshore spacing Δx and variable offshore spacings Δy_{ij} which will be simply denoted by Δy. Following the work of Perlin and Dean [3] an averaging approach over adjacent grid blocks is achieved using variable τ termed a dissipative interface parameter and has the role to improve the stability of the finite difference scheme ($\tau = 0.25$). The finite difference versions of the above two equations take the final forms:

$$\Theta_{ij} = \cos^{-1}(1/k_{ij} \{\tau (k \cos\Theta)_{i-1j+1} + (1 - 2\tau) (k \cos\Theta)_{ij+1}$$
$$+ \tau (k \cos\Theta)_{i+1j+1} - \Delta y/2\Delta x [(k \sin\Theta)_{i+1j} - (k \sin\Theta)_{i-1j}\}] \tag{8}$$

and

$$H_{ij} = (1/(c_g \sin\Theta)_{ij} \{\tau(H^2 c_g \sin\Theta)_{i-1j+1}$$
$$+ (1-2\tau) (H^2 c_g \sin\Theta)_{ij+1} + \tau (H^2 c_g \sin\Theta)_{i+1j+1}$$
$$+ (\Delta y/2\Delta x [(H^2 c_g \cos\Theta)_{i+1j} - (H^2 c_g \cos\Theta)_{i-1j}]\})^{\frac{1}{2}} \tag{9}$$

The wave transformation over the irregular bottom is achieved using the above finite difference equations for the computational grid. Each grid cell was assigned a mean depth h_{ij}, all major quantities H, Θ, k, are computed at the centroid of the cell. The boundary condition at the shoreward is simple that the wave angles are normal to the shoreline, and the deep water conditions are specified. For the side boundary conditions simple application of Snell's law supply the necessary information. The process is carried on in an iterative manner in the sense that the right hand side of Equations (8) and (9) involve some variables that are not yet known, these are assumed and the left hand side is computed. Newly obtained values are resubstituted in the right hand sides to obtain improved answers until convergence is achieved within an acceptable margin of error.

c) Sediment Transport and Bathymetric Variability Model

Coastal sediment movement can be broadly classified into two types according to the direction of motion: longshore transport and on-offshore transport. The transport rate of the former has been successfully formulated by several investigators -for example Komar and Inman [4], and Fulford [5] -a comprehensive review of this issue is given by Kamphuis [6]. As for on-offshore transport, much research ha attempted to determine relationships between the flow field and the on-offshore sand transport rate- for example Einstein [7], Bakker [8], and Sleath and Ahilan [9]. The work of Bakker seems to be of more practical side from computational point of view. However, no single formula seems to attain the agreement of all researchers in this field.

The governing equations that are implemented for simulating the sediment transport and bathymetry changes due to the prevailing wave field are:
-The continuity equation of sediment:

$$\frac{\partial y}{\partial t} + \frac{1}{m}\left(\frac{\partial q_x}{\partial x} + \frac{\partial q_y}{\partial y} \right) = 0 \tag{10}$$

where m is the beach slope, and x and y are the longshore and offshore directions respectively.
The longshore sediment transport per unit length q_x as a function of the offshore distance:

$$q_x(y) = 3/(1.25\ y_b)^3\ (y + a)^2\ \exp\{-\ [(y+a)/(1.25y_b)]^3\} \tag{11}$$

This equation is due to Fulford [5], where a is a constant to allow sediment transport above mean water line, i.e. swash transport or transport in region of wave setup, to be accounted for. And y is the distance to the point of breaking. It was determined that, (Perlin and Dean [2].

$$a = \frac{h_b}{\dfrac{\partial h}{\partial y}} \tag{12}$$

where h_b is the water depth at breaking.

Equations (11) and (12) are used to determine the fraction of longshore sediment transport between two y locations, y_1 and y_2 say,   by

$$Q_x \bigg|_{y_1}^{y_2} = C' H_b^{\frac{5}{2}} \sin 2\alpha_b \int_{y_1}^{y_2} q_x(y)\, dy \tag{13}$$

where

$\quad\quad H_b \quad$ is the wave height at breaking.

$\quad\quad \alpha_b \quad$ the orientation $\tan^{-1}$ (dy/dx) of mean water level contour at breaking

The constant C' - see Perlin and Dean [2] - is given by

$$C' = \frac{K \rho g^{\frac{1}{2}}}{16(\rho_s - \rho)(1 - p) K_s} \tag{14}$$

Where

$\quad\quad K \quad\quad$ 0.77 (Komar and Inman [4])

$\quad\quad \rho_s \quad\quad$ mass density of sediment

$\quad\quad p \quad\quad$ porosity

$\quad\quad K_s \quad\quad$ spilling breaker coefficient ($H = K_s h$) taken as 0.78

It should be mentioned that in the present study we are interested in simulating the changes in contour position, thus the above formulas are transformed into functions in h by multiplying by the corresponding Jacobin $\Delta y/\Delta h$.

Furthermore, the above equations are derived for the case of nearly parallel contours, thus to compensate for nonparallel nature of the contours the variables in Equation (13) namely α_b and H_b, are replaced by their corresponding local values.

The on-offshore sediment transport per unit length q_y (y) based on the work by Bakker [8] is related to the on-offshore transport Q_y between y locations by

$$Q_y = \int_{y_1}^{y_2} q_y(y)\, dy \qquad (15)$$

And if i, j denote values at particular x_i longshore location and y_j contour line then.

$$Qy_{i,j} = \Delta x\ C_{off}\ (y_{i,j-1} - y_{i,j} + W_{EQi,j}) \qquad (16)$$
where

C_{off} is an activity factor (= $3 \times 10_{-6}$ m/sec inside the surf zone)
$W_{EQi,j}$ is the positive equilibrium profile distance between the contour lines $y_{i,j}$ and $y_{i,j-1}$

The equilibrium profile used in the model is the one suggested by Dean [10] given by:

$$h = A.y^{2/3} \qquad (17)$$

Where   A is a scale parameter in the equilibrium beach profile to be supplied.
Outside the surf zone the concept of on-offshore transport is generalized based on the fact that bottom friction is the dominant factor of energy dissipation, as opposed to wave breaking in the surf zone. This observation leads to a variant formulation for $Qy_{i,j}$ outside the surf zone; namely

$$Qy_{ij} = Const_{i,j}\ [y_{i,j-1} - y_{i,j} + W_{EQi,j}] \qquad (18)$$
where

$Const_{i,j} = \Delta x.\ C_{off}\ (i,j)$
$C_{off}(i,j) = (4/5\ \Gamma)(C'\ \rho\ \sigma^3)/(g^{3/2}\ K^2_S\ A^{3/2}\ h)\ (H/\sinh kh)^3 x\ 10^{-5}$
Γ is a parameter relating the efficiency with which breaking wave energy mobilizes the sediment bottom ($0 < \Gamma < 1$).

COMPUTATIONAL PROCEDURE AND RESULTS

The numerical model of the above governing proceeds by implementing an explicit finite difference formulation for the refraction and diffraction scheme. This is subsequently followed by a totally implicit formulation of the sediment transport model. Evaluation of the magnitude of the coefficient in the on-offshore sediment transport formula is given special attention so that contour lines do not cross at any particular location. This is assisted by the fully implicit formulation of the sediment transport model. As a result of such formulation we obtain a system of linear equations of the form AY = B where A is a square matrix of size depending on the number of unknown new contour positions represented by the vector Y, and B is the corresponding right hand side of the obtained equations which includes both the boundary conditions specified at the domain boundaries as well as the information at previous time step.

The computer program starts by establishing the deep water wave characteristics and reads in bathymetric configurations as well as the computational grid. The next step is to determine the wave climate within the study area. Subsequently the sediment transport equation is implemented using both longshore and onshore-offshore transport. This results in a system of linear simultaneous equations whose unknowns are the new offshore distance of the specified N-contour lines. Upon solution of this system the changes in these contour lines are obtained and the process is repeated for another set of deep water wave conditions. Figure (6) represents the variability of a number of contours at the indicated locations as anticipated by the model.

REFERENCES

1. Fleming, C.A and J.N Hunt. Applications of a sediment transport model, Proc. of 15 the conf. on Coastal Eng., Vol 11, pp. 1184-1202, 1976.
2. Perlin, M. and R.G. Dean. 3-D model of bathymetric response to structures, Journal of Waterway, port, Coastal and Ocean Engineering Vol. 111, no.2 pp. 153-170, 1985.
3. Dobson, R.S. Some application of a digital computer to hydraulic Engineering problems, Tech. Rep. 80, dept.of Civil Eng. Stanford University, Stanford California,1967.
4. Komar, P.D. and D.L. Inman. Longshore Sand Transport on Beaches, J. of Geophy. Res., Vol 75, pp. 5914-5927, 1970.
5. Fulford, E. Sediment Transport Distribution Across the Surf Zone. Thesis presented to the University of Delaware, at Newark, Del.,1983.

6. Kamphuis, J.W., M.H. Davies, R.B. Narim and O.J. Sayao. Calculation of Littoral Transport Rate; Coastal Engineering, Vol 10, pp. 1-21, 1986.

7. Einstein, H.A. Sediment Transport by Wave Action. Proc. 13th, Conf. on Coastal Engineering, Vancouver, ASCE, pp. 933-952, 1972.

8. Bakker, W.T. The Dynamics of a coast with a Groyne System. Proc. of the 11th, Conf. on Coastal Engineering ASCE, pp.492-517, 1968.

9. Sleath, J.F.A. and R.V. Ahilan. Sediment Transport in Ascillalory Flow Beds, J, of Hyd. Eng. 113 (3), pp. 291-307, 1987.

10. Dean, R.G. Equilibrium Beach Profiles, U.S. Atlantic and Gulf Coast; Ocean Eng. Report No. 12, University of Delaware, Newark, Del., 1977.

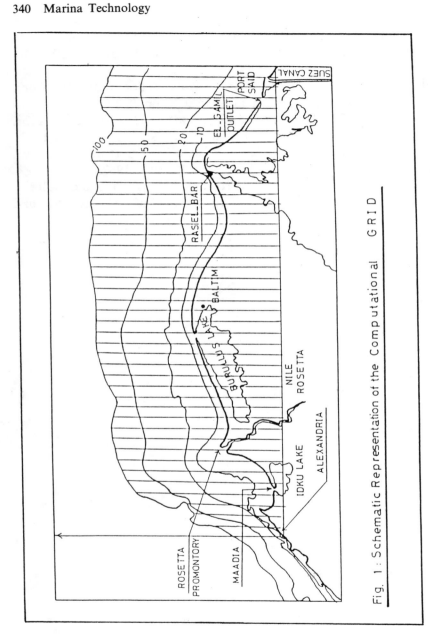

Fig. 1 : Schematic Representation of the Computational   G R I D

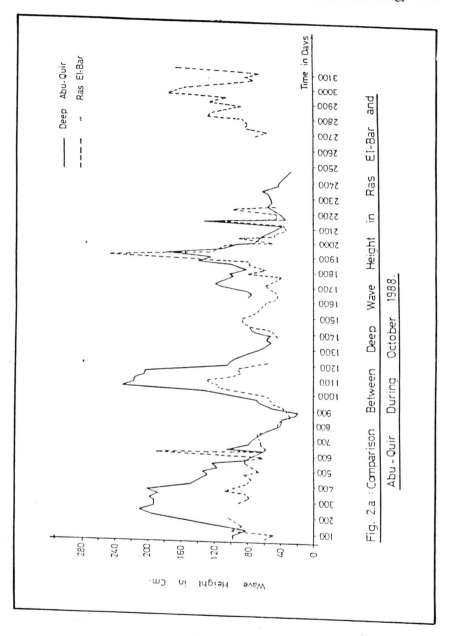

Fig. 2.a : Comparison Between Deep Wave Height in Ras El-Bar and
Abu-Quir During October 1988.

M

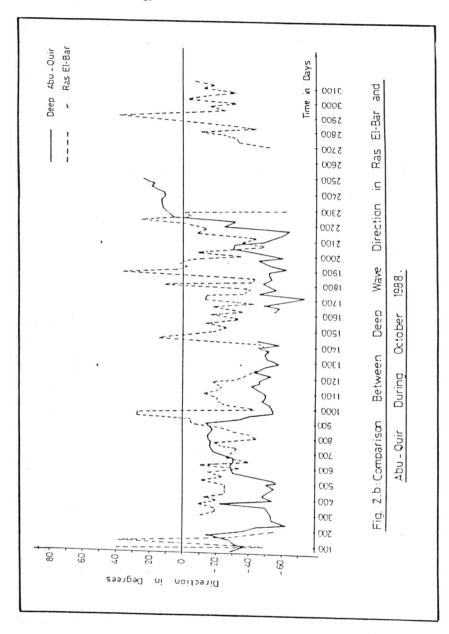

Fig. 2.b: Comparison Between Deep Wave Direction in Ras El-Bar and Abu-Quir During October 1988.

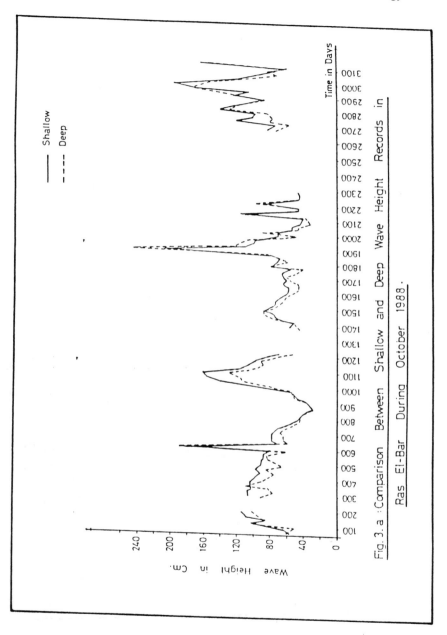

Fig. 3. a : Comparison Between Shallow and Deep Wave Height Records in

Ras El-Bar During October 1988.

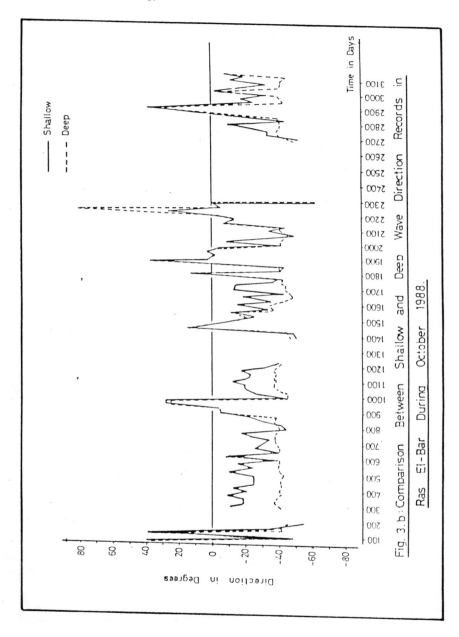

Fig. 3. b: Comparison Between Shallow and Deep Wave Direction Records in
Ras El-Bar During October 1988.

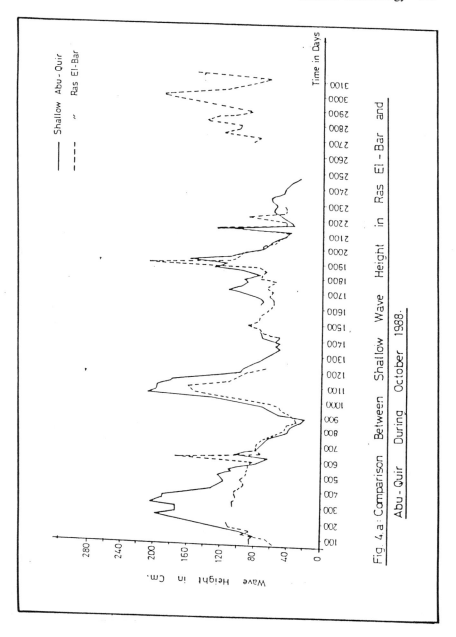

Fig. 4. a: Comparison Between Shallow Wave Height in Ras El - Bar and
Abu - Quir During October 1988.

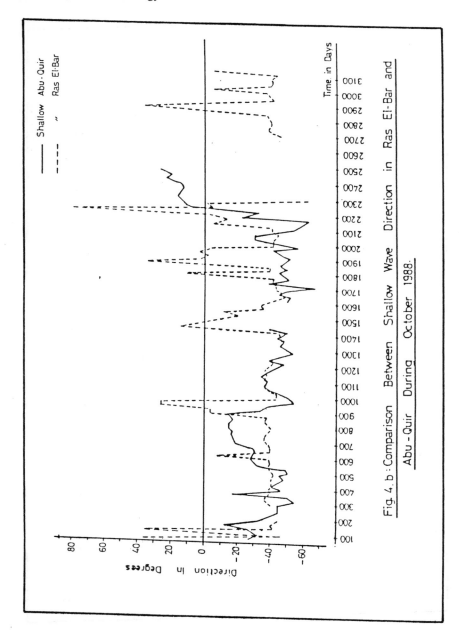

Fig. 4. b : Comparison Between Shallow Wave Direction in Ras El-Bar and

Abu - Quir During October 1988.

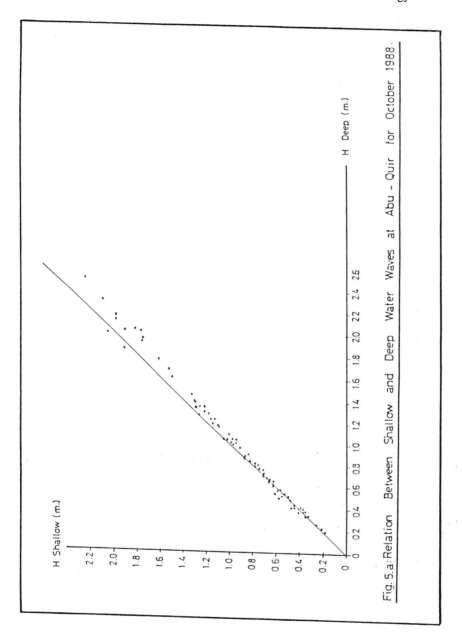

Fig. 5.a: Relation Between Shallow and Deep Water. Waves at Abu - Quir for October 1988.

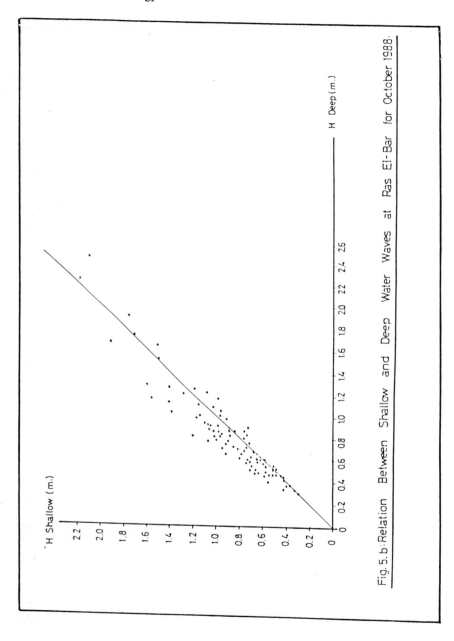

Fig. 5. b: Relation Between Shallow and Deep Water Waves at Ras El-Bar for October 1988.

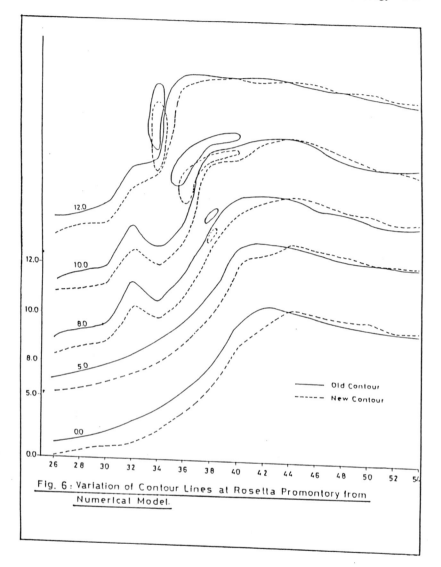

Fig. 6 : Variation of Contour Lines at Rosetta Promontory from
Numerical Model.

Improvements to Marina Design by Physical Modelling

D.H. Cooper (*), A.W. Grinyer (**)

(*) ABP Research and Consultancy Ltd., Hayes Road, Southall, Middlesex, U.K.

(**) R.J. Watkinson and Partners Ltd., Union Castle House, Canute Road, Southampton, Hampshire, U.K.

ABSTRACT

The evolvement of the Marina Village concept into dockyard re-development resulted in the construction of a residential marina inside Millbay Dock, Plymouth in 1987, with an open bottom wave screen to partially complete the fourth side of the retangular yacht basis.

Random wave tests on a 1:64 scale model of Millbay Dock following storm damage in 1989 showed surprisingly that only 14% improvement was effected on the overall wave height by closing the gap beneath the screen. The most significant improvement was achieved by narrowing the entrance and/or moving it to the middle of the screen, enabling a 50% improvement in wave conditions to be demonstrated.

A similar marina development on a larger scale is planned at Southampton's Town Quay. The site is relatively unprotected for yacht berths particularly from the south and west and wave protection will therefore be required.

Earlier numerical model studies have highlighted the problem of wave reflection within the proposed marina and thus a physical model has been commissioned to evaluate the effectiveness of the proposed breakwater and to optimise the form of construction to absorb wave reflections.

1. INTRODUCTION

The development of the Marina Village concept was fully reported in the Marina '89 Conference Proceedings. One of the most important factors is the use of pontoon finger berths for alongside mooring.

Two more recent very similar Marina Village Developments are at Ocean Village, Southampton, previously Princess Alexandra Dock and Millbay Marina in Millbay Dock, Plymouth, both old ferry terminals now obsolete and both with massive totally reflecting old stone walls.

These marinas are in relatively deep tidal water so that quayside housing has been constructed with adjacent pontoon marina berthing and linkspan bridges giving single point access to the yacht berths.

The entrance to Princess Alexandra Dock is on the River Itchen and well protected. However, at Millbay a wavescreen was considered necessary to provide additional protection for the marina.

Another marina has been proposed for Southampton in the area between the old eastern docks and Town Quay, for many years the only public access with the adjacent Royal Pier to the Southampton Waterfront.

Although relatively well protected, the requirement for a harbour to act as a Yacht Marina is very different from that for vessels in excess of 50 metres overall. For yachts in the normal 10 to 20 metre length range, much greater protection is required even from the wash of passing ferries. The maximum wave height, which is considered to be the acceptable limit for pontoon finger berthing is 0.3m (ie. 300mm or approximately 1 foot).

Previous wave studies for both Millbay Dock and Town Quay Marina have indicated the problems of hard boundaries and reflected waves which no doubt resulted in the storm damage of 1989 in Millbay, mainly due to long period wave reflections within the marina.

Consequently physical wave models for these two important marinas have been constructed at ABP Research and Consultancy in order to evaluate the engineering proposals put forward jointly by the Consulting Engineer and the Reasearch Centre, to alleviate these problems.

The results described in this paper indicate vastly different solutions for the two marinas to obtain the optimum benefit. As marina development in the future will no doubt continue to be in difficult sites or old shipyards and docks, then increasing use of physical model techniques will be necessary to fine tune designs so as to ensure that the optimum conditions for finger berthing are obtained at minimum cost.

2.   MILLBAY DOCK MARINA

2.1   Site Location and Condition.

Millbay Dock lies to the west of Plymouth Hoe at the head of the Sound and just to the north of Drake's Island, see Figure 1. The marina basin is immediately to the east of the wide dock entrance and protected from the south by Millbay Pier, as shown in Figure 6.

The vertical wave screen extends from Trinity Pier, the entrance to the marina being in the lee of Millbay Pier. The screen does not extend to the sea bed and therefore wave energy can be transmitted into the marina under the screen as well as through the entrance. In severe storm conditions on 16th December 1989 long period waves were reported to have overtopped the screen and penetrated the marina causing damage to the pontoons and walkways.

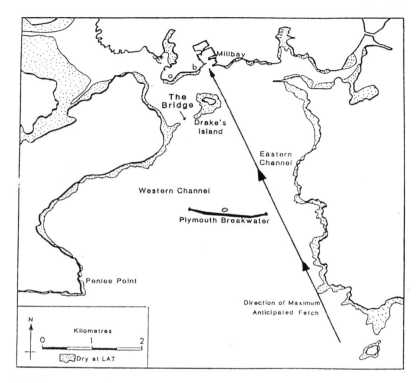

FIGURE 1   MILLBAY LOCATION

The severity of the storm was such that a length of adjacent sea wall was demolished and other waterside structures were extensively damaged. Due to the evidence of long period wave motion recorded on video, which coincided with a tidal surge in excess of 1.2m, improvement works needed to be evaluated and thus a physical model was commissioned based on earlier studies undertaken by ABP Research.

Millbay is exposed to wave activity from between southeast and southwest. However, a combination of the outer breakwater, Drake's Island and the Bridge, see Figure 1, effectively reduces the exposure to long fetch generated waves to between southeast and south. Throughout the tide minimum depths on this approach exceed 6m below Chart Datum (CD) allowing storm waves of greater than 6 second period to pass directly into Millbay. Waves entering Plymouth Sound around the western side of the outer breakwater from between south and southwest side are dissipated and the wave period reduced by the Bridge and Drake's Island where effective controlling depths are about 1m below CD. At high water and during surge conditions some higher energy wave activity may however pass over the Bridge and possibly enter Millbay. Thus wave refraction studies were undertaken to confirm this understanding and to determine test conditions for the physical model studies.

2.2   Refraction Studies.

A numerical wave refraction model of part of Plymouth Sound was designed to establish the amount of wave energy which could cross the Bridge and penetrate into Millbay under high water conditions. The model was mainly run at two water levels (6.3m and 7.0m above CD) with three wave periods of 4, 7 and 10 seconds. Wave orthogonals were computed for intial wave directions at 10° intervals from 180° to 225° true with an intial spacing of 25m. From each of the plots 'windows' were derived through which waves could possibly penetrate into Millbay. The model was re-run concentrating on these windows with wave orthogonals being started with a spacing as small as 1m. The wave height as a proportion of the initial deep water wave height has been presented as the waves approach Millbay.

An analysis of the wave orthogonal diagrams (Figure 2 is a typical example for the 7 second waves) shows that with waves from between south and southwest most of the wave energy which enters Millbay comes from the east side of Drake's Island and through a very narrow window. The shallow depths over the Bridge cause most 7 and 10 second waves to break against Drake's Island and the west side of Plymouth Sound. The few waves which do penetrate generally diffract around the north side of Drake's Island or strike the shoreline to the west of Millbay. The waves entering Millbay, having passed over the Bridge, are only 10% or less of the initial wave height.

7 Second , 7.0 m tide , 225 Degree

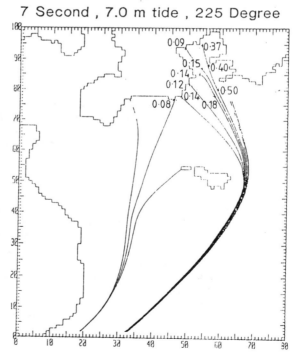

FIGURE 2

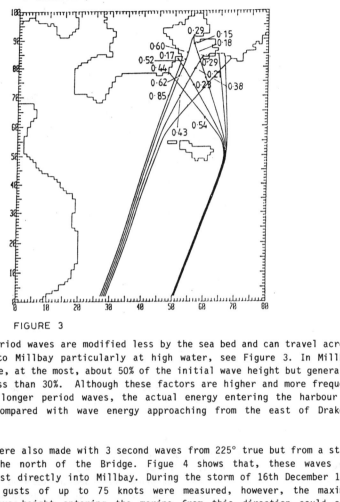

4 Second , 6.3 m Tide , 200 Degree

FIGURE 3

The shorter period waves are modified less by the sea bed and can travel across the Bridge into Millbay particularly at high water, see Figure 3. In Millbay these waves are, at the most, about 50% of the initial wave height but generally tend to be less than 30%. Although these factors are higher and more frequent than for the longer period waves, the actual energy entering the harbour is still small compared with wave energy approaching from the east of Drake's Island.

Computations were also made with 3 second waves from 225° true but from a start position to the north of the Bridge. Figue 4 shows that, these waves can penetrate almost directly into Millbay. During the storm of 16th December 1989 southwesterly gusts of up to 75 knots were measured, however, the maximum significant wave height entering the marina from this direction could only have been about 0.6m. Assuming that the average wind speed was in excess of 45 knots for a period of at least 20 minutes, the wave period would have been less than 3 seconds.

The refraction study has shown that only local waves of up to about 0.6m with a period of about 3 seconds can be generated from the southwest even though the storm which occurred on 16th December 1989 came, in principle, from this direction. Clearly the video recording made some hours after the peak of the storm shows that the damage was caused by waves with heights of approaching 2m and periods longer than 7 seconds. This refraction study indicates that such longer period waves can only approach from the southeast, that is east of Drake's Island. In view of these findings it was decided to concentrate the physical model tests on waves from this direction.

3 Second , 6.3 m Tide , 225 Degree

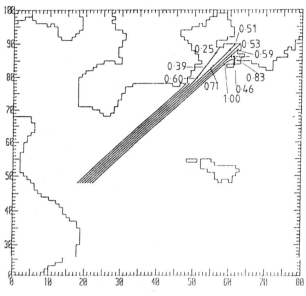

FIGURE 4

2.3   The Physical Model.

A physical hydraulic model was constructed to an undistorted scale of 1:64 on order to evaluate schemes for reducing wave disturbance in the Millbay Marina. The area covered by the model is shown in Figure 5. The bed of the model within Millbay was moulded to the survey dated June 1989 and in the approaches was moulded down to the -20m CD contour from data extracted from Admiralty Chart dated 1985. Random waves were generated by a bank of vertical generators positioned outside Millbay in a depth 20m below CD. The wave refraction study indicated that the predominant sector for wave energy entering Millbay is southeast; the wave generator was therefore orientated to produce waves from 145° true which is the direction of maximum fetch.

Wave measurments were made in Millbay off the No. 2 RO/RO dolphin during 1983 and 1984. Analysis of the data demonstrated that the site was subject to waves with periods between 3 and 180 seconds, 97% being below 10 seconds. The maximum significant wave height recorded during 1983 and 1984 was 1.64m, however, absolute maximum wave heights exceed 2m. Long term predictions from the site measurements indicate the 1 in 50 year significant wave height to be about 1.85m at the No. 2 RO/RO dolphin and the 200 hours significant wave height to be about 0.55m.

The tests with storm waves were undertaken at both MHWST (5.54m above CD) and HAT (6.25m above CD). Tests on selected schemes using the less severe wave conditions were only carried out at MHWST.

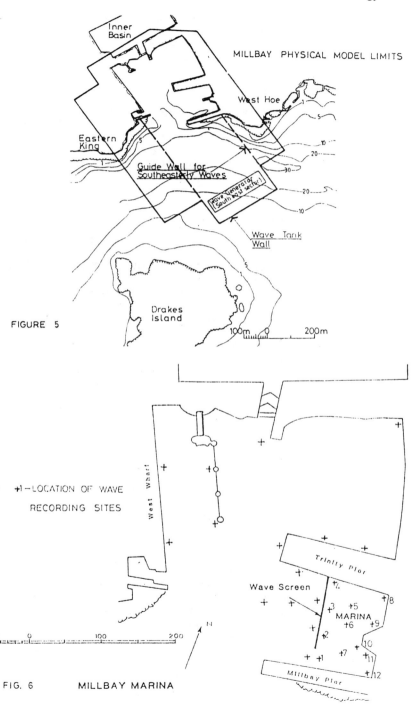

FIGURE 5

MILLBAY PHYSICAL MODEL LIMITS

Inner Basin

West Hoe

Eastern King

Guide Wall for Southeasterly Waves

Wave Generator South East sector

Wave Tank Wall

Drakes Island

100m   200m

+1 - LOCATION OF WAVE RECORDING SITES

West Wharf

Trinity Pier

Wave Screen

MARINA

Millbay Pier

50   0   100   200

N

FIG. 6   MILLBAY MARINA

Wave heights within the marina and the outer harbour were measured at 27 sites, the results being recorded directly by computer for subsequent analysis to determine standard wave parameters. The recording sites are shown in Figure 6.

2.4   Schemes Tested in the Model.

Waves enter the present marina during storms under and occasionally over the wave screen and through the entrance. The schemes tested in the model were thus aimed at reducing this wave penetration and are listed in Table 1 below.

Table 1 - Schemes Tested in the Model Using Storm Conditions.

SCHEME	DESCRIPTION
A*	Existing screen.
B	Existing screen extended down to the bed.
C	Existing screen extended down to the bed and raised to prevent overtopping.
E	Scheme C with a stub screen extending out from Millbay Pier, reducing entrance width to 12m.
F*	Scheme C with existing screen extended to reduce entrance width to 12m.
H	Scheme C with existing screen extended to reduce entrance width to 8m.
J	Scheme F with entrance depth reduced to -2m CD
I	Scheme H with entrance depth reduced to -2m CD
G*	Scheme F with 18m long internal overlapping screen (6m overlap).
D	Scheme C with 36m long external overlapping screen (12m overlap).
L*	Scheme C present entrance closed, new central 12m entrance.
K	Scheme L with an 8m entrance.
N	Scheme L underwater bank removed.
O	Scheme L with angled internal screen at the north side of the entrance.

* Schemes selected for $H_{200 \text{ hours}}$ tests.

The wave heights in the marina have been compared, in all cases, with that recorded with the present harbour configuration, Scheme A. The average significant wave heights (storm conditions) in the marina with the various schemes are listed in Table 2. Since the difference between HAT and MHWST (0.71m) is relatively small compared with the depth of the water in the marina (about 11m at HAT), wave heights recorded with both water levels were very similar. Of the schemes tested with storm waves Schemes F and G (present entrance location) and Scheme L (central entrance location) appeared to give better results and were therefore also tested using the less severe ($H_{200 \text{ hours}}$) wave conditions. The average significant wave heights recorded in the marina with these conditions are listed in Table 3.

Wave heights in the outer harbour (immediately outside the marina, at West Wharf and Trinity Pier Berths) were recorded with all schemes; average significant wave heights in these areas are also given in Table 3.

Table 2 - Average Significant Wave Heights (Storm Conditions) in the Marina.

SCHEME	AVERAGE SIGNIFICANT WAVE HEIGHT (m)		
	MHWST	HAT	
A*	1.13	1.12	Existing entrance
B	0.96	0.98	
C	0.95	0.96	
E	0.86	0.82	Reduced width
F*	0.82	0.78	
H	0.72	0.70	
J	0.80	0.76	Plus reduced depth
I	0.71	0.67	
G*	0.73	0.71	Overlapping screen
D	0.67	0.64	
L*	0.56	0.52	Central entrance
K	0.50	0.47	
N	0.61	0.59	
O	0.57	0.54	

Table 3 - Average Significant Wave Heights ($H_{200 \text{ hours}}$) in the Marina and Outer Harbour.

SCHEME	AVERAGE SIGNIFICANT $H_{200 \text{ hours}}$ at MHWST (m).			
	MARINA	OUTSIDE THE MARINA	WEST WHARF	TRINITY PIER BERTHS	
A	0.26	0.43	0.50	0.42	Existing
F	0.22	0.43	0.48	0.42	Narrow entrance
G	0.19	0.45	0.47	0.41	Overlapping
L	0.18	0.46	0.51	0.40	Central entrance

2.5 Storm Wave Tests.

The results of the storm wave tests are indicated on Figures 7, 8 and 9 and can be summarised as follows:-

1. The hydraulic model tests indicate that most of the wave energy in the marina penetrates through the entrance, rather than under and/or over the present wavescreen. Closing the gap under the screen, Scheme B, would reduce present storm wave heights in the marina by about 14%; raising the crest level of the screen, Scheme C, however would only give a further 1% improvement to about 15%.

2. Narrowing the present entrance reduced present storm wave heights in the marina by about 26% to 0.8m with Scheme E and Scheme F. Also by about 37% to 0.7m, with Scheme H. The narrower 8m entrance, Scheme H, although more beneficial in terms of wave height reduction, tended to produce higher velocities and unacceptable turbulence in the entrance.

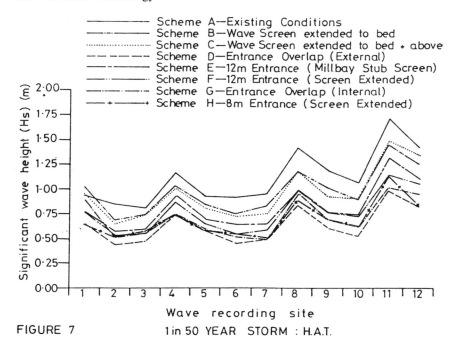

FIGURE 7

1 in 50 YEAR STORM : H.A.T.

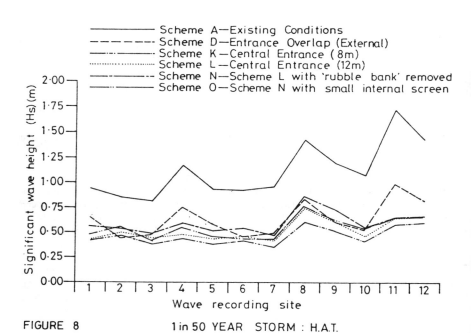

FIGURE 8

1 in 50 YEAR STORM : H.A.T.

3. Reducing the width and depth at the entrance, Scheme I, (8m) and J (12m), reduced present storm wave heights in the marina by about 38 and 30% respectively. In view of the 26% reduction already obtained with Scheme E the additional 4% reduction achieved by raising the bed level would probably not be cost effective.

4. The effect of the overlapping screens was to reduce present significant wave heights in the marina by about 42% to 0.6m, with Scheme D. Also by about 36% to 0.7m, with Scheme G.

5. With the relocated central 12m entrance, Scheme L, the present significant wave heights in the marina were reduced by about 52% to 0.5m. The further improvement in wave heights with the 8m entrance, Scheme K was 5% but high velocities and unacceptable turbulence were again observed with the central entrance.

6. The effect of removing the fissured limestone bank in front of the 12m central entrance, Scheme N was to increase the average significant wave heights observed with Scheme L by 6%, even so present average sigificant wave heights were still reduced by about 46%. The addition of an angled screen northside of the entrance, Scheme O, overcame the adverse effect of removing the underwater bank, reducing present average significant wave heights by about 51%.

7. In the present configuration storm wave heights were highest along the east wall and at the corners of the marina. The effect of reducing the entrance width and/or providing overlapping screens was to produce a more uniform wave height distribution in the marina. With the "central" entrance there was less wave reflection inside the marina and, therefore, a much more uniform distribution of wave heights than with the present entrance.

2.6 Conclusions.

The wave refraction study has shown that most of the wave energy entering Millbay originates from the east side of Drake's Island. Only local short period waves of up to 0.6m can be generated from the southwest, that is north of the Bridge, even in storm conditions.

The hydraulic model tests have shown that in the present harbour configuration most of the wave energy in the marina enters through the entrance rather than under and/or over the present screen.

Reducing the present entrance width to 12m would reduce average significant storm wave heights in the marina by 25 to 30% from about 1.1m to 0.8m. Reducing the entrance width further would reduce wave heights in the marina but there would be undesirable turbulance and current velocities in the entrance.

Relocation of the marina entrance near to the centre of the wave screen and reducing the width to 12m would produce a reduction in average significant wave heights in the marina by about 52% to 0.5m. With this entrance wave heights around the marina would be more uniform than at present.

Reduction in the central entrance width to 8m reduced average significant wave heights by a further 5% but tended to create undesirable flow conditions in the marina. None of the schemes tested produced a wave climate under severe storm conditions within the normal maximum acceptable wave height of 0.3m for finger berthing.

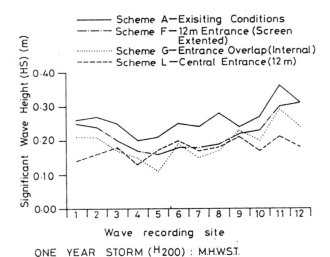

ONE YEAR STORM (H200) : M.H.W.S.T.

FIGURE 9

3.   TOWN QUAY MARINA

3.1 Location and Conditions.

The Port of Southampton lies some 10 miles up the estuary of Southampton Water which is further protected by the Isle of Wight. It is well known for its double high water and is one of the best protected natural harbours along the south coast of England for ocean going ships. The only direction from which Southampton Port is exposed to storm winds and waves is from the southeast although even then, because of the narrowing effect of Southampton Water, long swell ocean waves do not reach the Port, see Figure 10.

The location of Town Quay Marina is susceptable to winds and waves from the southeast, southwest and northwest. The longest fetch and the worst storm wave condition is from the southeast, while the northwest fetch is also significant. The southwest direction is relatively short and was not originally thought to be important. However, the results of the previous studies indicate that for the type of layouts proposed, the southwest waves are likely to lead to more severe conditions inside the marina.

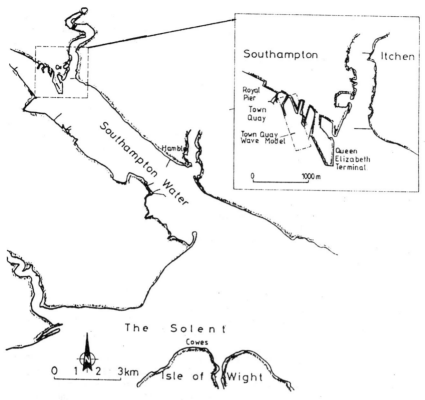

FIGURE 10   LOCATION PLAN

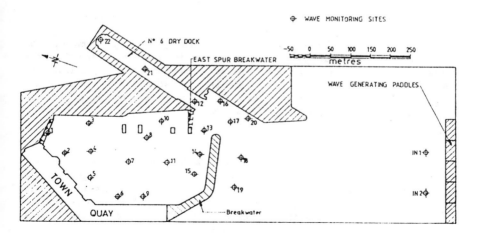

FIGURE 11   MODEL TEST LAYOUT

Both the 1987 and 1990 wave studies by Hydraulics Research indicate that even with a generous overlap and with careful orientation of the Marina entrance, there were still considerable problems within the marina area due to the reflection of waves from the vertical walls. The conclusion was that for all the layouts tested, wave heights in the marina area were in excess of 0.3m significant wave height for a 1 in 50 year return period event.

Wave periods are also critical for marina design and predictions of wave conditions at Town Quay have been made by Hydraulic Research at Wallingford and are given in their reports of 1987 and 1990. For southeast, southwest and northwest directions, these vary from 0.8m to 1.3m in height and 2.5 to 3.5 seconds duration.

Waves of around 6 second duration are often experienced in in-shore waters so that the 3 second wave predictions for Town Quay are relatively slight and would not normally be a problem to cope with. However, it is the particular configuration of the proposed Town Quay Marina that gives rise to reflected waves that require careful design and attention to detail.

3.2 Previous Studies and Objectives.

During 1986 the developers commissioned Hydraulics Research to undertake numerical wave disturbance and hindcast modelling together with flume testing of various marina layouts incorporating a 'skirt breakwater'. These tests were, however, terminated before an optimum scheme had been determined.

Outline planning permission was granted in March 1988 for the construction of a 450 berth marina with residential, shopping and leisure facilities at Town Quay. Schemes were designed to include an area for car parking on top of a wave protection barrier. These schemes were also tested on the HR numerical model but did not provide an acceptable solution as wave heights within the marina were still too high.

The results obtained indicated that it should be possible to provide a marina where significant storm wave heights are less than 0.3m, but for this to be achieved it would be necessary to prevent wave energy entering the marina from under Town Quay, particularly from waves approaching from the sector southwest through northwest. In 1991 ABP Research and Consultancy Limited were commissioned to review the results of the previous studies and then construct a physical model to test a further series of design layouts incorporating solid breakwaters with car parking facilities on the top.

3.3 Model Design.

Since the previous studies showed that wave energy should be prevented from entering the marina under Town Quay, the physical model was designed with a vertical face on the west (Royal Pier side) of the quay. This will enable different slopes, types and positions of 'solid blockage' to be investigated at the marina side, should this be required to further reduce wave heights in the marina.

By closing Town Quay to wave transmission the most important wave direction that will directly affect the marina will be 153°. The orientation of Southampton Water and the need to consider diffraction around the Dock Head suggested the worst wave conditions would be generated by winds/waves from the southeast. Wave directions between south and west would also be important if the marina entrance were located nearest the Town Quay. However, the previous studies suggest that entrances in this position were unlikely to give acceptable wave heights within the marina.

Taking the above situations into account, together with the layout of schemes to be tested, the effects of the possibility of opening up the No. 6 dry dock and the effects on the RO/RO terminal at 49 berth, an undistorted physical model covering the area shown in Figure 11 was constructed to an undistorted scale of 1:50, which yields a time scale of approximately 1:7.

Random waves were produced in the model from a 'bank' of five electrically operated vertical piston paddle wave generators from a direction approximately south-southeast as shown in Figure 11. The required input spectrum to the generating system was checked to ensure that a sufficiently representative spectrum of waves against those required was achieved in the model, see Figure 12.

Data was recorded by computer using twim wire resistance probes sited at locations agreed with the Client. Two sites were continuously monitored in front of the wave paddles to monitor the continuity of the input data during the test programme. These sites were also used to calibrate the model against the required spectra.

No further wave data has been collected from the site, therefore, the same significant wave heights and peak periods of the spectra have been used as derived by HR for previous studies and are set out in Table 4.

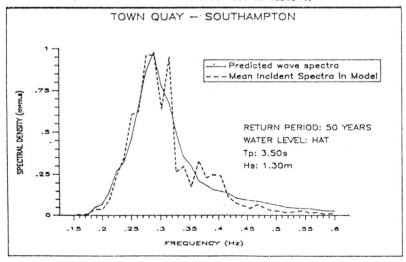

COMPARISON OF PREDICTED AND MODEL WAVE SPECTRA
FIGURE 12

Table 4 - Model Test Conditions.

RETURN PERIOD YEARS	SIGNIFICANT WAVE HEIGHT	PEAK ENGERY PERIOD	SPECTRUM	WATER LEVEL	DIRECTION
1	0.92m	3.2s	JONSWAP	MHWST	150°
50	1.30m	3.5s	JONSWAP	HAT	150°

Two waves conditions were used in the study, the first was representative of a 1 in 50 year storm and the second a 1 in 1 year storm. The 1 in 50 year return period wave condition was run at a water level of HAT (highest astronomical tide +4.9m CD) whereas the 1 in 1 year condition was run at MHWST (+4.5m CD) to give a more representative condition.

Water surface elevations were monitored at 22 sites in the proposed marina area, No. 6 dry dock and No. 49 RO/RO terminal for conditions representing the existing layout and then the schemes to be tested. The measuring sites are shown on Figure 11.

At each site the data was recorded directly by computer for subsequent analysis to determine the standard wave parameters. Each set of data was collected twice to ensure the accuracy of the results which are reproduced graphically of Figures 13 and 14.

3.4 Description of Test Schemes.

The prime objective of the model tests was to evaluate two schemes involving a solid breakwater from the end of the existing Town Quay with a marina entrance close to the No. 6 dry dock. After testing the two modified designs, three further schemes were derived for testing. The individual schemes are described below.

'A' - Existing layout of Town Quay area.
'B' - Solid main 'car park breakwater' with vertical sides and a small sub-surface toe revetment on all sides and a 'spur breakwater' on the east side between the dolphin and the No. 6 dry dock wall.
'C' - As above with a 1:3 sloping rubble revetment from the bed to the cope level placed along the outer face of the main breakwater and around the head of the breakwater and to the spur breakwater.
'D' - As above except with the revetments steepened to 1:2 and the entrance to No. 6 dry dock open.
'E' - As above with the length of the main breakwater reduced by 25m.
'F' - As above except a 1:2 revetment was placed all around the spur breakwater and outermost dolphin.

In the model the revetments were constructed using limestone chippings which were scaled from results obtained from the numerical software package 'BREAKWAT'. This package aids the design of breakwaters but for this study was used to calculate the size of stone required to withstand a 1 in 50 year return period storm for revetment slopes of 1:2 and 1:3.

3.5 Results and Conclusions.

For the purpose of simple comparison of individual schemes, average significant wave heights have been calculated for specified areas for both 1 in 1 and 1 in 50 year return period storms. These heights are summerised for the three areas in the marina for all schemes tested in Tables 5 and 6 together with the overall average.

The tables show that all schemes considerably reduce wave heights. However, only Schemes C and D produce overall average significant wave heights for the 1 in 50 year return period below 0.3m, the generally accepted standard for pontoon finger berthing.

The model tests indicate that the optimum solution for providing an acceptable small boat marina at Town Quay will involve a long breakwater with an entrance as narrow and as far east as possible, consistent with providing adequate access to the RO/RO berth and the dry dock. The breakwater should incorporate a wave absorbing sloping revetment which reduces wave activity in the vicinity of the marina entrance and the RO/RO berth. It will also be essential to include an overlapping east spur breakwater.

Table 5 - Average Significant Wave Heights (Hs) in the Marina with 1 in 1 Year
Return Period at MHWST

1 in 1 Year Return Period   Marina Area

SCHEME	NORTH (m)	MIDDLE (m)	SOUTH (m)	WHOLE (m)
A	0.51	0.53	0.59	0.54
B	0.19	0.18	0.13	0.17
C	0.15	0.16	0.11	0.14
D	0.14	0.16	0.10	0.13
E	0.26	0.24	0.19	0.23
F	0.14	0.19	0.13	0.15

A shorter breakwater opens the 'effective' width of direct exposure to wave activity in the vicinity of the entrance allowing greater penetration of wave energy into the marina. To provide a suitable marina with a shorter main breakwater, increased overlap of the east spur breakwater would be required and would 'cancel out' the cost benefit of shortening the main breakwater. Moreover, the car park area would be reduced and the reduction in wave heights at the RO/RO berth will also be less beneficial.

Table 6 - Average Significant Wave Heights (Hs) in the Marina Area with a 1
in 50 Year Return Period at HAT

SCHEME	NORTH (m)	MIDDLE (m)	SOUTH (m)	WHOLE (m)
A	0.97	1.02	1.03	1.00
B	0.37	0.39	0.33	0.36
C	0.24	0.26	0.24	0.25
D	0.25	0.28	0.23	0.25
E	0.33	0.37	0.31	0.34
F	0.37	0.32	0.25	0.32

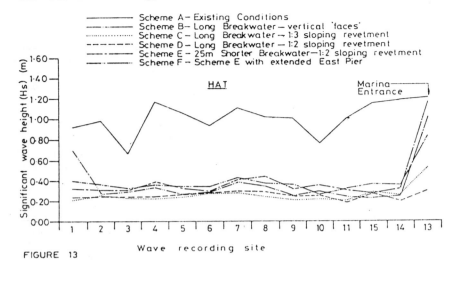

FIGURE 13

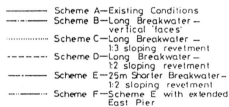

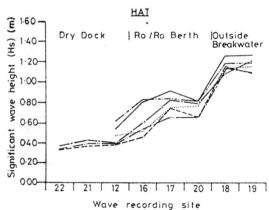

FIGURE 14

Two sites (18 and 19) were measured outside the breakwater and significant wave heights were marginally reduced for all schemes compared with the existing conditions whereas the wave periods were unaffected. Visually, however, the revetment schemes showed calmer conditions. With the vertical breakwater the higher waves overtopped the breakwater whereas the slopping revetments stopped all overtopping even under the 1 in 50 storm conditions.

4. SUMMARY

Physical hydraulic model studies have been carried our to evaluate alternative ways of improving wave conditions in the marina at Millbay, Plymouth, and to evaluate the optimum design for a breakwater at Town Quay, Southampton.

The conclusion of wave refraction studies for Plymouth was that most of the wave energy which penetrates Millbay comes around the east side of Drake's Island and that only local short period waves can be generated from between the south and southwest.

The present marina appears to be well protected from storm waves by Millbay Pier and the vertical wave screen although storm damage has occured. The model showed that raising the crest level of the screen to prevent overtopping and by closing the gap to the bed underneath would give some improvement, however, a narrower entrance reduced storm wave heights much further.

The effect of relocating the 12m wide entrance towards the centre of the screen further reduced storm wave heights in the marina to about 0.5m. With less severe wave conditions wave heights in the marina were on average less than 0.2m with all schemes, the central entrance giving a more uniform distribution of wave height around the marina.

Several different proposals for a breakwater at Town Quay had previously been studied but none were found to give acceptable conditions. The aim of the present physical model was to produce a design that would provide an acceptable wave climate within the marina commensurate with minimum cost and safe navigational requirements.

The model demonstrated that this can be achieved provided that the existing Town Quay structure is closed to prevent wave energy passing underneath it, an overlapping entrance is constructed as narrow and as far to the east of the 'opening' as possible and the outer face of the breakwater is constructed as a wave absorbing structure with slopes not steeper than 1:2.

The effectiveness of wave modelling techniques and in particular physical models has been amply demonstrated. The ease of modifying proposed designs and evaluating their effect almost instantly in the micro environment makes the use of physical modelling as a design tool a practical and cost effective adjunct to the preliminary design.

Even if only a few minor amendments to the design are necessary, the model costs are still only a few percent of the total capital value of a project. Unsatisfactory untested designs can cost considerably more to modify retrospectively.

Model Tests of Floating Breakwaters

J. Rytkönen, P. Broas

Technical Research Centre of Finland, Ship Laboratory, Tekniikantie 12, SF-02150, Espoo, Finland

ABSTRACT

This paper describes the model tests of floating-type breakwaters for marina applications. Several cross-section types for floating bodies were tested in a flow flume in order to study the behaviour of the bodies and movements in waves. The main goal was to determine the limits where a certain body can be used as a berthing pontoon or as a floating breakwater. Thus the wave transmission coefficients of selected pontoons were measured in regular waves. The motion responses of floating bodies were also determined using irregular waves. The tests were arranged in a wave flume having a cross-sectional area of 1.1 m x 1.5 m, the test configuration thus following the two-dimensional principles. Together with the flume tests an aerodynamic model was constructed for wind force measurements. The total wind force affecting a marina full of or without boats having different rig arrangements was measured for different wind angles.

1 INTRODUCTION

A small-craft harbour can be protected against waves by arranging breakwater structures, these being of two principal types: fixed and floating breakwaters. Conditions where the structure is situated, construction and maintenance costs as well as the degree of protection required all define limits on the use of the structures in question. Floating structures have many advantages over fixed structures, though they are not suitable for very exposed locations. In particular, ice forces, the anchoring of structures, the dampening ability as well as the assessment of wind loadings and movements create difficulties in planning.

This paper describes the experimental flume and wind tunnel tests for small-craft marina applications conducted at the VTT Ship Laboratory. The tests presented in this paper are a part of a more extensive research project concerning information collectioning on the properties of floating pontoon jetties and breakwaters [1]. The most important questions in this respect were the various loading factors, the types of floating breakwaters and their wave-dampening characteristics.

The goal of the research project was to provide information on the movements of various types of floating structures, the environmental loads on them as well as their wave-dampening characteristics in order for the planners of small-craft harbours to make decisions on the suitability of a certain structure as either a breakwater or mooring jetty for small craft.

2 WAVE TESTS

2.1 Test procedure

The wave tests were performed in a large class-walled flow flume having a cross-sectional area of 1.5 x 1.1 m^2. The model scale of 1 to 8 was selected for the measurements, thus the full-scale water depth of 8.0 m corresponded to 1.0 m in the model scale. The Froudian similitude principle was followed in the modelling.

A total of 10 different cross-sectional floating constructions were selected for the measurements (Fig. 1). The selected test types well represent the floating pontoon types constructed in Finland, thus the pontoon contractors and designers can easily use the results. The full-scale data for the test pontoons are shown in Table 1. The cross-sections 1 to 5 represent rectangular-shaped pontoons, which are usually constructed with a concrete box filled with a certain flotation material, for example styrox. The bottom of the box is normally left opened, without a concrete cover. However, the cross-section no. 4 represents a heavier pontoon type with concrete surfaces and concrete inside partitions.

Models 6 - 10 represent catamaran-type pontoons, which are also commonly used in Finland both as mooring jetties and breakwaters. Models 6 and 8 have a similar basic construction, but model 8 has double the amount of circular flotation parts. Cross-section 7 had underwater keels below the flotation parts. Cross-sections 9A and 9B were modified versions of model 10 made by inserting a horizontal plate below the construction. The vertical distances between the plate and the construction were one (model 9A) and two metres (model 9B) in the tests. The cross-sectional area of the plate followed the size of model 10. The connection between the plate and the construction was stiff.

The anchoring principle of the test configuration followed catenary-anchoring system. The lengths of anchor lines used in this test procedure were approximately 4 times the water depth. Later a more extensive study on the effects of various anchoring methods will be carried out. The measurements will be extended to shallower water depths, too. The different wave propagation angles against the structure will also be studied numerically after the required hydrodynamic coefficients from the flume tests have been analyzed.

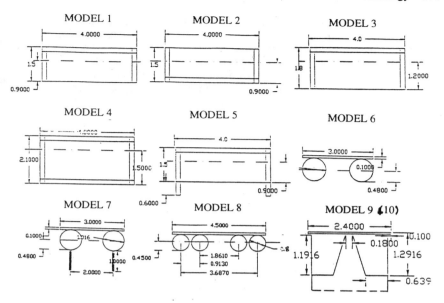

Fig. 1. The cross-sections of the test pontoons.

The measuring system of the wave test consisted of a microcomputer-based data collection system and the control of a plunger-type wave maker of the flume. Eight measuring channels were used for the tests. However for this paper only wave measurements and wave dampening ability were analyzed.

Table 1. The full-scale design parameters of the test pontoons.

M	BREADTH [m]	HEIGHT [m]	DRAUGHT [m]	VERTICAL CENTER OF GRAVITY *) [m]	RADIUS OF INERTIA [m]	MASS [kg/m]
1	4.0	2.1	0.9	1.09	1.59	3600
2	4.0	2.1	0.9	0.41	1.59	3600
3	4.0	2.1	1.2	1.27	1.63	4803
4	4.0	2.1	1.5	1.056	1.57	5995
5	4.0	2.1	0.92	1.06	1.6	741
6	3.0	1.2	0.478	0.808	1.02	880
7	3.0	1.24	0.478	0.62	1.12	961
8	4.5	0.99	0.45	0.58	1.46	972
9A	2.4	1.29	0.66	0.68	1.06	972
9B	2.4	1.29	0.66	0.62	1.2	972
10	2.4	1.29	0.62	0.79	1.01	906

*) Upwards from the bottom of the construction.

Both regular and irregular waves were used in the tests. The period range of the sinusoidal regular waves was 0.5 - 5.0 s and their steepnesses were 1 in 30 or 1 in 50. Two kind of irregular wave spectra were used: Jonswap-type spectra were calculated for the wind speeds of 10 m/s, 20 m/s and 30 m/s. The maximum fetch was limited to below 5 km. Because the Jonswap-type spectrum is narrow-banded ITTC-spectra for wind speeds 10 and 20 m/s were also performed. Fig. 2 shows an example of the measured time history and energy spectrum of a Jonswap-type spectrum.

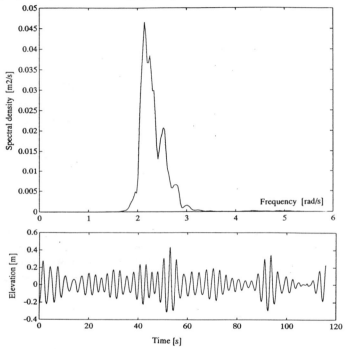

Fig. 2.   The measured time history of an irregular wave train and corresponding energy spectrum.

2.2 Results

The results for the wave dampening characteristics in regular waves are shown in Figs. 3 - 5 as a transfer function. The vertical axis shows the value of the transmission coefficient, ie. the relationship between the dampened and incoming wave height. The horizontal axis is the wave length versus the breadth of construction.

All the wave data of Figs. 3 - 5 are based on tests using regular waves with the steepness of 1:50. The transmission coefficients for waves having the steepness 1:30, however, equal well the results shown here. Only insignificant differences were observed.

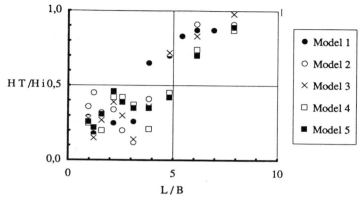

Fig. 3. The measured transmission coefficients, models 1 - 5.

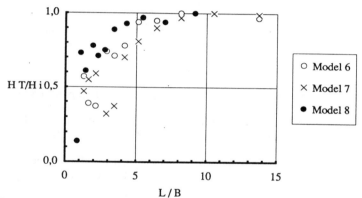

Fig. 4. The measured transmission coefficients, models 6 - 8.

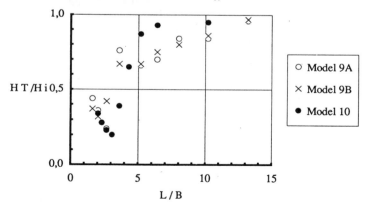

Fig. 5. The measured transmission coefficients, models 9A, 9B and 10.

2.3 Discussion

The wave dampening results of models 1 - 5 have all quite similar character. All cross-sections tested seemed to have the best dampening ability near the value L/B = 3. For longer waves the wave dampening ability decreased rapidly.

The observed minima of transmission coefficients were near the natural period of heaving motion of the structures. This indicates the heaving motion will generate waves against the incoming waves. Thus the incoming waves will be reduced, and less wave energy will be transferred to the leeward side of the structure.

The tested structures did not roll significantly in short waves and even in longer waves the natural period of rolling motion affected not significantly on the results of wave transmission coefficients. Similar conclusions will also be drawn from the results of Figs. 4 and 5.

The underwater keels of model 5 do not increase the wave dampening ability. However the heavier pontoons (or the decrease of the breadth/draught ratio) had smaller transmission coefficients for longer waves, which means better wave breaking characterictics and slower motions in shorter waves.

The underwater keels of model 7 increased the wave dampening ability compared to the model 6. Due to the increased added mass the natural period of heaving motion was higher. The model 8 was observed to follow the incoming waves too easily, thus the transmission coefficients were in high level too.

The catamaran models, 9A, 9B and 10, functioned well below the value L/B < 0.4 m. The inserted horizontal plates showed better wave dampening ability for longer waves, which can be noted from Fig. 6.

The tests were all carried out in a water depth of 8.0 m. Thus the measured motions and anchor forces will do differ from shallow-water conditions. However the basic aim was to determine data for the breakwater-design, thus the results are well suited for these conditions. Later the measurements will be extended for shallow ie. 2-4-m water depth, in order to study the limit when the structure's function is more a mooring-orientated than wavebreaking one. In this respect the various modes of body motions, measured accelerations and anchor forces will be reported.

3 WIND TESTS

3.1 Test Procedure

The tests were carried out in the VTT closed-circuit low speed wind tunnel using scale models (1:35). The wind tunnel is designed for aeronautical research and the test section is octagonal; 2 m high, 2 m wide and 4.5 m long. A separate ground board is normally used when testing ships or other vehicles to allow the use of the whole width of the test section. The boundary layer on the ground board is thin and because of the restricted length of the board the simulation of the atmospheric boundary layer is very difficult. In this case a sawtooth plate and roughness elements were used to create some turbulence and to thicken the boundary layer. The velocity and turbulence intensity profiles were measured using a hot-wire anemometer in three lateral positions. The measured mean velocity is compared in Fig. 6 to the power law curve with an exponent of 0.16. The boats reach a height of about 5 cm (without mast). In this region the turbulence intensity is 10-20 %. From the height of 200 mm (7m on full-scale) the velocity distribution, however, differs clearly from the theoretical curve of the mentioned atmospheric boundary layer.

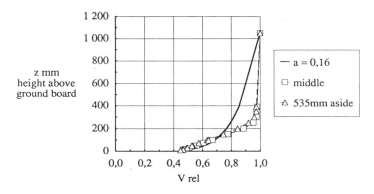

Fig. 6.   The measured boundary layer profile on the ground board with roughness and without model.

The models were fixed to the turntable using aluminium plates. The 1.2 m diameter turntable was connected to a three-component wind tunnel balance lying under the test section. The velocity head of the air flow was measured with a pitot-static tube in free stream 1.05 m above the turntable. Signals from the force transducers of the balance and from the pressure transducer were transmitted to the data-acquisition board of a microcomputer which calculates the force coefficients as an average of 15 000 individual measurements.

The models were designed to represent the most common sized boats in these kinds of marina. About 40 waterline models of a 33' sailing-boat and a 24' family power-boat were made of polystyrene for the tests. The sailing boat was equipped with a 16 m mast with a boom and a sailbag on it (Fig. 7). The boats had sitting boxes and some smaller details, but ropes and railings were omitted.

The measurements were made to find out the effect of different marina layouts, the length of walkways and the number of boats (Fig. 8). The effect of various boat types were investigated using both sailing- and motor-boats and also different boat arrangements in the marina. Three walkway lengths with 12, 16 and 18 boats were tested. The T-shaped walkway represented a slightly more complicated layout. In Finland, large boats are often moored using buoys at the stern, and therefore the forces from the boats on the windward side of the walkway are not transmitted to the pier. Instead these boats provide wind shelter for the leeward side boats. This situation was examined too, as well as the wind shelter of a row of boats at various distances.

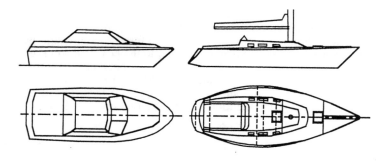

Fig. 7. Schematic views of the tested boat types.

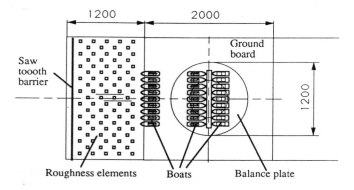

Fig. 8. Schematic view of the test arrangement.

3.2 Results

The aerodynamic force and moment coefficients for various test versions with tested yaw angles were calculated from measurements using the following equations:

$$C_x = \frac{X}{qA}; \qquad C_y = \frac{Y}{qA}; \qquad C_n = \frac{N}{qAl}$$

where C_x is normal force coefficient (perpendicular to the walkway)
C_y is side force coefficient (parallel to the walkway)
C_n is yawing moment coefficient
X is normal force, Y is side force, N is yawing moment
q is velocity head of free stream at a height of 1.05 m
A is reference area = 10 m² (1:1)
l is length of sailing-boat = 10 m (1:1).

These coefficients can be used to compare various versions with one other. Full-scale forces can be simply calculated using these coefficients with different wind velocities. In Figure 9 an example of results is presented, where for example the difference between straight and T-shaped marinas can be seen.

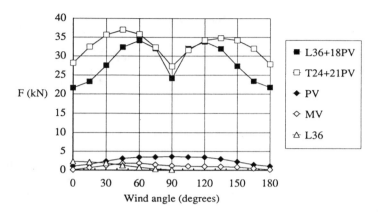

Fig. 9. The total wind forces of two marina layouts, one sailing boat, one motor boat and the walkway with yaw angle.

3.3 Discussion of results

At the moment of writing, the wind tunnel tests have just been completed and the analysis of the results is under way. Hence only a brief discussion is presented here.

The total wind force increases almost linearly with the number boat pairs. The wind force is 40 - 50 % smaller if there are boats only on one side of the walkway, which shows that the wind shelter of the first row is quite small. The wind force of sailing-boats is almost four times higher than the wind force of motor-boats. It should be mentioned here, that the drag of the sailing-boat masts is not yet corrected for the Reynolds number effect, which can decrease the wind forces of sailing-boats.

The results were compared to wind forces calculated using equations from Burn[2] and Tobiasson[3]. The measured forces of marinas with sailing-boats are about half of the calculated forces. The difference is probably due to the difference in reference velocity. Owing to the poor boundary layer simulation, the reference velocity (as well as the velocity at 10 m height) is about 70 % higher than the velocity at the boat-deck level. In practise, the velocity at 10 m height is only about 25 % higher than the deck level velocity. The increase in the total wind force with yaw angle is according to the literature somewhat larger than the measured increase.

Additional testing on the effect of the atmospheric boundary layer is planned and a more profound treatment of wind tunnel test results will be presented at the conference.

4 CONCLUSIONS

The wave transmission measurements showed a floating pontoon will dampen the short waves effectively, while the dampening of long waves seem to be rather ineffective. The draught of the rectangular cross-section seemed affect only slightly on the wave dampening characteristics. The increase mass of the body together of the construction width decreased more clearly the transmission coefficient.

More tests will be carried out in shallower water to find out the performance limits of the floating jetty or breakwater. After the test programme of anchor forces has been completed, the motions, anchor forces and an evaluation of the environmental total forces against the floating jetty or breakwater will be reported.

These two-dimensional tests will be extended numerically for other wave propagation angles. Additional full scale measurements are also needed to solve the real three-dimensional wave-structure interaction near the shore line.

Wind tunnel tests showed significant differences between various boat types and various marina lay-outs. The total force affecting on the certain type marina lay-out will later studied numerically.

Additional tests are needed to determine the effect of atmospheric boundary layer.

5 REFERENCES

1. Rytkönen, J. & Pakarinen, E., The properties of floating pontoon jetties and breakwaters - a preliminary survey. VTT Notes, Espoo. 1991. 95 p + 6 app.

2. Burn H. Load Sharing Benefits of Composite Pile Systems. World Marina 91. L.A.California, U.S.A. p. 202 - 215.

3. Tobiasson, B. O. Marina Layout Parameters, Vessel Characteristics and Design. Proc. of the International Conference on Marinas, Southampton, UK, September 1989. P. 235 - 246.

Applications of Single Layer Armour Units in the Construction of Breakwaters

S. Hettiarchchi

Maritime Engineering Group, Ove Arup & Partners, London

Abstract

This paper refers to the performance of single layer hollow block armour units used in the construction of breakwaters in ports and marinas. These units have been produced in different shapes of which the cubic (SHEDs and COBs) and hexagonal (SEABEEs) shapes are two examples. A characteristic feature of such units is their placement in a pre-determined manner which together with their fixed geometry allows close control of the voids matrix of the primary armour layer. The single layer placement and the high stability of the layout can produce significant cost savings. This paper focuses attention on experimental investigations which have been conducted to assess the hydraulic performance, including wave induced loading and the overall performance of these units in service.

1. Introduction

Various shapes of artificial armour units have been developed by breakwater designers in order to obtain a high degree of hydraulic stability at a relatively small armour block weight. It was expected that these armour units would withstand the design wave height without significant damage to individual armour units or to the breakwater as a whole. The different types of artificial armour units used in practice can be broadly classified into three types, namely,

(a) Bulky
(b) Slender interlocking
(c) Single layer hollow block

Bulky armour units rely mainly on their weight for stability and are usually placed at random. The slender interlocking type of units have the advantage of greater hydraulic stability due to interlocking effects. However, armour units of this type develop greater static and dynamic forces under wave action. These armour units, which have a relatively reduced block weight, are usually placed at random. There are instances where pre-determined laying arrangements are used. It is important to note that in the case of both bulky and slender interlocking types of units, the voids which contribute to the dissipation of wave energy are established between the armour units in a random manner. On most occasions, at least two layers of units are used for the primary armour.

Hollow block armour units are of more recent origin and are somewhat different to the other two types in that the voids are built into the individual units in the required form. Armour units belonging to this type are usually placed as a single layer to a predetermined form. Thus the resulting voids matrix of the primary armour is geometrically well-defined in contrast to that of the other two types.

2. Characteristics of the Hollow Block Armour Unit

The unique characteristic of the hollow block concept is the systematic analysis of the voids matrix of the primary armour layer. This allows absolute control of the geometry of the voids within the confined boundaries of an individual armour unit or a group of units to produce a cost effective primary armour layer which is very efficient with respect to wave energy dissipation. The stability of a breakwater consisting of hollow block armour units does not depend on the degree of interlocking between the units and as a result the weight of the individual armour units can be reduced considerably. A characteristic feature of a hollow block armour unit is the presence of a large volume of void in the unit relative to the volume of solid material These units have been produced in various external shapes of which the cubic form has been more popular. Figure 1. illustrates some of the units used in practice.

Hollow block armour units can be broadly classified into three types, as given below, based on the presence of lateral porosity and the method of placing.

(1) Armour units without lateral porosity
(2) Armour units with lateral porosity
(3) Armour units with lateral porosity and placed in sets of two or more units

In the case of the first group when the armour units are placed in the prescribed manner on the breakwater slope, the resulting voids matrix is not interconnected laterally although individual armour units have a void in the direction normal to the slope (e.g. SEABEE).

In contrast to the first group, units belonging to the second group generate a laterally interconnected voids matrix. By adopting a method of placement in which alternate rows are staggered by a length of half an armour unit it is possible to generate a voids matrix of equal porosity but with increased tortuosity (e.g. COB, SHED).

Armour units categorised into the third group are very similar to those belonging to the second group but are placed in sets of two or more armour units. They have been designed on the basis that three-dimensional symmetry of individual units is not an essential requirement in relation to the top surface of the armour layer and that having parallel edges of adjacent blocks is uneconomical with regard to material usage. It is evident that the armour units belonging to this group will have a rather complex geometry in contrast to those belonging to the first and second groups (e.g. DIODE, REEF).

Water entering a hollow block unit having a lateral porosity spills in four directions within the unit and out into the adjacent units where it encounters water moving, generally in opposite directions. Wave energy is thus dissipated in turbulence within the block. Since these units are placed close to each other other forces of appreciable magnitude do not act for force apart adjoining blocks and this is assisted by the relatively small surface area of the blocks. When the armour units are correctly placed they exhibit hardly any movement, with contact forces uniformly distributed. A typical cross section of a breakwater constructed with COB units is illustrated in Figure 2.

3. Factors which influence the design of armour units

3.1 Hydraulic performance of armour units

The emphasis in most research work on rubble mound breakwaters had been on the hydraulic stability of armour units and to a lesser extent on dynamic forces under wave attack, material properties of the units and the hydraulics of wave motion within the porous structure. Breakwater cross-sections constructed with model armour units are subjected to design wave conditions in the laboratory in order to justify their use in the prototype. A major design consideration is the stability of individual model armour units with reference to their displacements from the original position. This enables the identification of different levels of damage and the definition of stability coefficients (KD) based on Hudson's formula for different types of armour units (ref. 1). Refined formulae have been presented more recently by Van der Meer (ref. 2).

3.2 Wave induced loads on armour units

An increased number of breakwater failures recently has made it necessary to consider in detail the structural integrity of armour units together with other design factors. Until this stage, designs were based mainly on hydraulic stability tests for which the strength of armour units was not scaled. This led to the development and use of comparatively large interlocking type of armour units. On most occasions these large concrete armour units have been used without reinforcement material. Recent failures indicate that the limits of applicability of these units have been exceeded mainly because due consideration was not given to other aspects, particularly the influence of dynamic forces and the capability of armour units to withstand such loads in a hostile marine environment.

3.2.1 Types of loads acting on armour units

For a given armour unit there are three important phases that can be identified in relation to its overall performance

(1) Manufacture and transport for storage
(2) Transport to site and placing
(3) In service

The load conditions corresponding to the first and second phases are reasonably well defined and are mainly influenced by the static weight of the unit. Thermal stresses due to temperature differences during the hardening process may also be present. Under normal circumstances these phases do not produced critical loading conditions. The load conditions corresponding to the service state are complex and demand closer examination. The types of load encountered in this phase include static, dynamic, abrasive, thermal and chemical loads.

3.2.2 Static loads on armour units

The static loads mainly consists of the weight of units and stresses due to settlement of underlayers. Sometimes, armour units, particularly those of an interlocking type, may become wedged between other units during wave action resulting in an additional static load. It should be noted that the settlement of the underlayers generally contributes to increased static loads. Although static loads are not usually critical when considered in isolation, they may prove to be decisive when acting in combination with other types of loads.

3.2.3 Dynamic hydraulic loads on armour units and their influence

Dynamic hydraulic loads acting on armour units are essentially of two types. The first are oscillatory forces which are gradually varying or quasi-static loads due to wave action on the slope. These oscillatory forces are usually exerted during uprush and downrush of waves. The second type of dynamic hydraulic loads are impact forces due to direct wave impact. The presence of these forces and their magnitude will be very much dependent on the type of wave profile at the point of wave impact. A typical schematization of time-dependent wave impact forces by periodic water waves is illustrated in Fig. 3. Impact forces due to direct wave action, influence the armour unit in two ways. Firstly, the impost hydraulic impact loads of high magnitude acting over a very short time interval. This is of particular relevance to armour units placed in the vicinity of the still water level. Secondly, they cause the movement of a given armour unit which in turn will strike neighbouring armour units, thus imparting structural impact loads. Rocking, rolling and collisions between armour units and parts of one broken units striking another units are some of the main effects of this type of load.

4. Design criteria for hollow block armour units

4.1 Hydraulic stabililty

One of the main difficulties in using hollow block armour units is establishing an appropriate design criterion. Unlike other types of armour units they have proved to be extremely stable during hydraulic model tests and the definition of a stability coefficient on the basis of Hudson's approach is not applicable. The external geometry of the unit and the predetermined packing arrangement restricts the movement of individual armour units to a minimum. Hence it is not surprising that values of K^D greater than 80 have been observed for these units which have been found to be more stable om steeper slopes rather than less stable (ref. 3, 4).

When attempts were made to identify the failure mechanism it was observed that excessive overtopping of a breakwater having a relatively mild slope and with an unsupported crest, dislodged several units. However, this problem was overcome by using appropriate restraining measures (refs. 4 & 5). A comparatively loose laying pattern also resulted in the movement of a few units mainly by rocking or lifting at high incident wave amplitudes. Once a unit is extracted from the armour assembly due to the wave induced forces, there exists an opportunity for other units to fall over or to be lifted from their positions. The resulting instability will be characterized by lifting, rocking and rolling of armour units. This state corresponds to one of the possible failure mechanisms for hollow block armour units provided that the crest and the toe wall of the breakwater remains stable.

4.2 Wave loadings

In the case of hollow block armour units which are placed to a predetermined layout, each unit is in contact with the neighbouring unit such that the contact surfaces are well defined and controlled. Hence the influences of rocking, rolling and collisions are reduced to a great extent. In addition, the slope corresponding to the upper surface of the armour units is aligned throughout the breakwater and it is very unlikely that parts of the units will be removed and displaced over a considerable distance. Hence, for this type of unit, forces due to direct wave impact play a vital role. They impost impact components in the directions parallel and perpendicular to the slope. It is also evident that the applied static loads are well defined.

For hollow block armour units of which most have rectangular vertical foaces, the study of lift and along-slope forces is adequate to understand the forces acting on such a unit. The assessment of the upward lift component and the impact components of both lift and along-slope force is of particular relevance to the long term durability of materials and the understanding of possible failure modes for the breakwater.

Force measurements were made on model SHED and COB armour units by adopting a specially designed cantilever type strain gauged transducer. The instrumented unit was located centrally and positioned such that it was not in contact with the neighbouring units or with the underlayer (ref. 6). The study revealed that the critical aspects of the force traces were the upward normal component of the lift force and the impact components of both lift and along-slope force.

For the experimental conditions used for this study, it was observed that the upward normal force which tends to lift the armour unit out of the primary armour assembly occurred during the run down phase when the water level on the breakwater slope was between the maximum value of the run-up and the still water depth.

The upward normal forces are restricted by the corresponding component of submerged weight and the frictional forces between the armour units. Removal of armour units by lifting is one of the possible failure modes for breakwaters consisting of a hollow block armour units and as a consequence a proper estimate of the lift force is necessary. During the service state partial lifting of armour units will lead to the relative magnitude of the positive lift force was found to be within acceptable limits for the incident wave conditions used and the hydrodynamic forces were not high enough to extract the unit from the armour slope.

Results from tests using regular waves indicated that for a given armour unit - depending on its relative position and incident wave conditions - impact loads were superimposed on gradually varying or quasi-static loads. These loads were characterized by the presence of peak forces in the positive along-slope and the negative normal (negative lift) directions corresponding to the instant of impact of the wave front (Figure. 4). Of these two force components the positive along- slope force was found to be the dominant loading force. The results from the study provided an assessment of the order of magnitude of the respective force components acting on a typical hollow block unit under different incident wave conditions. Although forces corresponding to critical states of instability would not be achieved due to limited wave heights under the available laboratory conditions, the results would correspond to service loads encountered by the armour units. Detailed results are presented in reference 6.

As the wave steepness decreased, the force traces gradually became free from sharp peaks corresponding to impact loads and were characterized only by the gradually varying type of dynamic loads (Figure 5). For armour units located in the immediate vicinity of the still water level impact loads were observed in both along-slope and normal components corresponding to the point of impact, with impact increasing with increasing wave steepness. The intensity of the impact force was reduced as the degree of submergence increased. This was mainly due to the fact that at greater levels of sub-mergence most of the impact energy was absorbed by the row of armour units positioned above that of the row containing the instrumented armour unit. Hence under these conditions the instrument armour units was not fully exposed to direct wave impact forces

The importance of the downward along-slope force acting on a hollow block armour unit can be assessed in relation to the design of the toe beam of the breakwater. From the force records it was observed that impact forces were not present for this component. In the absence of such components in that direction it was evident that minimum damage would occur from pos-sible collisions between armour units. For a breakwater constructed with hollow block armour units, the toe should be able to withstand the down-ward component of the static weight of armour units and additional wave induced forces. In comparison with the static weight corresponding to a typical breakwater section having eight to ten rows of armour units, the wave-induced forces in that direction are small. The influence of wave in-duced downward forces can be incorporated in the estimation of the overall load acting on the toe beam.

Research on the hydraulic performance of and wave induced loading on single layer hollow block armour units have been conducted at Imperial College (ref. 6 &7), Hydraulic Research Limited (ref. 5) and Wimpey Lab-oratories (ref. 4). A major research project on this subject is currently in progress in the UK. Preliminary results from this study are available (ref. 8) and detailed results will be published in the near future.

The tests performed at Imperial College also indicated that a rectan-gular block consisting of COB or SHED units (with or without a vertical impermeable rear face) exhibited a high degree of hydraulic efficiency with respect to both wave reflection and transmission. Similar structures built with porous stacked blocks are frequently used in coastal and harbour works in Japan (ref. 12). A vertical porous screen of COBs and SHEDSs, as sin-gle unit in width, also proved to be very effective with respect to energy dissipation. This type of structure, identified as a porous wave absorber, has been used frequently in harbour and marina works. Test have also indi-cated that COB and SHED armouring could also be used effectively in front of vertical walls which are subjected to toe erosion due to high reflections from the vertical face. This type of armouring develops reduced levels of reflection which over a long period will contribute to the long term stability of the structure. The investigations clearly indicated the importance of the hydraulic efficiency of voids matrix of the COB and SHED unit under different conditions of application.

5. Performance in service

Single layer armour units have now been used effectively on various projects in different parts of the world. The designs are based on incorporating the fundamental characteristics of the hollow block concept outlined earlier. Relevant details of the projects are outlined in reference 9. It is recommended that high quality concrete is used to provide early high strength and durability. Considerable use has been made of chopped polypropylene fibre as reinforcement in the porportion of 0.2% by weight to improve impact and handling stress resistance.

When constructing breakwaters with hollow block armour units attention has to be focused in the preparation of the underlayer, the toe structure and the placement of units. Important aspects relating to the construction are described in references 3, 10 and 11.

Adequate support should be provided during handling, transport and placing of units. A certain amount of abrasion amd impact may occur when placing one unit beside another. Sufficient precautions should be exercised when doing so and in particular when placing on a predetermined layout. If hollow block units are not placed with care it is possible that long narrow spaces may be generated between the units. This will encourage the movement of armour units and also create regions in which wave pressure can concentrate. Under wave loading, these conditions contribute to the potential occurrence of structural impact loads, particularly in the upward direction, and abrasive forces. It should be noted that even if precautionary measures are adopted. It is quite possible that narrow crevices, however small, are formed within acceptable standards of construction. These may also develop due to the settlement of the underlayer. Attention should also be focused on armour units placed on the roundhead of the breakwater. Due to the external curvature of the structure it is difficult to achieve perfect alignment between the units in these regions, thus allowing movement under wave attack. The above observations are of particular importance to the performance of armour units in the near vicinity of the water-line where high impact loads can be expected.

Hollow block armour units have performed efficiently in service. No sign of significant movement of units has been observed. The underlayers have remained stable although some of the underlying stone is considerably smaller than the apertures in the units, but it is not dragged out by the waves. On none of the breakwaters constructed so far has any unit been removed from the primary armour layer due to wave action on the slope and in no case has the cracking of units caused further disintegration of affected the stability of the overall structure.

Cracking of units is mainly due to point loading between units arising from toe and bank settlement and is not considered a problem for long term stability. The importance of accurate placing of armour units with particular reference to the developement of structural impact loads was discussed earlier. The presence of these loads also contribute to the formation of cracks. It is in this respect that in the design of armour units it is important to understand the types of loads acting on them due to wave action (refer to section 3.2).

The principle advantages of the hollow block design in revetments and breakwaters can be summarised as follows:-

1. The use of steeper slopes on the seaward face (generally 1:1 1/3) with a consequent saving in the volumes of the underlying materials.

2. Only one layer of hollow blocks is required for the principal armour.

3. The weight of the hollow block units is considerably lighter than alternative armour units for a given design condition. Consequently, the overall weight of the armouring units is reduced substantially, thus providing an economic solution.

4. On most occasions the layer of material underlying the hollow block armour units serves as a filter layer rather than as a secondary armour layer. Consequently, the layer thicknesses is relatively small with the required grading being generally available.

5. The geometry and the reduced weight of the blocks are such that they permit easy production, storage and greater flexibility with regard to site transportation.

6. Closure

This paper identified the important characteristics of single layer hollow block armour units used for the construction of breakwaters. Reference was made to experimental investigations which have been conducted to assess the hydraulic performance. These investigations clearly illustrate that single layer hollow block units exhibit a very high overall hydraulic efficiency. The performance of these units in sevice was discussed and the principal advantages of using this type of armour design were summarised.

References

1. **HUDSON, R.Y.** Laboratory Investigation of Rubble Mound Breakwaters. Journal of the Waterways and Harbours Division. ASCE, 1969. Vol. 85, WW3, pp.93-119.

2. **VAN DER MEER, J.W.** Stability of Breakwater Armour Layers - Design Formulae Journal of Coastal Engineering. 1987. Vol. 11, pp. 219-239.

3. **WILKINSON, A.R. and ALLSOP, N.W.H.** Hollow Block Units. Proc. ASCE Conference Coastal Structures '83 Virginia, USA. 1983, pp. 208-221.

4. **STICKLAND, I.W.** COB Units - Report on Hydraulic Model Research Wimpey Laboratory. Ref. No. H/334, 1969.

5. **HYDRAULICS RESEARCH STATION** The SHED Breakwater Armour Unit, Model Tests Using Random Waves. Report EX 1124. 1983

6. **HETTIARACHCHI, S.S.L** The influence of Geometry on the Performance of Breakwater Armour Units PhD Thesis, Imperial College, University of London, 1988.

7. **HETTIARACHCHI, S.S.L and HOLMES. P.** Performance of Single Layer Hollow Block Armour Units Proc. Conf. on Design of Breakwaters, ICE, London 1988.

8. **DAVIS, J.P., WALDRON, P., EDWARDS, D.J. and STEPHENS, R.V.** Acquisition of Data from Single Layer Armour Units in Breakwaters using Radio Telemetry. Proc. IASBE Colloquium on Monitoring of Large Structures and Assessment of their Safety, 1987.

9. **DUNSTER, J.P., WILKINSON, A.R. and ALLSOP, N.W.H.** Single Layer Armour Units. Proc. Conf. on Design of Breakwaters, ICE, London, 1988.

10. **WILKINSON, A.R.** St. Helier Consulting Engineer. Vol. 42, No. 10, October 1987.

11. **COODE and PARTNERS** Artificial Armouring of Marine Structures Dock and Harbour Authority. Vol. 51, No. 601, November 1970.

12. **KAIYO RESEARCH INSTITUTE INC** Illustration of Coastal Projects Using Neptune Caisson Research Report

			SIZE	WEIGHT	POROSITY
COB 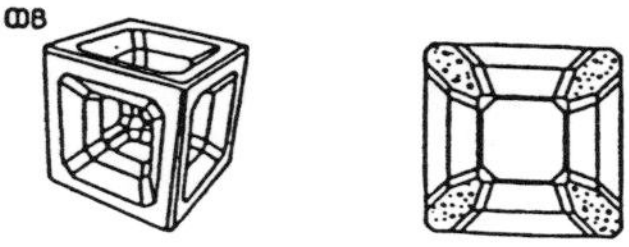			1.3 m cube	2 tonnes	60%
SHED 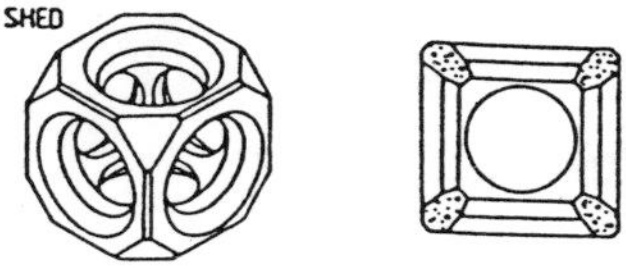			1.3 m cube	2 tonnes	60%
DIODE 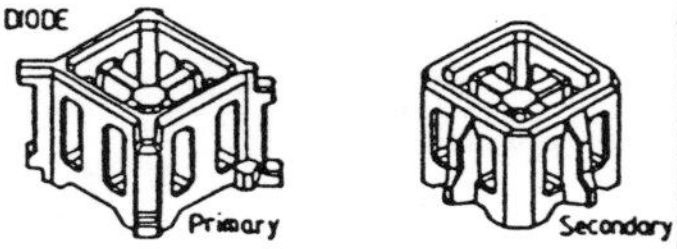 Primary		Secondary	varies typically 2 2 1.1 m	4 2 tonnes	Typically 60%
SEABEE 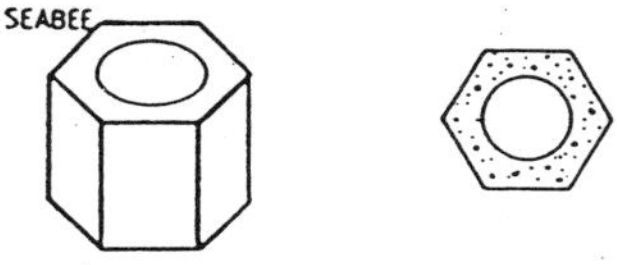			varies	3-12 tonnes	Normal to slope only

Note:- Mild steel reinforcement has been used for the DIODE. Chopped polypropylene fibres or stainless steel loops has been used for the SHED.

Fig. 1: Hollow block armour units

(ref. 9)

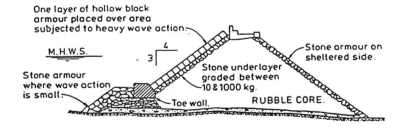

Fig. 2: Typical crossection of a breakwater in Jersey using COBS

(ref. 3)

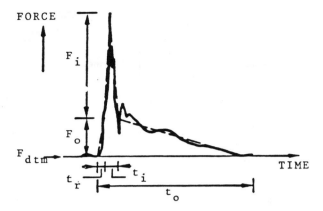

F_{dtm} = force corresponding to still water depth (datum force)

F_o = gradually varying or quasi-static force

F_i = impact force (in excess of F_o)

t_r = rising time of F_o

t_o = duration of the gradually varying force

t_i = impact duration

Fig. 3: Typical schematization of time-dependent wave impact
forces by periodic water waves

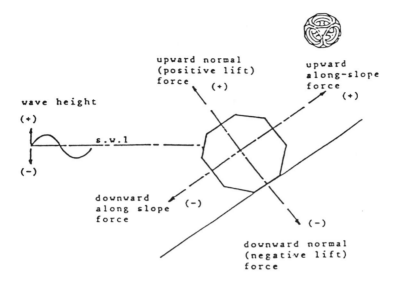

Sign convention for wave height and force measurements

(ref. 6)

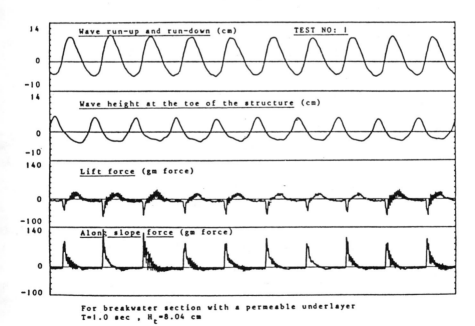

For breakwater section with a permeable underlayer
T=1.0 sec , H$_t$=8.04 cm

Fig. 4: Wave height and force measurements (measured values)

(ref. 6)

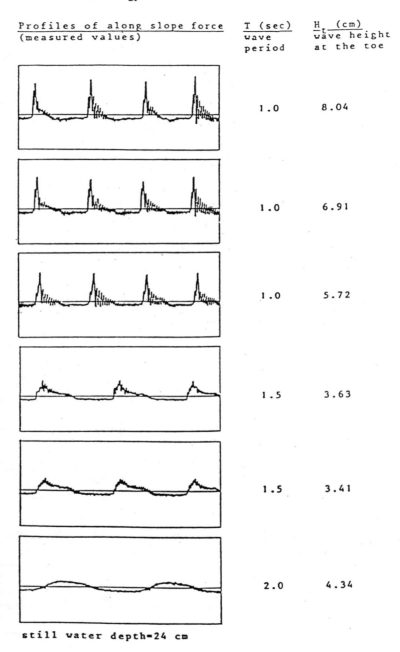

Profiles of along slope force (measured values)	T (sec) wave period	H_t (cm) wave height at the toe
	1.0	8.04
	1.0	6.91
	1.0	5.72
	1.5	3.63
	1.5	3.41
	2.0	4.34

still water depth=24 cm

Fig. 5: Variation of along slope force (measured values) for different incident wave conditions

(ref. 6)

SECTION 8: MARINA DEVELOPMENTS

The Development of the National System of Leisure Harbours in Greece

A.J. Rogan

Rogan Associates,

Consulting Engineers-Architects, 9 Valetta Street,

GR-15771 Zografos, Athens, Greece

ABSTRACT

The regional planning of leisure harbours in Greece, by developing a system of navigational zones, each one offering a selection of sites suitable for Marinas (big leisure harbours), Skalas (medium-small leisure harbours) and Anchorages for refuge, is presented in this paper. Details of the 1st Phase of development in 1981-82 are given, followed by an Intermediate Period of unexpected evolution which led to a 2nd Phase of development in 1990-91, according to which the selection of sites in four navigational zones is finalized, thus doubling Greece's capacity in berths.

INTRODUCTION

Greece, with some 16.000 kms of coastline, more than 1600 bays and gulfs and about 3000 islands, has, since the ancient times, been a predominantly maritime nation. It is mainly by sea and through the country's ports that communication has taken place and there are at present 127 ports varying both in scale and range of activities. It is, therefore, obvious that any further development in yachting is inevitably related to the classical sea routes and to the majority of the existing ports.

The Greek National Tourism Organization (GNTO) has realized the need for leisure harbours and the possibilities offered by Greece's natural and cultural attractions. On the other hand, it had become obvious that the some 65 yacht supply stations, which are now out of date, had to be made suitable for present-day needs, parts of existing ports had to be transformed into yacht harbours, as well as creating new ones. In view of this goal, at the beginning of 1981, the GNTO asked the then Chair of Harbour Works of the National Technical University of Athens (NTUA) to study the 1st Phase of a National System of Navigational Zones for Leisure (NZL). This study was completed under the Coordination of the Writer (Professor of the Chair of Harbour Works at that time) in the beginning of 1982, it was approved by the GNTO and is presently being realized step by step.

In a 2nd Phase, mid 1990, the GNTO appointed a Working Group (WG), consisting of 5 GNTO distinguished Officials (Messrs A.Bratacos, C.Papageorgiou, C.Kostoyannou, A.Markatou and N.Pagonis), 2 Experts (Prof.H.Coccossis and Ass.Prof.J.Golias and the writer as Coordinator of the WG.

The GNTO has asked the WG to actualize four navigational zones (NZ) of the NZL of 1982:

(a) Ionian Sea (NZ1)
(b) Eastern Aegean Sea - Dodecanisos Islands (NZ3)
(c) Island of Crete (NZ4)
(d) Gulf of Corinthiacos-Saronicos-Argolicos (NZ6)

More specifically, the WG had to collect all possible information concerning works or approved designs of leisure harbours in the four NZ, inspect the sites, in order to verify the new elements, investigate the possible sites in each broader area, considered suitable for the creation of a leisure harbour, present and discuss GNTO'S proposals with the authorized representatives of the Local Authorities (in Greece: the Prefecture Councils) and form GNTO'S final proposals.

The NZL, when completed, will offer the possibility of leisure cruising with either owned or chartered boats, for a period of 10-15 days. It has to be pointed out that such a type of yachting is based more on navigating-sight seeing and less on navigating-fishing, as it may happen elsewhere in the world (i.e. west coasts of U.S.A.). The Eastern Mediterranean Basin, with the variety of coasts and islands, as well as the continuous alterations of cultural sites, is particularly suitable for this type of nautical tourism.

THE CONCEPT OF A LEISURE HARBOUR

A yachtsman expects to find some or all of the following facilities as berthing, direct services, indirect services and auxiliary services

But one should keep in mind that the creation of a leisure harbour produces very often a complexity of problems, which have to be examined very thoroughly. If this is not done, the investor may be taking a real financial risk.

Greece aims at producing a new type of leisure harbour development, more appropriate to the conditions of the Eastern Mediterranean Basin.

Some of the main characteristics of this type of development are listed below, namely:

(a) a special emphasis is given to the protection of the natural and cultural environment. Greece possesses a nature of an extreme variety and a multitude of archeological sites, witnesses of numerous past civilizations. Therefore,

any new development should be well integrated into the
above historical framework, paying a special attention to
the architectural design of the onland structures, the use
of local materials etc.

(b)   Antipollution measures must be carefully studied. Sea
currents are not very strong in Greece and waters in the
numerous gulfs, bays and harbours must be meticulously
protected from pollution. This can be achieved either by
inciting people living on boats to use the well-equipped
onland facilities, which will be provided at strategic
points throughout a marina, or by obliging the owners to
fit sewage storage tanks in their boats. Since October 1990
the new Common Ministerial Decree (69269/5387/90) has been
introduced (following the directive 85/337/CEC of27.06.85
and Law 1650/86) obliging the competent bodies of any new
proposal for investment or new structure to fill a
questionnaire and thereafter, according to a specification
of the Directorate General of Environment of the Ministry
of Physical Planning, Environment and Public Works, to
prepare the Environmental Impact Study (Assessment and
protective measures).

(c)   The majority of the leisure harbours of a NZ have to be
developed within the human scale. Apart from the few
"Marinas" (as they are specified in this chapter), which
will necessarily have a big reception capacity and all
possible services, and which are very often brand new
creations, it is hopeful that the remaining smaller units
("Skala") of a NZ would be integrated into the medieval
ports or into the smaller island harbours (former
commercial harbours). Nevertheless, any leisure harbour has
to be examined as an autonomous entity too and any
proposed solution must be accompanied by a detailed
examination concerning at least the site selection, the
port structures, the reception capacity, the master plan,
the classification and size distribution of the yachts.

After the site selection and definition of the necessary
port structures, a predimensioning can be produced according to
the Final Report of PIANC'S "International Commity for Sport and
Leisure Navigation" (1975), which is still very useful.

Thus,for instance the reception capacity, length of berthing
front, basin's surface, onland surface can easily be evaluated.

Finally, it is worth pointing out that the conceptual design
is completed only when the master Plan of the onshore area is
prepared. Such a Master Plan has to foresee the full development
and at the same time a phasing of the creation of the leisure
harbour.

For the moment, the few Marinas in the Athens Greater Area
are filled by assault, even before construction and equipping
are completed (a recent example has been GNTO' s Marina in
Alimos).

It is also worth pointing out that in parallel and in coordination with the NZL,the GNTO for the first time, prepared and completed in July 1991 ten (10) Master Plans prepared by the WG together with two expert Architects: Ass.Prof. Spyros J. Rogan (NTUA) and Andronicos Tzigounis.

THE CONCEPT OF NAVIGATIONAL ZONES FOR LEISURE (NZ)

As the number of yachtsmen has considerably increased during the last decades and cruising is no longer the sport of some privileged and experienced "sailors", the "regional planning" of the coasts had to be organised on a State level.

The scope of such a regional plan was multiple. Here are some objectives:

(a) The development of maritime tourism and yachting
(b) The creation of safe "navigational zones" for yachting
(c) The systematic supply to the yachtsmen of refuge, fuel, water, food, possibilities for repairing, etc.

Such a NZ is a well organised system with inter-working MARINAS, SKALAS and ANCHORAGES OF REFUGE. The existing 65 yacht supply stations could be incorporated into the Marinas and Skalas.

Marina (in greek "MARINA")

Marinas are usually the largest specially constructed leisure ports for receiving yachts. They offer, among others, well equipped berths, gardiennage or valet service, on-shore boat storage, parking for berth owners or visitors, boat repair facilities, shopping center (foods,shipchandler, etc), catering, organised refuse disposal and sewage treatment, an administration and reception center (Capitainerie, customs, port police, meteorological service etc), ablution blocks, etc.

The Marina is more viable if it is the (or one among few) center of a well organised NZ. Thus, it can offer to its yachtsmen a series of interchangeable cruises in a system of "Skalas" with adequate "Anchorages of Refuge". The site where to create a Marina must be carefully selected, if possible in a well protected gulf, in order to avoid the excessive cost of breakwaters, and close to an international airport or road axis.

Skala (in greek "ΣΚΑΛΑ")

Skala is, practically, a small Marina, with restricted facilities, which is not necessarily created from scratch. It is often created in a part of an existing commercial, fishing or historical port. It offers cultural and touristic interest, being close to antiquities, festival places, fishing areas, as well as a relaxing environment.

Apart from fuel, a Skala offers at least: well equipped berths, an administration and reception center, possibility for small repairs of boats and purchase of nautical material (shipchandler), some food stores, guaranteed refuse disposal etc.

Yachtsmen call at a Skala and remain from one up to a few days according to the general interest of the site.

An optimum distance between "Skalas" can be 20-30 nautical miles, which allows the yachtsmen to sail the above distance in, approximately, one day.

Anchorages of refuge (in greek "ΚΑΤΑΦΥΓΙΟ")

Yachtsmen must be protected from sudden storms, which are usual in the Mediterranean Sea. In order to protect inexperienced sailors, such refuges must not be farther than 10 nautical miles from each other.

There are generally two types of refuge, which can be selected along a cruise: the natural one, in an anchorage or gulf, well protected from the winds (a signal and some simple buoys are the only necessary investment) on the one hand, and the existing ports of any type, on the other.

THE NATIONAL SYSTEM FOR LEISURE HARBOURS IN GREECE

1st Phase

According to the Study prepared in 1981-82, Greece has been divided into eight (8) subsystems, or NZ, presented in Table 1 which also shows the number of sites selected during the study and considered suitable for the creation of "Marinas", "Skalas" or "Anchorages for Refuge". Figure 1 shows Greece's Navigational Zones for Leisure Harbours.

Table 1: Sites suitable for creation of a Marina, Skala or an Anchorage of Refuge (NZL/GNTO 1982)

NZ	NAMES	Selected sites for		
		MARINAS	SKALAS	ANCHORAGES
1	Ionian Sea	5	14	36
2	Thracian Sea-Thermaicos	4	11	31
3	Eastern Aegean Sea -Dodecanisos	4	22	30
4	Island of Crete	3	13	21
5	Southern Peloponnisos	4	9	17
6	Corinth.-Saronic.-Argolic. Gulfs	3	8	25
7	Southern Eubean Gulf-S.Aegean Sea	2	18	59
8	Northern Eubean and Pagasiticos Gulfs - Sporades Islands	1	11	29
	Totals	26	106	248

Immediate Period

Meanwhile, the typology introduced in the NZL has not been always applied and, very often, the word "Marina" has been misused, meaning at the end, everything offering berthing facilities, even to a few yachts! Futhermore the word "Skala" has not proved successful.

Last but not least, internationally, the word "Marina" corresponds to a very specific, by size, leisure harbour, with strong housing development.

During the same period 1980-90 it proved that, due to the greek administrative system, several Authorities decided the creation of leisure harbours before applying to the GNTO for approval. Thus, the NZL has not been always applied in both parameters: site and size, according to the approved regional planning (NZL, 1982).

The NZL of 1981-82 has selected 27 sites suitable for Marinas, allover Greece but the 2nd Phase did not follow immediately after, in order to select the specific place for the development of each leisure harbour, in its broader area, obtaining in parallel the approval of the Local Authorities concerned.

The Rodos Marina has been a typical example of what could happen:

During NZL's 1st Phase the broader area of the Town of Rodos had been selected for the creation of a Marina. GNTO selected a specialised Design Office which investigated five different sites suggested by the local section of the National Chamber of Greece. After a professionally perfect comparative study, the Designer suggested the area of Phaliraki which was approved by the GNTO, the Port Treasury and the Prefecture of Dodecanisos Islands. The construction design was prepared, an invitation for tender published, but, due to the reaction of the inhabitants of Phaliraki, the Tender was repeated three times and finally cancelled. The Prefecture Council of Dodecanisos Islands then asked the GTNO to reconsider the site selection and present new proposals. As a result, the creation of the Marina was postponed for 5 years!

Facts like the above example and the common Ministerial Decree 69269/5387/90 contributed to the official decision that the advice of the Local Authorities has to be taken for each site selection.

2nd Phase

The 2nd Phase of the development of the NZL in Greece concerning the four NZ (Ionian Sea, Eastern Aegean Sea -Dodecanisos Islands, Island of Crete and the Gulf of Corinthiakos - Saronicos - Argolicos) was completed in April 1991. During the

elaboration of the 2nd Phase the WG suggested and the GNTO
approved, to use the wording "LEISURE HARBOUR" and distinguish
various   sizes,   according to harbour's capacity "c", for
instance c>500 or c = 150-500 etc. In this case, the capacity
"c" totalizes all berths in the water basins and yachts dry
stored.

During the 2nd Phase,the WG inspected 29 sites and presented
to the relevant Prefecture Councils, written proposals including
urban - regional, transportation and port analysis.

The opinions of the Prefecture Councils were taken into
account for the formulation by the WG of the final proposals,
for twelve (12) Leisure Harbours. The concise description of
actions undertaken for each NZ follows.

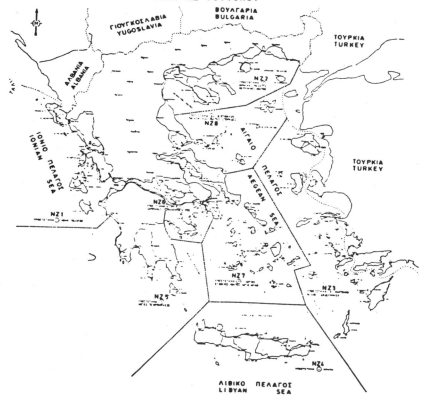

Figure 1 : Greece's Navigational Zones for Leisure Harbours
           (1981-1982)

NZ1 Ionian Sea: During the 1st Phase,5 sites for Marinas, 14 for
Skalas and 36 for Anchorages of refuge were selected as suitable
for the development of   the relevant NZ. Table 2 shows the
situation, as it was in April 1991, after the completion of the
2nd Phase. The Leisure Harbour of Pylos, although belonging to

NZ5, was included in NZ1, due to its vicinity to NZ1 and the progress of its construction works.

Table 2: NZ1 (Ionian Sea), April 1991

Leisure Harbour		Capacity			Phase of materiali-sation	Entity	Inclu-sion in the M.I.P.	Type accor-ding to the 1st Phase	Master Plan of Onshore Instal-lations
Area	Site	No of Berths	No of yachts dry-stored	Total [c]					
GOUVIA	Southern part of the Gulf Gouvia	1000	-	1000	Tender for preselection (deadline: (18.12.91)	GNTO	Yes	Marina	Yes
PAXOI	Ozias	150	-	150	Prelim. Design	GNTO	No	Skala	No
PREVEZA	Commercial Harbour	200	-	200	Final Design	GNTO	No	Inter-mediate	No
ARGOSTOLI	Commercial Harbour	200	-	200	Final Design	GNTO	No	Inter-mediate	No
LIXOURI	Lixouri	200	-	200	Final Design	GNTO	No	Inter-mediate	No
LEFKADA	Town of Lefkada	500	-	500	Acceptance of the site	GNTO	No	Marina	No
PATRA	NE of the commercial harbour	250	-	250	Prelim. Design	GNTO	No	Inter-mediate	No
ZAKYNTHOS	N of the commercial harbour	250	-	250	under construc-tion	GNTO	Yes	Inter-mediate	Yes
PYLOS	N of the commercial harbour	250	-	250	under construc-tion	GNTO	Yes	Inter-mediate	No

NZ3 - Eastern Aegean Sea-Dodecanisos Islands: During the 1st Phase, 4 sites for Marinas, 22 for Skalas and 30 for Auchorages of refuge were selected as suitable for the development of the relevant NZ. Table 3 shows the situation as it was in April 1991, for NZ3, after the completion of the 2nd Phase.

Table 3: NZ3 (Eastern Aegean Sea-Dodecanisos Islands),April 1991

Leisure Harbour		Capacity			Phase of materiali-sation	Entity	Inclu-sion in the M.I.P.	Type accor-ding to the 1st Phase	Master Plan of Onshore Instal-lations
Area	Site	No of Berths	No of yachts dry-stored	Total [c]					
CHIOS	N.Commercial Harbour	120	-	120	under construc-tion	GNTO	Yes	Skala	No
SAMOS	Bay of Karpovo-los	250	-	250	begin of construction	GNTO	Yes	Inter-mediate	No
IKARIA	Aghios Kyrikos	250	-	250	Tender	GNTO	Yes	Inter-mediate	No
KOS	E.of commercial harbour	250	-	250	under construc-tion	GNTO	Yes	Inter-mediate	No
RODOS	NE of Akadia commercial harbour	500-1000	-	500-1000	Conceptual Design	GNTO	Yes	Marina	No
MYTILINI	Makris Yalos	200	-	200	site approved	GNTO	No	Skala	No

NZ4 - Island of Crete: During the 1st Phase, 3 sites for Marinas, 13 for Skalas and 21 for Anchorages of refuge were selected as suitable for the development of the relevant NZ. Table 4 shows the situation as it was in April 1991, for NZ3, after the completion of the 2nd Phase.

Table 4: NZ4 (Island of Crete), April 1991

Leisure Harbour		Capacity			Phase of materiali- sation	Entity	Inclu- sion in the M.I.P.	Type accor- ding to the 1st Phase	Master Plan of Onshore Instal- lations
Area	Site	No of Berths	No of yachts dry- stored	Total [c]					
CHANIA	Venetian Harbour	100	-	100	under constru- ction	GNTO	Yes	Skala	No
RETHYMNO	Commercial Harbour	250	-	250	under constru- ction	GNTO	Yes	Inter- mediate	Yes
RETHYMNO	Aghios Nicolaos -Koube	500	300	800	Prelim. Design	GNTO	Yes	Marina	Yes
HERAKLEION	Bay of Dermata	400	300	700	Prelim. Design	GNTO	No	Marina	Yes
AGHIOS NIKOLAOS	SE of the com- mercial port	250	-	250	under constru- ction	GNTO	Yes	Inter- mediate	No
AGHIA GALINI	Commercial Harbour	100	-	100	under constru- ction	GNTO	Yes	Skala	No
PALAIOXORA	Fishing port	150	-	150	under constru- ction	Chania Port Treasury	Yes	Skala	No

NZ6-Gulfs of Corinthiakos - Saronicos - Argolicos: During the 1st Phase, 3 sites for Marinas, 8 for Skalas and 24 for Ancho- rages of refuge were selected, as suitable for the development of the relevant NZ. Table 5 shows the situation as it was in April 1991, for NZ6, after completion of the 2nd Phase. Figure 2 shows the situation, at present in NZ6.

Table 5: NZ6 (Korinthiakos-Saronikos-Argolikos Gulfs),April 1991

Leisure Harbour		Capacity			Phase of materiali- sation	Entity	Inclu- sion in the M.I.P.	Type accor- ding to the 1st Phase	Master Plan of Onshore Instal- lations
Area	Site	No of Berths	No of yachts dry- stored	Total [c]					
PIRAEUS	Zea	365	-	365	in function	GNTO	-	Marina	No
NEO PHALIRO	Flisvos A'	173	-	173	under constru- ction	GNTO	Yes	Marina	Yes
	Flisvos B'	569		569	Conceptual Des.	GNTO	-	Marina	Yes
ALIMOS	Alimos	677	-	677	in function	GNTO	-	Marina	No
GLYFADA	Glyfada	780	-	780	in function	Township	-	Marina	No
BAY OF VOULIAGMENI	Vouliagmeni	115	-	115	in function	GNTO	-	Inter- mediate	Yes
METHANA	Methana	70	-	70	in function	Township	-	Skala	No
LOUTRAKI	Poseidonia	420	280	700	Prelim. Design	Township	No	Marina	Yes
POROS	Greek Navy	-	-	500	site approved	GNTO	No	Marina	No
PORTO HELI	W of the bay	-	-	250	site approved	GNTO	No	Inter-	No
NAFPLIO	Commerc.Harbour	-	-	500	site approved	GNTO	No	Marina	No
ELEUSIS	E to the com- mercial harbour	-	-	250	under constru- ction	Township	No	Skala	No
AGHIOS KOSMAS	Hellenikon (airport)	-	-	1000	site approved	GNTO	No	Marina	No
ANAVYSSOS	Aghios Nikolaos	-	-	1000	site approved	GNTO/Sports Secret.Gen.	No	Marina	No
LAVRIO	Panormos Bay (Gaidouromadra)	-	-	>1000	site approved	Olympic Marine S.A.	No	Marina	No
PORTO RAFTI	NW of the bay	-	-	<500	site approved	GNTO	No	Marina	No

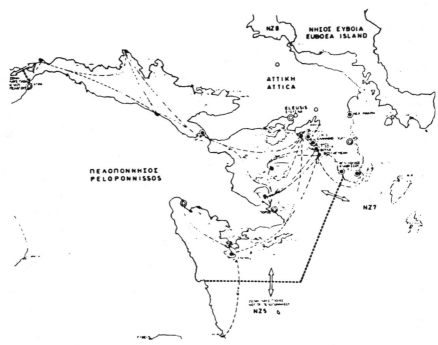

Figure 2 : Navigational Zone 6 (Gulfs of Saronikos -Corinthiakos -Argolikos) as it was in April 1991

If all Leisure Harbours presented in Tables 2 to 5 are created, the Official National Greek capacity will exceed 15000 units (around 7000 at present)

It is expected that the GNTO will proceed with the 2nd Phase of the remaining four NZ (i.e. Thracean Sea - Gulf of Thermaicos, Southern Peloponnisos, Southern Eubean Gulf -Southern Aegean Sea, Northern Eubean Gulf - Pagasiticos Gulf and Sporades Islands).

Victoria & Alfred Waterfront - from Concept to Reality

A.H. Burggraaf

Victoria & Alfred Waterfront (Pty) Ltd., Cape Town and Portnet, Johannesburg, South Africa

INTRODUCTION

Cape Town is probably regarded as the South African city with the most onerous requirements for any property development. Apart from the fact that it houses the repository of the nation's heritage, it has a natural beauty and grandeur that almost defies contemporary man-made artefacts.

Its citizens have become very vocal over the past 20 years, guarding the value of their historic and natural assets. Its municipal rating system penalises property owners, since 'liberal' Cape Town always attempted affirmative action, paying more than lip-service to the needs of the disadvantaged.

Nevertheless, 20 months after Victoria & Alfred Waterfront began operating, it was able to open the first phase of one of the most talked-about developments in South Africa.

It was achieved through:

• the benefit of timing (most battles had already been fought);
• management assembling one of the most comprehensive teams of consultants seen in South Africa and focusing on mutual benefits to all parties;
• City Council (responding to the public mood) created new procedures to realise the project;
• the shareholder's willingness to finance initial phases.

HISTORIC OVERVIEW

Midshipman HRH Prince Alfred, Queen Victoria's second son, tipped the first load of stone to start construction of the breakwater for Cape Town's harbour on 17 September 1860. This formed the beginning of what was to become the Port of Cape Town of today.

Alfred Basin, the first of many basins providing shelter for shipping, was completed ten years later, but was already too small for the growing fleets and increasing size of ships. Steam ships were replacing sail; gold and diamonds had been discovered and the development of the hinterland had begun. A second basin, the **Victoria Basin** was completed 35 years later and served as the gateway to Southern Africa until Duncan Dock opened in 1944.

To achieve the depth required for the new ships arriving at the Cape, Duncan Dock was built nearly a kilometre out to sea, and Cape Town gained nearly 200 hectares of reclaimed land. For the rapidly growing city between Table Mountain and Table Bay, this was fortuitous – it created a harbour appropriately known as the 'Gateway to Southern Africa,' and much-needed land for city development.

Nobody anticipated that Cape Town would lose its gateway status to Johannesburg with the growth of air transport, and that the city's new Foreshore would, in effect, separate the old city from the sea.

Although the Suez crisis of 1966 saw a resurgence in traffic around the Cape, with shipping queuing in the roadstead, this did not last and with the advent of containerisation in sea transport, the number of ships began to decline dramatically. A new, purpose-built dock was constructed on the seaward side of Duncan Dock and, since this attracted the majority of traffic, Duncan Dock appeared frequently and noticeably dormant.

With commercial shipping accommodated elsewhere, the historic Victoria and Alfred Basins became the centre for the fishing industry and smaller scale ship repair work. Whilst Duncan Dock frequently lay idle, the historic docklands were always a hive of activity.

With the growth of yachting and other maritime leisure activities, a new focus fell on small boat facilities and public quayside amenities. While Capetonians started yearning to 're-instate the city's lost attributes as 'Tavern of the Seas,' the boating fraternity focused their attention on obtaining the Victoria and Alfred Basins for leisure activities.

In 1979, the Department of Sport and Recreation commissioned a study into the feasibility of a small boat marina at Granger Bay, on the seaward side to the north of the historic harbour. The City's planners suggested that this development should be complemented by residential and tourist/leisure-orientated development.

At the same time, prominent architect, conservationist and yachtsman, Gawie Fagan, publicised his proposals for redevelopment of the Victoria and Alfred Basins as a small boat marina.

Capetonians and the media took sides in the debate, but port authorities saw no reason to disrupt activities in one of their busiest precincts.

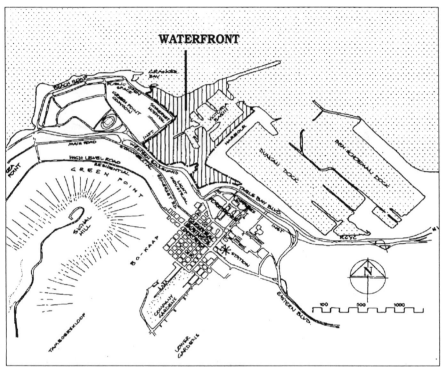

The Waterfront is an extension of both the central business district as well as the desirable Atlantic seaboard suburbs.

While use of the historic docklands was denied, neighbours to Granger Bay also opposed any new development at the alternative site.

In 1984, to break the deadlock, and encouraged by members of the Granger Bay Planning Committee and the city's planners, Cape Town's mayor formed a Waterfront Steering Committee and started lobbying to re-establish the city's links with the sea. He also used his office to obtain permission for a festival in the historic Pier Head precinct, which was organised by a group of waterfront enthusiasts under the chairmanship of prominent businessman, Mr Harold Gorvy, to focus public attention on the area. The festival was visited by several cabinet ministers and, at the conclusion of the 16-day event, the Minister of Environment Affairs and Tourism gave Mr Gorvy an undertaking to liaise with his colleagues to achieve rapid progress.

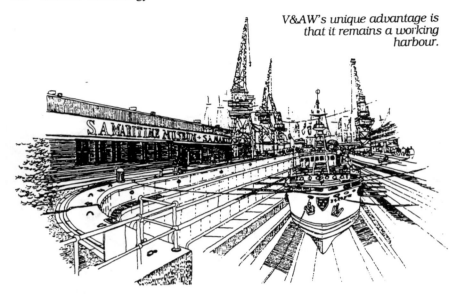

V&AW's unique advantage is that it remains a working harbour.

THE COMPANY

It was as a result thereof that a committee was established in 1985 by the Minister of Transport Affairs, under whose jurisdiction the SA Harbours resorted, to investigate their potential for greater public use. This committee was under the chairmanship of the author, then Inspecting Engineer – Harbours. The Committee reported on Cape Town harbour in 1987, proposing that the historic docklands around Victoria & Alfred basins be redeveloped as a mixed-use area focusing on tourism and residential development, with the continuing operation of a working harbour. The recommendations were accepted in full in May 1988.

The formation of the Victoria & Alfred Waterfront company (V&AW) was announced in November 1988. The board of directors comprises three representatives from Transnet, a representative from the Cape Town City Council , and is chaired by Prof. Brian Kantor, a well-known champion of the free enterprise system from the School of Economics at the University of Cape Town, while the managing director, Mr David Jack, was Cape Town's city planner prior to this appointment.

The company is a wholly-owned subsidiary of Transnet Limited, which itself was formed from the former SA Transport Services (comprising the national railways, harbours, airways, pipelines and road transportation) as a precursor to possible privatisation. In this reorganisation, the separate business unit of SA Harbours was renamed Portnet.

Relevant to V&AW, as a property company, were the provisions introduced in the South African Transport Services Act of 1981 and the Legal Succession to South African Transport Services Act of 1989 which also determined V&AW's relationship to the local authority insofar as agreements for any change in land use.

V&AW's principle asset was a land area of approximately 83 hectares on long leasehold from Transnet. At that stage activities in the area were running at an operational loss, and the challenge was to add both value to the site and to realise a positive cash flow.

The new company defined its mission as the following:

V&AW manages and develops Cape Town's historic docklands to maximise the long-term benefit to its shareholders, Capetonians and visitors.
V&AW is committed to:

- *enhancing its maritime image;*
- *retaining working harbour activities;*

creating a:

- *quality environment;*
- *desirable place to work, live and play;*
- *preferred location to invest;*

satisfying:

- *the aspirations of Capetonians and visitors*

MUNICIPAL AREA

Negotiations for development rights were complicated by the fact that the municipal boundary runs through the site. Some 40% of potentially rateable land (– reclaimed land beyond the original high water mark) falls outside the municipal area.

State and parastatal organisations had historically been exempt from payment of municipal rates, but this was phased out with the Rating of State Properties Act of 1984. Although the areas falling within the municipal boundary were subject to rates, an agreement was needed to cover areas outside but adjacent to the municipal area that required servicing by the municipality.

Council interpreted the legislation in such a way as to link planning agreement to all servicing and financial agreements, thus finalising all agreements simultaneously.

Both parties started at the extremes, with V&AW trying to exclude as much as possible, and the local authority wanting everything incorporated. The compromise lay mid-way in the sense of reaching approval on a phased

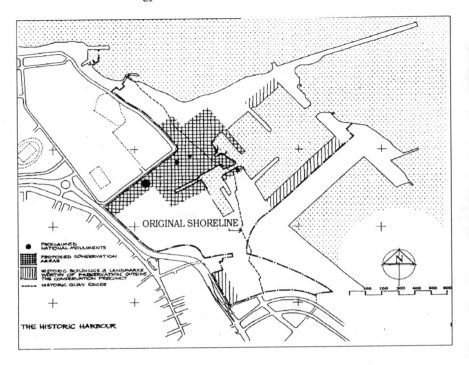

incorporation of the area outside the municipal area.

Payment of municipal rates on certain parts of the project had to be balanced against the provision of municipal services in the areas outside. This was resolved by agreeing that V&AW would pay contributions in lieu of rates for the area outside the municipal boundary.

There were, however, two exceptions. The first related to Council's agreement that no rates (or contributions in lieu of rates) on land and improvements used for working harbour activities (inclusive of the fishing industry) should be paid.

Secondly, not all public-access areas would be handed over to the local authority because of the perceived inadequacies of municipal budgets, management, etc. V&AW was allowed to retain management control and in return receive 'agency payments' for services rendered, such as cleansing, horticulture, etc.

HEADS OF AGREEMENT

Heads of agreement between Cape Town City Council, V&AW and Transnet covering servicing, financial and planning aspects were decided on in August 1990 and signed in August 1991. These were based on the 'Package of Plans' approach as the basis for the future and detailed planning and development, as well as zoning in terms of the Legal Succession to the South African Transport Services Act of 1989.

The package of plans is analogous to the Planned Urban Development (PUD) approach developed by the Urban Land Institute in the USA.

The Development Framework Report (which was submitted to Council in September 1989) was regarded as constituting the basis for development of the area. The entire area was divided into 14 precincts in terms of existing physical characteristics and primary land uses. A range of activities was envisaged within each precinct, with the aim of avoiding the sterile monofunctional use of land that results from conventional town planning schemes. The report also set out the proposed development rights for the entire area although, in terms of the flexible approach to accommodate changing scenarios, this was made subject to review by the Council at five yearly intervals.

ZONING

Development at the Waterfront is not subject to a rezoning in terms of the Land Use Planning Ordinance, but is administered under the Legal Succession to South African Transport Services Act which became effective in April 1990.

This statute allows Transnet to develop land for purposes other than transportation, but requires that '... such development shall only take place after agreement has been reached with the relevant local authority and after such consultation as the local authority may deem necessary, and if such agreement is not reached with the permission of the relevant Administrator (head of provincial tier of government), on such condition as he may deem fit.. .'

In obtaining agreement for an agreed upon usage it was incumbent upon the local authority to record a formal zoning within three years of the Act becoming law.

This did not negate the need to obtain in terms of the Physical Planning Act the Administrator's approval for an amendment to the applicable Guide Plan where such was required, as was the case in point.

Council has interpreted this, as previously mentioned, not only in a planning sense, but in the broad context covering planning, services and financial/rating issues.

Planning is pursued by means of a hierarchy or 'package' of plans. The package is a multi-tiered system consisting of the following levels:

• Contextual framework
• Development framework
• Precinct plans
• Development Area or Site Development Plans
• Building Plans

The plans encompass very broad city and district-wide considerations at the Contextual Framework level, becoming more and more specific as plans move from context down to individual sites and buildings plans. Responsibilities for plan preparation were allocated, the contents of each plan level was described in terms of the role each was intended to play. Only the Contextual Framework plan – in effect, Council's policy statement for the entire area – was prepared by Council, and all other planning phases are the responsibility of V&AW. (See diagram 1)

DIAGRAM I: PACKAGE OF PLANS

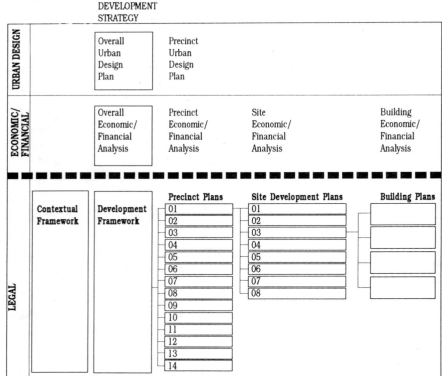

DEVELOPMENT RIGHTS

Although planning agreement had been very speedily reached with Council at the end of 1989 to allow preliminary redevelopment of the Pierhead precinct to begin, the Heads of Agreement were required before the site effectively was granted its full development rights.

These rights which were negotiated on the basis of the Development Framework report are as follows:

Retail	$30000m^2$
Offices	$150000m^2$
Entertainment	$7000m^2$
Museums	$6000m^2$
Hotels	$60000m^2$
Residential	$1500units$

The rights cover the entire area and are not broken down to be specific to individual precincts and thus offer the scope to maximise precinct development. Council's planning approval initially related to only $90000m^2$ of office development, but through the judicious purchases of two adjacent properties, V&AW obtained a further $58000m^2$ of development rights, enhancing substantially its development opportunities.

DEVELOPMENT PROCESS

A major advantage, which enabled V&AW to get off the ground so quickly was that Council, in approving the overall Development Framework, delegated approval of Precinct Plans to its Town Planning Committee, Site Development Plans to its City Planner, while in terms of the three year moratorium, Council did not consider it to fall within its ambit to formally approve building plans.

Council officials are, however, consulted on a pre-scrutiny basis, where they make a very positive contribution, especially on key public safety issues.

Plans are approved by the company on certification by registered professional architects and engineers that all by-laws, etc, have been complied with.

LIAISON

To avoid the controversy which had dogged proposals in the early 1980's, the Minister of Transport Affairs appointed a Liaison Committee at the same time the company was established to ensure that cognisance was taken of public concerns, and that the public participated in the development

*One of the first refurbished buildings at
the Waterfront was the Information
Centre, which opened nine months after
the company started work.*

process.   Representatives of various organisations, which amongst others includes a number of affected ratepayers' associations, serve on this committee which meets with V&AW management and consultants on a regular basis.

Media liaison was facilitated by public interest in and support for the project, and contributed greatly to the process of development.

Further contact with interest groups was implemented by addressing all associations and other bodies which requested presentations, while regular meetings were also held with neighbouring ratepayers' associations. This is an ongoing process.

More recently, liaison with Council has been extended to include regular meetings to discuss development proposals received by the Company and programming, as well as proposals which Council has received for other development neighbouring and impacting on the project area.

COMPANY STRUCTURE

Though it was originally intended at inception to extend a 75 year lease to the company, the creation of Transnet Limited as a public limited liability company necessitated a different structure.

As a result, the company will manage the development while the various precincts have been sub-divided further for transfer and registration into separate companies.

The latter arrangement offers the opportunity to adopt a very flexible approach to the development responding to market demands and financial restraints. The subsidiary companies draw the distinction between properties held for investment purposes, and those for trading purposes – which draw a higher rate of tax. Holding investment and trading activities under the same umbrella would have attracted a tax penalty.

In establishing subsidiary companies, the following options are catered for:

1. Development by V&AW and retained.
2. Development by V&AW and disposed of to create roll-over capital
3. Development with partners and retain
4. Development with partners and dispose
5. Provide infrastructure and dispose
6. Dispose of without any new infrastructure

DEVELOPMENT TO DATE

A substantial amount of soul searching went into the approach to development.

The primary approach to development was naturally one of generating adequate returns. No redevelopment would be contemplated if the expected returns could not support this.

In this regard the consultants developed an exceptionally user-friendly costing model which assists in rapid decision making by a sensitivity analysis of the optimum mix of development.

A total of 24 building and engineering contracts started in February 1990 and ran simultaneously within Pierhead's 1,5 hectare area.

By December 1990, some 20 months after the company started operating, the following amenities were operating:

Victoria & Alfred Hotel – a 68 room hotel in a converted quayside warehouse.
V&A Arcade – 19 speciality shops in the same warehouse.
Mitchells Waterfront Brewery and Ferrymans Tavern.
Bertie's Landing – a tavern and restaurant.
Quay Four – a tavern and restaurant.
Green Dolphin – a restaurant featuring jazz musicians.
Dock Road complex – a theatre, restaurant and multi-purpose venue.
Union Castle House – a jewellery centre.
SA Maritime Museum - with both indoor and floating exhibits.

Organisational diagram illustrating the separate companies. Note that precinct numbers refer to the new business units rather than the initial planning precincts.

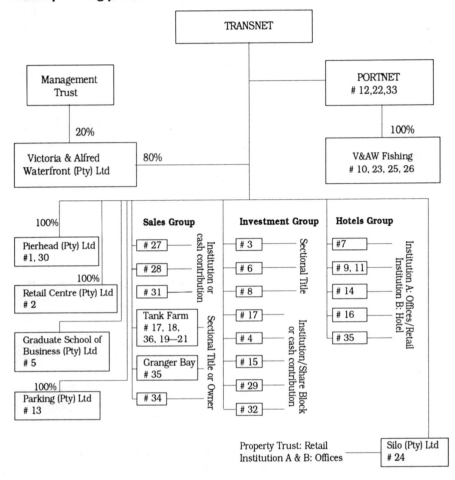

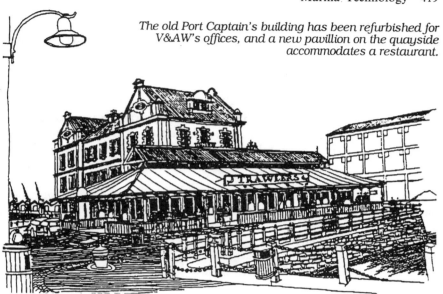

The old Port Captain's building has been refurbished for V&AW's offices, and a new pavillion on the quayside accommodates a restaurant.

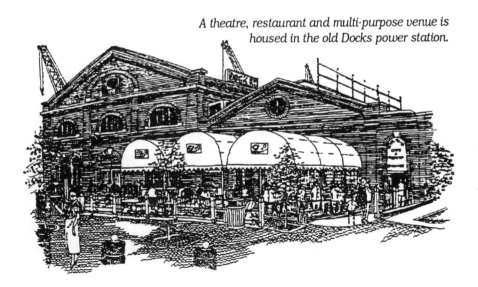

A theatre, restaurant and multi-purpose venue is housed in the old Docks power station.

During 1991, these were complemented by a jazz club, three further restaurants, a wine bar, additional speciality shops, and a boathouse and operations centre for the National Sea Rescue Institute.

Existing service infrastructure was old and totally inadequate for the Waterfront as a whole. Virtually all services had to be replaced. Of the R63 million spent on the first phase, R35 million was for infrastructure – roads, water, stormwater and sewerage reticulation, hard and soft landscaping and street furniture, as well as the relocation of existing services.

Responding to existing tenants, assisting with relocation in certain instances, and tying this into a very tight construction programme provided one of the biggest challenges in the first phase.

The largest single contract to date has been for the construction of the Waterfront's speciality marketplace, named Victoria Wharf, which is scheduled to open in October 1992. The contract was awarded in February 1991.

This is a $25000m^2$ complex overlooking Victoria Basin which will accommodate 11 cinemas, 16 restaurants and coffee shops, fish and fresh produce markets and some 80 speciality traders, with a predominance of fashion.

Redevelopment of the historic Portswood Ridge business park is underway. Several of the Victorian buildings have been refurbished as offices, and the University of Cape Town's Graduate School of Business is under construction and opens at the Waterfront at the beginning of 1992.

DEVELOPMENT: 1992 – 1997

The major items planned over the next five years include the following:

200 room hotel linking to Victoria Wharf at Quay Six.
An aquarium
1, 000 bay parking garage and $18000m^2$ office space on Portswood Ridge
Flooding the old quarry to form Alfred Marina
First two phases of Alfred Marina residential development
Offices & hotel alongside a new waterway at Amsterdam Battery
Construction of breakwaters for Granger Bay residential area

Preparatory work for the flooding of the old quarry is moving ahead rapidly to meet a target date of Easter 1993 and presents one of the most complex challenges in the entire development.

Existing access to the Waterfront from the Foreshore will be upgraded and the road relocated. Existing services need to be relocated to allow the

1. *GRANGER BAY: Residential area of 700 units around a small craft marina with direct access to Table Bay.*

2. *PORTSWOOD RIDGE: Business park adjacent to the old breakwater Gaol, which is now University of Cape Town's Graduate School of Business.*

3. *PIERHEAD PRECINCT: Commercial and entertainment core of the Waterfront in the heart of the historic working harbour. This comprises hotels, restaurants, speciality shopping and museums.*

4. *YACHT MARINA: Marina with mooring for 400 yachts inside Table bay's main breakwater.*

5. *ALFRED MARINA: Small craft basin in the old quarry. An aquarium, hotels and restaurants overlook the marina.*

6. *NEW BASIN: New basin and waterway which stretches towards the CBD, surrounded by housing and offices.*

7. *AMSTERDAM BATTERY: Landscaped office park alongside the new waterway.*

8. *SILO PRECINCT: Occupied mainly by long leases. The only new activity is Bertie's Landing.*

9. *OLD POWER STATION & ICS SITES: Outside the Waterfront site, but redevelopment will improve the link to the CBD.*

REVISED PROGRAMME

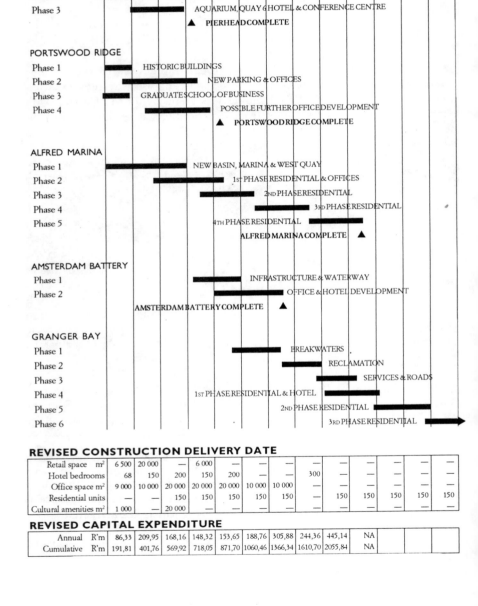

	1991	1992	1993	1994	1995	1996	1997	1998	1999	2000	2001	2002	2003

PIERHEAD
Phase 1 — ▲ INITIAL PHASE OF DEVELOPMENT COMPLETE
Phase 2 — VICTORIA WHARF & MARKET SQUARE
Phase 3 — AQUARIUM, QUAY 6 HOTEL & CONFERENCE CENTRE
▲ PIERHEAD COMPLETE

PORTSWOOD RIDGE
Phase 1 — HISTORIC BUILDINGS
Phase 2 — NEW PARKING & OFFICES
Phase 3 — GRADUATE SCHOOL OF BUSINESS
Phase 4 — POSSIBLE FURTHER OFFICE DEVELOPMENT
▲ PORTSWOOD RIDGE COMPLETE

ALFRED MARINA
Phase 1 — NEW BASIN, MARINA & WEST QUAY
Phase 2 — 1st PHASE RESIDENTIAL & OFFICES
Phase 3 — 2ND PHASE RESIDENTIAL
Phase 4 — 3RD PHASE RESIDENTIAL
Phase 5 — 4TH PHASE RESIDENTIAL
ALFRED MARINA COMPLETE ▲

AMSTERDAM BATTERY
Phase 1 — INFRASTRUCTURE & WATERWAY
Phase 2 — OFFICE & HOTEL DEVELOPMENT
AMSTERDAM BATTERY COMPLETE ▲

GRANGER BAY
Phase 1 — BREAKWATERS
Phase 2 — RECLAMATION
Phase 3 — SERVICES & ROADS
Phase 4 — 1ST PHASE RESIDENTIAL & HOTEL
Phase 5 — 2ND PHASE RESIDENTIAL
Phase 6 — 3RD PHASE RESIDENTIAL ➔

REVISED CONSTRUCTION DELIVERY DATE

	1991	1992	1993	1994	1995	1996	1997	1998	1999	2000	2001	2002	2003
Retail space m²	6 500	20 000	—	6 000	—	—	—	—	—	—	—	—	—
Hotel bedrooms	68	150	200	150	200	—	—	300	—	—	—	—	—
Office space m²	9 000	10 000	20 000	20 000	20 000	10 000	10 000	—	—	—	—	—	—
Residential units	—	—	150	150	150	150	150	—	150	150	150	150	150
Cultural amenities m²	1 000	—	20 000	—	—	—	—	—	—	—	—	—	—

REVISED CAPITAL EXPENDITURE

	1991	1992	1993	1994	1995	1996	1997	1998	1999	2000	2001	2002	2003
Annual R'm	86,33	209,95	168,16	148,32	153,65	188,76	305,88	244,36	445,14	NA			
Cumulative R'm	191,81	401,76	569,92	718,05	871,70	1060,46	1366,34	1610,70	2055,84	NA			

cut between Alfred Basin and the new marina. Provision for a rail terminus for special-event steam trains is also planned in this location.

Apart from planning issues, engineering aspects relate to water quality, the quayside revetments, the marina jetty design and lock systems, the physical implementation of the cut and introducing water into the waterways that stretch through the residential areas towards the CBD.

The anticipated problem of surge through the cut which the flooding of the old quarry might have introduced was analysed by mathematical modelling at the time Bertie's Landing was being designed. Cape Town's harbour has historically suffered from this problem, especially during the north-westerly swells in winter, although it has diminished as a result of additional harbour works which have reduced wave heights.

What the model demonstrated was that a fixed jetty in front of Bertie's Landing would have exacerbated the problem (which led to the option of a floating jetty), and that the flooding of the old quarry would almost eliminate surge completely. The depth of the quarry needs to be reduced by 9m to ensure that once it is flooded, there will be adequate tidal flushing to maintain water purity and avoid stagnation. The biggest challenge is the issue of water purity, especially with the location of the aquarium in the new basin.

The new canal extends from the Waterfront to the site of an old electrical power station, which was supplied with ducted cooling water from Duncan Dock. This facility has been retained and will assist in flushing both the canal and the new basin.

Regarding fill for the basin, V&AW has been stockpiling suitable rubble and soil for this purpose over the past two years. The remainder of the fill required will come from the excavation for the new Dock Road parking garage (built into Portswood Ridge) and the new basin and waterway on the CBD side of Alfred Marina.

There can be little doubt that the flooding of the quarry is the single event at the Waterfront which the public looks forward to most, since it marks the beginning of residential development which has attracted widespread interest.

LESSONS LEARNT

Cape Town's Waterfront has learnt from the lessons available elsewhere, and a concerted effort was made to benefit from the perceived opportunities and pitfalls which developers had identified at other waterfronts around the world.

The historic Time Ball Tower — one of four national monuments at the waterfront.

V&AW's directors, senior management and its consultants have acquainted themselves with overseas trends and met with many of their foreign counterparts. If Cape Town's Waterfront provides any lessons for other developers, V&AW's design review process must be singled out as a key factor in the success to date.

The Waterfront is a sensitive area with a large number of historic buildings. The restoration and re-use of these buildings, and the demolition of some structures, required very careful consideration. All actions fell under the spotlight and there was an awareness of the criticism which has historically been levelled at any development in Cape Town, and especially at the Waterfront.

To achieve the widest possible basis for the decision making which would take place, specialist advisors were appointed, some of whom had been outspoken adversaries of earlier proposals for the Waterfront. All decisions were taken with input from these advisors, each having acknowledged expertise.

From the outset as many different project teams as possible were established.

But all had to work together, understanding what the others were doing.

To achieve this design review meetings were instituted which involved the advisors, the core consultants and the project teams.

From initial concepts through to sketch plan and final designs, these were reviewed as a group, much the same as 'crit' sessions at university.

These were always demanding, lively, and sometimes very tense, as prominent architects challenged each other. To a large extent, the success and appropriateness of the new and restored buildings stemmed from this process.

From an operational point of view, the importance attached to cleanliness, security and personal safety is a key factor in the Waterfront's success. The biggest challenge which is receiving concerted attention is public access and the availability of sufficient, convenient parking.

Preliminary Design for a Marina in Arrecife Lanzarote

M. Poole Pérez-Pardo

Nautilus Servicios Técnicos, S.A. Consulting Engineers, Camino Alto, 26, La Moraleja, Alcobendas, Madrid, Spain

INTRODUCTION

The Spanish Recreational Fleet and Marinas

Spain is situated in the extreme south-west of Europe. It is composed of the Iberian Peninsula (except Portugal), the Balearic Islands, the cities of Ceuta and Melilla on the northern coast of Africa and the Canary Islands in he Atlantic Ocean.

It is a country with a maritime history, with a coastline of more than 7900 Km.

Until the invention of steam-powered shipping, the Iberian Peninsula was one major point of embarkation of Western Europe for all the other continents, and the Canary Islands were the last European port on the route of the Trade Winds.

The development of recreational sailing has been late but extremely intense in Spain and these circumstances have resulted in the yachting fleet being very modern and the resultant marinas developed being amply suited to whatever requirements the modern yachting community may have. These facilities and the fleet rank among the best in Europe.

At present the Spanish fleet is composed of 150 000 boats and the number of yachting marinas stands at 280, with five able to accommodate over 1000 moorings, 29 can contain between 500 and 1000, 59 between 250 and 500, leaving 187 with a capacity less than 250 berths.

The greatest development has naturally occurred in the Mediterranean sea, the 'Sea of the Gods' and the cradle of Western Civilization.

The Canary Islands

Crossing the Atlantic between the Americas and Europe, the Canary Islands constitute a natural landfall, being situated 1000 miles from Cádiz and 3000 miles from the Caribbean Sea and the Gulf of México.

Five centuries have passed since the first small fleet, captained by Christopher Columbus,used the Canary Islands as a stopover on their voyage to the New World: a port of call that has been repeated by all the sailing boats that cross the Atlantic from East to West.

Steam-powered shipping and aviation in the last 50 years have considerably reduced commercial traffic but recreational navigation keeps crossing the Atlantic along this 'route, and the number of yachts that dare to jump the Ocean increases every year. Most of them utilize one of the ports in the Canary Islands as a stepping stone in preparation for the last stage to America.

The distance between the individual Canary Islands is between 40 to 60 miles, the wind is constant day and night and the weather is a continued European summer. These circumstances, along with the tourist development of the archipelago, have resulted in a considerable growth in the local fleet and therefore an increasing demand for mooring space which is combined with the demand for moorings by the trans-Atlantic visitors.

In this year, 1992, is the 'Regatta of Discovery' in which more than 1000 will participate in following Columbus' route with a resultant stopover in the Canary Islands.

At present in the Canary Islands there are eighteen ports with facilities catering for yachts. Twelve of these ports were constructed solely for this purpose and a few more are projected.

The Island of Lanzarote and the Port of Arrecife

The Island of Lanzarote is the first landfall from the Iberian Peninsula, and was incorporated into the Castilian Crown in 1402 by the Caballero Bethencourt, during the reign of Enrique III. The city of Arrecife has been the capital of the island since 1618.

Lanzarote, known as the Island of the Volcanos, has up to 100 identified volcanic cones in its 800 Km2 area.

The contrast between the navy-blue of the sea, the areas of volcanic rock and the tropical vegetation in the rest of the island, make it an exotic place, unique in Europe. Presently tourism is the major income of the Island. The new airport and the new 'los Mármoles' dock in the harbour of Arrecife receive visitors from all countries, making it frequent to see mooring at its wharves the biggest cruising boats of the world, including the Queen Elizabeth II - the flagship of the British cruising fleet.

THE PRELIMINARY DESIGN FOR THE SPORTING/RECREATIONAL MARINA IN ARRECIFE

This section summarizes the proposal of 'Sporting Port of Arrecife, Lanzarote', put forward with the 'Technical Assistance of NAUTILUS S.A.' under the direction of the Spanish National Port Authority in Las Palmas.

The objective is to define the facilities necessary to meet the demands placed by the tourist and private sporting boats on the Island of Lanzarote, as well as to contribute to the improvement in the quality of tourist and leisure services of the city of Arrecife.

These are as follows:

. Growth in demand for the moorings and services by the local tourism and sporting fleet on the island of Lanzarote.

. Increase in the number of visiting boats based in other sporting ports of Lanzarote and other islands of the Archipelago.

. A rise in the number of boats that are based in the Iberian Peninsula and other European countries that desire a berth in the Archipelago of the Canary Islands.

. Growth in the Trans-Atlantic traffic that stopover mid-crossing.

. The possibility of increasing the service capacity of the sheltered water within the Port of Lanzarote for smaller boats, improving utilization and profitability.

. The landscaping of the waterfront within the harbour and the development of the tourism services of the city of Arrecife.

A description of the existing facilities

The harbour facilities of the Port of Arrecife, at present, are situated within three basins:

1. The old commercial harbour known as the Port of Arrecife.

2. The new basin of 'Naos', which primarily is dedicated to the fishing fleet.

3. The 'Muelle de los Mármoles', a basin built recently for commercial traffic.

The construction of the new basins of 'Naos'and 'los Mármoles'with their adequate installations has permitted the transfer to them of all the commercial and fishing activity of the harbour; leaving the basin of the old harbour.

Reasons for the adoption of this solution

Because of its natural orientation and protection, its site in relation to the city and for its existing maritime installations, the old port of Arrecife is the most appropriate area to build a nautical, tourist and sporting complex of maximum quality.

The proposed marina in its finished form will consist of a basin able to accommodate up to 520 yachts, and also have the potential for extension at a later date.

These following existing factors have been taken into consideration in this proposed development:

. A natural harbour exists that has been the origin of the city of Arrecife, and has been the traditional and most important point of arrival and departure of the Island.

. The harbour has been rendered obsolete by the size and number of the merchant fleet in service, and their moorings have been transferred to a new basin.

. The facilities for the fishing fleet have also been transferred t a harbour built for that purpose - leaving free the small island of 'San Gabriel'and the wharf of 'La Pescadería'.

. The most important part of the city historically, residentially, commercially and in tourism has developed along the waterfront of the old harbour.

. The existing configuration of the breakwaters of the harbour protects the basin against all inclement winds except the South-East.

. The bed of the harbour is composed of volcanic rock which is extremely expensive to dredge. Blasting with dynamite has been rejected for any attempt to do so would have severe ecological consequences thus placing restrictions on such an option.

. The tourist/recreational development situated on the 'Isla del Amor'is practically finished and can be integrated within the overall planned urbanization of the area of the proposed marina.

. The waterfront unifies the commercial streets of Arrecife and also is the main thoroughfare connecting the east and west of the city, which is frequently congested.

. A great demand for parking spaces for cars exists within the area adjacent to the piers.

The socio-political characteristics that determined the nature of the proposal

. Conservation of the character of the group of small islands and their linking causeways that make up the harbour.

. Maintenance, as far as possible, of the waterfront of the city to all practical purposes.

. Minimal dredging of volcanic rock for economic and ecological reasons.

. Inclusion of a pedestrian passage encircling the harbour (at a later stage. Stage II).

. Inclusion of a selective commercial zone with emphasis on tourist services.

. Maintenance of the direct link of the existing Sailing Club to the sea with its possible transfer to a new site within the marina at a later date.

Outline of the Preliminary Design and its general characteristics
The proposed marina is a sporting/recreational tourism harbour situated in the urban centre of the capital of the island. Tourism is the main source of economic growth and development of Lanzarote and any improvement in the standard of living of its inhabitants has to be the final objective of this proposal.

The proposal's aim is to satisfy the needs of the local tourism and recreational fleet in the immediate future, to promote interest and involvement in the sea and nautical sports and to draw the 'custom' of a larger number of sailors from the archipelago, the Iberian Peninsula and other countries who come as visitors, providing the incentive for them to base themselves with their boats in Lanzarote for a short term or permanently.

The rationale behind this new marina is based on amplifying the contact between the city and the sea through the construction of an area of walks - cultural, commercial and recreational - of which all the locals as well as the visitors to the island can enjoy. At the same time, parallel to this rational, there is the aim to provide facilities making the Port of Arrecife the most attractive and exotic marina of Europe.

One main limitation imposed upon the development is the minimization of dredging of volcanic rock which has meant the adaptation of the wharves and filling to the existing bathymetrics. This has resulted in the shape of the basins following the natural topography of the immediate environment.

Further, because of its historical value, the proposal has been careful to preserve in its present state the form of 'Isla de San Gabriel'or also called the 'Island of the English/Isla de los Ingleses'with its accesses, as well as the access to the 'Muelle Antiguo' in its eastern part and of the existing dyke.

The design of the supporting breakwater has been defined by following the bathymetric level minus 7 in the mouth of the port and in siting the launching ramp on the natural angle and support of the volcanic rock base, with the protection of the 'Isla del Amor'.

The physical interconnection of the small islands will be maintained through the further use of bridges (as well as the two already existing). These small islands "The Archipelago of the Reefs 'Arrecifes'" has been the origin of the port and of the name of the city.

Within the proposal landscaping forms an important facet, making the marina an artificial park with small islands, passages lined with palms and with extensive areas of gardens.

The landscaped areas will contain buildings and facilities of a cultural nature. This provision for a cultural area includes the positioning of an open air theatre and other social/cultural buildings, as well as a school of sailing and navigation.

These facilities will give rise to a demand for parking spaces, which can not foreseeably be contained by the present urban structure of the city; so the provision of several controlled parking areas are provided for in the design. These areas will be bordered by low hedges and gardens, obscuring them from the view of the pedestrian passages and the wharves.

The circulation of vehicles will be limited only to the access road to the parking areas and the maintenance areas, and on the wharf of the 'Muelle Antiguo' only under selective conditions.

The proposal also includes the establishing of beaches fronting the islands of 'San Gabriel' and 'del Amor'.

The northern frontage along the service road on the 'Muelle Antiguo'will not be modified, excepting the rehabilitation of the beach adjoining the Castle, leaving the island of 'Juan Rejón' in its present state, with the only variation being the construction on its extreme of the lighthouse designed by the

Lanzarote artist César Manrique.

In its second stage, the construction of a maritime passage is foreseen, continuing the theme of a palm lined passage for pedestrians along the waterfront of the city, thus encircling the inner harbour.

The characteristics of the harbour are as follows:

. The area of the basins: 82 000 m²
. The area of the land: 80 000 m²
. Average dimension per mooring: 44.0 m²
. Total length of wharves: 1300 m
. Length of floating pontoons: 730 m
. Area of maintenance area: 6500 m²
. Area of light sailing boats and School of Sailing: 3200 m²
. Number of berths and their dimension:

Table I: Berths

Dimensions	m²	Number	Total m²
25 x 6	150.0	4	600
20 x 5.5	110.0	23	2530
15 x 5	75.0	26	1950
12 x 4	48.0	271	13008
8 x 3	24.0	196	4704
		520	22792

. Number of car park spaces: 500 units
. Area of parks and gardens: 10 000 m²
. Area of amphitheatre and cultural buildings: 2000 m²
. Area of commercial buildings: 4000 m²

CONCLUSIONS

The location of the harbour of Arrecife is the best in the Island of Lanzarote and the most appropriate for the arrival of leisure ships navigating between the islands and for European crossings.

The development of the Marina/tourist centre favours the increase in the number an quality of visitors, an objective being common to the whole of Spain.

The construction of a marina/recreational park of the category and dimensions that is outlined in this paper, is the investment which will allow a better commercial and urban development of the city of Arrecife.

The solution described above is the most appropriate for the utilization of the existing port facilities resulting in the development of the economic, commercial, recreational and social standards of the city of Arrecife.

The construction of this port will increase the use of other marinas on the Island of Lanzarote, as well as on the Canary Islands themselves, attracting a greater number of European and American visitors to the islands.

ISLA DEL AMOR

MUELLE ANTIGUO

ISLOTE DE JUAN REJON

CLUB NAUTICO

MUELLE DE LA PESCADERIA

ISLOTE DE SAN GABRIEL

CASTILLO DE SAN GABRIEL

THE OLD PORT OF ARRECIFE   FIGURE 1

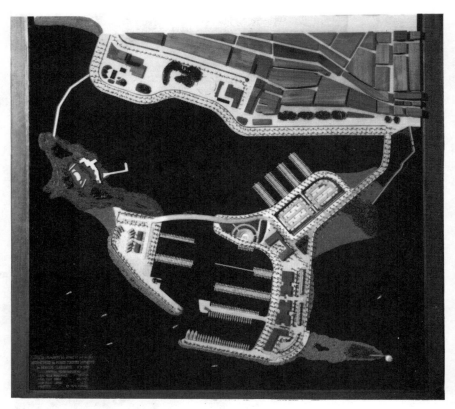

GENERAL LAYOUT OF THE MARINA   FIGURE 2

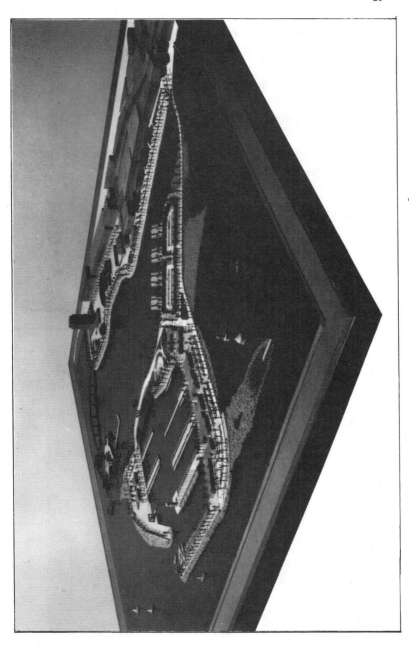

EAST VIEW   FIGURE 3

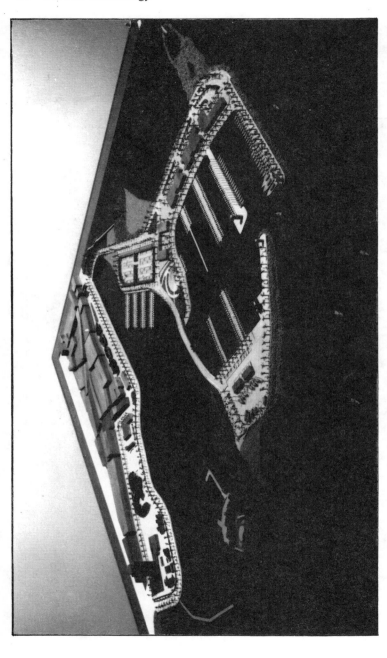

SOUTH WEST VIEW

FIGURE   4

SECTION 9: DESIGN AND CONSTRUCTION

The Use of Gabions in Marinas, Docks and Canals

A.D. Crowhurst

River & Sea Gabions Ltd., Tricorn House,
51-53 Hagley Road, Edgbaston, Birmingham,
B16 8TP U.K.

ABSTRACT

Both for inland and tidal marinas and small boat harbours the choice of construction materials has to be made balancing aesthetics, costs, constructability and efficiency in service with the need for longevity.

Similar considerations apply to the approaches to marinas whether through a natural or artificial channel or between breakwaters protecting the entrances.

In order to develop certain aspects of design relating to the use of gabion mattresses in navigational channels such as those leading into small boat harbours or marinas as well as into commercial docks research work has been carried out for the Maccaferri Group by Delft and by Sogreah.

This has centred on the effects of wave action on mattress linings due to ship generated waves and comparisons with rip-rap or armour stone and the stability of these mattress linings.

Within an enclosed area the use of vertical gabion walls has advantages in that the open nature of the stone fill permits dissipation of some of the wave energy with resulting reduction in wave reflection, an obvious advantage to the enjoyment of a peaceful mooring.

In more open water gabions should normally be constructed to slopes not exceeding 1:2 or 2.5 in order to withstand wave action. When so used, as with all hydraulic applications for gabions or wire enclosed rip-rap, a high standard of construction is essential coupled with the use of materials of proven strength and durability.

INTRODUCTION

For the smaller marinas and harbours built for and in some cases by the members of clubs or the developers of a project the need to ensure that the most economic scheme is used does not reduce the need for longevity.

In earlier times construction of inland waterway facilities was carried out using masonry or brick to form substantial structures which remain as monuments which, in many cases, are being rehabilitated to their former glory i.e. the Gas Street Basin in Birmingham.

For similar work currently economics dictate that alternatives be evaluated and these can include structures built with both gabions and Reno mattress. For tidal work also the gabion method of construction can be utilised for shoreline and harbour structures subject to constraints of wave climate and availability of hard durable stone.

Materials
The modern gabion, which can be traced back to wickerwork cages used for both civil works and military fortifications in medieval times, is a cage or basket made from steel wire, heavily galvanised and, for most inland and all coastal locations, additionally coated with a PVC sleeve.

The projects and applications mentioned in this paper incorporate mild steel wire woven into a double twist mesh which was found to give much greater strength and durability than the single or chain-link mesh type used in the last few years of the 19th century.

Zinc coating to BS443 has been used for many years being much heavier than the commercial grade coatings used for some wire products. To obtain a sound bonding, zinc coating has been found to be essential under the PVC sleeve.

The gabions are fabricated with thicker selvedge wires along all cut ends and with thick edge wires also so that forming the units into rectangular boxes on site is facilitated.

The grading and quality of the stone fill is critical in all marine and, indeed, most other structures. The fill must not pass through the mesh but it also must not be too large.

For walls of all types the greater the variation in stone size the more hand placing becomes necessary to minimise voids and to avoid settlement. Within a harbour or marina there are advantages in using a single size aggregate close to the smallest practicable size since this will give rise to a larger number of small voids which assist in dissipation of wave energy.

For mattress linings which are used, for example, to line river banks in the vicinity of marina entrances or to line the banks underneath piled structures the stone size has similar lower limits relative to mesh openings and has an upper limit such that there is at least a double thickness of stone within the pockets.

Geotextiles are also frequently specified as an underlayer to the mattress or as a lining behind gabion walls. Selection of type to be specified must take into account the need to avoid clogging of the interstices by the finer soil particles since one of the advantages of this type of construction is its ability to drain freely with a minimum increase in hydrostatic pressure.

Coastal Marinas and Small Boat Harbours   The use of a system of placing pre-filled gabions has made possible construction of harbour walls using, typically, 2m x 1m x 1m gabions. An early example was that of Ruweis in the State of Qatar in the Arabian Gulf where walls up to six metres deep were constructed by a local contractor. Stone collected from the desert was used and the gabions were filled within a "bottomless" box to give a good rectangular shape.

The approach channel at Ruweis was dredged through coral and was well sheltered with a very small tidal range and limited exposure to wave action.

The jetty was designed to handle dhows used
in coastal traffic and incorporated concrete
blocks to anchor the fendering system together
with a sufficiency of bollards.

In vastly different circumstances a small
boat harbour was built on the Island of Tristan da
Cunha, following the volcanic eruption some years
ago, for use by fisherman and for ferrying goods
and equipment from the regular supply shops. The
inner face of the breakwaters were lined with PVC
coated gabions, the basin itself being excavated
by explosives. No attempt was made to design the
gabion structures to permit mooring alongside, all
boats being brought up onto the beach.

More recently there have been a number of
attractive structures built to accommodate yachts
and cruisers rather than commercial vessels
although there have been applications in recent
years such as the wall built within Newlyn Harbour
in Cornwall.

In most coastal structures in the UK the use
of gabions or wire-enclosed rip-rap is restricted
to the perimeter walls since tidal variations are
such that floating pontoons are more suitable for
ease of access to boats moored alongside them.

Inland Waterways   The use of gabions for inland
waterways used for navigation is well documented
and ranges from the lengthy bank protection
carried out by the Manchester Ship Canal where
ocean going vessels pass many excellent examples
of gabion design and construction from its
downstream entrance for the Mersey right up to the
City of Manchester itself although few large
vessels now go so far. Unfortunately most of
these structures can only be seen from the decks
of freighters or tugs since the canal is not in
regular use by pleasure boats.

Across in the United States the Tennessee-
Tombigbee Waterway is probably the largest canal
or canalised waterway built in the last decade or
two. Following completion the US Army Corps of
Engineers found that it was necessary to protect a
length of the bank of the canal section. Reno
mattresses were used and were constructed partly
on a specially built platform extending from the
deck of a barge and partly on the bank itself. By

moving the barge at right angles to the bank the mattress was speedily and effectively laid underwater.

The Corps of Engineers also have used gabions and mattresses on one of the navigation channels leading into Lake Okeechobee, a very large body of water in central Florida. The design of the revetment included short lengths of walls with access ramps to facilitate access to the bank by disabled anglers.

Further west in the City of New Orleans stern wheelers berth at a jetty in Audoban Park, which has a long frontage to the Mississippi protected with PVC coated Reno Mattresses. Of course the water levels of the Mississippi which is deep enough for ocean going vessels at this point, varies widely and the revetment is of considerable width.

Inland Marinas For inland structures on canals where water levels do not vary by more than a few centimetres gabions may be used not only for perimeter walls but also for finger piers.

In Eire on the Shannon sytem gabions have been used for a small public marina or boat harbour at the delightful village of Terryglas on Lough Derg. Just a short distance away a new marina is being built at Gortmore Harbour by a private consortium, one of whose members is a professional engineer. The marina is being constructed in the dry in a basis excavated into the banks of the Lough. Armour stone has been used to protect the lake bank, the inner side of which is formed by a three metre high gabion wall which will have a 0.5 metre freeboard. Halfway along this wall is a spur or jetty, two metres wide. Eight Shannon barges will be moored normal to the main wall.

The inner half of the harbour will have a similar height of wall and will accommodate cabin cruisers. The gabion walls have been built by a two man crew using an excavator/loader. Although the construction period has been extended due to other contractual commitments a high standard of finish has been achieved.

<u>Research</u>   The preceeding paragraphs have concentrated on actual structures and practical considerations but the need to carry out research into aspects of canal, coastal and harbour works has been recognised.

From their initial uses in the closing decade of the last century many gabion uses have been developed purely based on experience. However, in the last twenty years considerable research has been carried out worldwide. Initial attention was given to the design of retaining walls and research studies were carried out by the University of Bologna, Italy, by STS Consultants, Chicago as part of research initiated by the Federal Highways Administration and by others in Australia and elsewhere.

Research into hydraulic applications included a major study by Simons Le & Partners of Fort Collins, Colorado which included full scale flume tests at Colorado State University.

Of direct relevance to this topic was a study carried out by Delft on the banks of the Haartel Canal in Rotterdam. This investigated the performance of various bank protection structures when subject to ship generated waves. As part of this study a 170mm thick Reno mattress was exposed to waves generated by barges navigated at various speeds and along varying tracks. The structures were very well instrumented and the results proved to be satisfactory with little or no movement of the mattress.

Research has also been carried out on gabion and mattress performance under various wave regimes, this research being handled by Sogreah at their laboratories in Grenoble, France. This research has enable design recommendations to be tabulated taking into account gabion thickness, the stone size and slope of protection relative to wave amplitude.

<u>Conclusions</u>   At the site listed and at many more, including several quite close to the Conference location, the gabion method of construction has been found to be a satisfactory and attractive solution to problems arising from development in locations that may have conditions ranging from deep estuarial silts to peat. The ability of a

gabion structure to settle without failure may be essential.

The ongoing research programme has already provided much useful data that can be used in future projects.

Zante Island: A Third Generation Marina - Design and Construction Problems

N. Panagopoulos, J. Sioris, V. Tsamis
Triton Consulting Engineers, 2 Hakedonon Street, Athens 11521, Greece

ABSTRACT

On the island of Zante in the Ionian Sea a third generation residential marina is constructed next to the commercial harbour. The works include the construction of three artificial islands connected among them as well as landscaping of the surrounding coastal area. This paper concentrates on the problems associated with the foundation and construction of these islands on the very soft clayey sea bottom, as well as the problems arising from the subsequent construction of buildings, taking into account the high seismicity of the area.

INTRODUCTION

For the last 30 years Europe has seen an unprecedented growth in tourism especially in the sea-born sector where a fundamental evolution has taken place. The Western European Mediterranean coastline is today saturated with old and new port facilities, marinas, fishing harbours, anchorages, e.t.c. due to the continuously increasing number of yachts and boats of pleasure.

In contrast, the Eastern Mediterranean basin and Greece in particular, has been relatively undeveloped in this field because of geographical, historical and economic reasons. Until now the seafarer tourist has found shelter, anchorage, food and provisions in the small picturesque fishery harbours of the Greek islands. Today these harbours are getting more and more congested and cannot fulfil the needs and requirements of the modern and affluent yachtsman.

The Zante Marina is the first project in Greece where the financial aspect of a marina, in providing not just yachting facilities, but a comprehensive set of services in the recreational and housing sectors, has been recognised.

Zante island lies in the Ionian Sea along the western sea route to Greece and therefore attracts large numbers of yachts all year round. The marina is planned in the south of the harbour of the Zante city, the capital of the island. Although the surrounding coastal area is in the rundown part of the city land prices remain relatively expensive. The design therefore, provides for the creation of a small urban center on artificial islands built along the coast, which will eventually offer the yachtsman full residential facilities, as well as complete mooring and boat maintenance services.

The marina will be built around 3 artificial islands (designated B,C, & D, see Fig. 1) and two coastal areas (designated A and E) with a total land area of 60.000 m² out of which approximately 30.000 m² will be reclaimed from the sea.

The project includes 1850 m. of quaywalls, 260 m. of floating piers and built up areas of approximately 36.500 m². Construction of the first stage of works is already underway while tendering for the second stage is scheduled for the near future.

This paper presents the problems associated with the foundation and construction of the artificial islands and the coastal reclamation works. The methods applied to make the very soft clayey seabed stable and capable to bear the significant loads associated with large residential building constructed next to the water front in an earthquake prone area, are of particular interest to both the ports and the geotechnical engineer.

SITE CONDITIONS

The marina is constructed south of the existing commercial harbour sharing with it the south leeside breakwater.
The sea bottom in the area is very mildly sloping and the main soft deposits encountered up to a depth of 41m are:

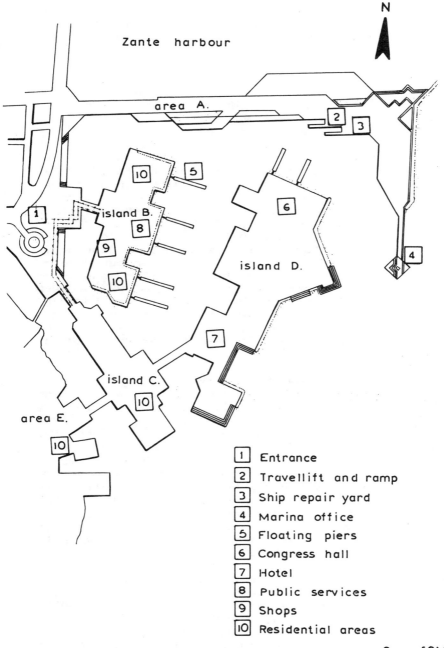

N

Zante harbour

area A.

island B.

island D.

island C.

area E.

1	Entrance
2	Travellift and ramp
3	Ship repair yard
4	Marina office
5	Floating piers
6	Congress hall
7	Hotel
8	Public services
9	Shops
10	Residential areas

0   50M

FIG. 1_ MARINA GENERAL LAY-OUT.

a. A layer of greenish-grey SILTY CLAY of medium to low plasticity [CL] 20 to 30 m thick. The number of SPT blows N ranges between 2 and 5, the plasticity index PI between 7 and 21, the undrained shear strength Cu between 5 and 35Kpa and the compression index Cc between 0.09 and 0.38.

b. Interbedded in the above layer at depths ranging between -11 and -22 below sea bottom, a layer of greenishgrey SANDY SILT [CL-ML] generally 2.5m thick (and locally up to 10m thick) is encountered. The N value of the layer, ranges between 2 and 5, the PI never exceeds 7 (in general the layer appears non plastic), the unconfined compression q_u ranges between 33 and 54kpa and the compression index at one location was found Cc=0.08.

c. Finally a layer of greenish to bluish grey CLAY of high plasticity [CH] is encountered at an elevation varying between -22m and -33m below sea bottom. The value of SPT blows N ranges between 4 and 9, the plasticity index PI between 23 and 31, the undrained shear strength Cu is approximately 20-25kpa while the compression index Cc varies between 0.20 and 0.33.

It is clear from the above that the site investigation indicated unusually low values of strength, quite high values of compressibility with a wide scattering of these values from place to place. This fact combined with the special nature of the project which includes not only conventional marina quaywalls and harbour works but also an extended building programme with two and three storey houses, hotels etc, resulted in the adoption of conservative values for the design soil parameters. Furthermore, as the island of Zante belongs to one of the most sensitive earthquakewise areas of Greece a high seismic coefficient was adopted.

Regarding the environmental conditions the marina area is naturally sheltered from the prevailing north winds and waves and it is only exposed to the east and north east sector waves albeit with very limited fetches due to the Peloponnesian coastline to the east. Thus, only limited external protection harbour works are required.

ALTERNATIVE FOUNDATION AND RETAINING TECHNIQUES

a. Artificial Islands

The architectural planning calls for buildings to be established on the artificial islands and the coastal zone E. Zone A was reserved for mooring facilities, the main entrance and the parking lots.

From the foundation point of view three particular design cases appear:
i.   Buildings founded on the interior of the islands away from the perimeter.
ii. Buildings founded upon or close to the quaywalled perimeter of the islands.
iii. Buildings founded close to the sloping perimeter of the islands.

A number of options for the improvement of the seabottom were examined by the designers.

Use of Geotextiles: Geotextile application in this case prevents formation of slip surfaces. Therefore high strength reinforcement geotextiles were examined but the problem of excessive settlements remained. For this reason, as well as the unacceptable high cost of the material for such a large area, their application was rejected.

Soil improvement by gravel piles (stone columns): The construction of gravel piles increases the mean shear strength of the seabottom improving the bearing capacity of the artificial islands. Furthermore they speed up the consolidation process, reducing the consolidation time and improving the compressibility of the underlying strata. Analysis indicated that such a solution would be effective if hard bearing strata could be reached at relatively small depths below the sea bottom. In such a case, the gravel piles could be founded on the harder strata and bear most of the loads. Unfortunately boreholes down to a depth of 40m from the sea bottom could not reach a bearing stratum and therefore this method was also rejected.

Preloading the marina islands: The idea is to increase the strength of the bearing strata and reduce their compressibility by means of preloading the seabed by constructing an artificial mound up to a higher level than the final level of the island. The excess fill material is removed subsequently. This option was finally chosen and the problems associated with it are described in detail herebelow.

b.Quaywalls

A similar analysis was also carried out for the selection of the optimum retaining structure for those sections of the perimeter of the islands where a vertical seafront was envisaged.

Sheet piling flexible solutions were examined and finally rejected because even with the improved undrained strength caused by preloading, development of sufficient passive earth pressures to resist the active ones and the lateral stresses from the nearby foundations could not be safely assumed, unless the upper 3 meters of the seabed layer were replaced by sandgravel immediately after driving the sheetpiles to the required depths. Moreover the single sheetpile wall is also rejected due to the large lateral deflections to which it would be subjected, while the solution of two sheetpile walls cross-connected with anchor bars creates significant constructional problems, among which of particular concern is the placement of the anchors below sea level.

Finally gravity type structures (quaywalls of concrete blocks) resting on a significantly improved soil by:
i. replacement of the top seabed strata around the quaywall foundation and
ii.subsequent preloading, proved to be the optimum technoeconomical solution.

Soil removal and replacement with layers of graded sand and rubbles for increased bearing capacity of the seabottom is restricted over the quaywall foundation zones only. Along these zones, a foundation trench will be excavated on the seabed and filled with suitably graded rubblestone.
Since preloading is required throughout the islands' area, an extensive preloading programme combined with a dense pattern of plastic drains to speed up consolidation was devised.

Soil removal and replacement with layers of graded
sand and rubbles for increased bearing capacity of
the sea bottom is restricted over the quaywall
foundation zones only. Along these zones a
foundation trench will be excavated on the seabed
and filled with suitably graded rubblestone.
Preloading will be applied by fill material
deposited up to the elevation of +3.0 m. above
Mean Water Level (M.W.L.). Subsequently this
mound will be removed down to +0.80 m M.W.L. i.e.
to the final general elevation of the islands'
superstructure. This overloading corresponds
approximately to the average foundation stress to
be imposed on the islands by the buildings.

Preloading will be applied in one stage lasting up
to twelve months over the areas where the total
height of the fill measured from crest to sea
bottom is less than 5 m.
On the remaining areas preloading will be applied
in two stages. In the first, preloading will be
applied for six months up to a height of 5 m above
the sea bottom. This period was computed to be
sufficient for the seabed's undrained strength to
be increased to such a value that bearing the
total preloading mound becomes possible.
In the second stage the preloading mound will be
raised to the final elevation of +3.00 m. above
MWL and will be maintained there for six more
months.
The whole preloading program was substantiated by
extensive slope stability analyses using both
Bishop and Fellenius methods with circular slip
surfaces, applied in both the preloading and the
final working stages, using in each step as input
the improved strength parameters obtained from the
previous stage.

Thereafter detailed bearing capacity analyses of
all types of quaywalls were carried out taking
into account both the various earth and seismic
forces as well as the loads from the adjoining
building complexes. Finally, bearing capacity and
settlement analyses were carried out for the
foundation beams of the two and three storey
buildings.

Material for preloading is specified with an
apparent unit weight of 17 KN/m3 . Even material
from dredging may be used provided it will be
approved by the supervising authority.

Due to the very limited permeability of the clayey strata of the seabottom the speeding up of the pore pressure dissipation is necessary in order to achieve consolidation within an acceptable time schedule. Adoption therefore of vertical plastic drains to be inserted down to depths of −22 m from the sea bottom came out as the only viable option.

Effective operation of plastic drains dictates the placement of their heads in a highly permeable granular material as sand, gravel, rubbles etc. Thus after dredging and filling the quaywall foundation trenches with a thin layer of sand and gravel, the plastic drains will be inserted through this first layer of foundation material.

For the "inland" areas of the artificial islands away from the quaywalled perimeter, the plastic drains will be inserted through a first layer of fill material that will initially be deposited directly on the sea bed. In all cases this first layer will not have a thickness more than 1.0 m.

To test the accuracy of the analysis, a "pilot" island 20 X 20 m. is to be constructed at depths ranging between −3.0 and −4.0 m. This pilot mound will reach a height of 1 m above the M.W.L. and from its crest "test" plastic drains will be embedded. Five settlement marker plates one in each corner and one in the center, along with five piezometers at depths of −5, −10, −20, −30 and −40m will be installed to monitor the development of settlements and pressure dissipation.

Simultaneously from this pilot structure two deep sampling boreholes and three static penetrometer tests will be carried out. The boreholes will be terminated at a depth of 100 m provided refusal in SPT tests is not observed earlier. High quality undisturbed samples by piston sampler will be taken and field vane tests along with the corresponding lab tests will be carried out.
Settlements and pore pressures recording will be carried out during the first six months and the results will be analysed in order to check and readjust if necessary, the preloading programme. The supervising authority may accordingly alter the pattern or even the application period of the preloading.

During construction, settlement marker plates and piezometers will be installed in 20 X 20 m and 30 X 30 m patterns respectively, covering all areas of the islands. The piezometers will be installed in depths -5, -10, -20, -30 and -40 m, while the settlement marker plates will be placed at elevation $+0.20$ m, their stems reaching an elevation of $+3.20$m.

These instruments (their output to be constantly computer recorded) will monitor settlement and pressure during each phase of construction and will remain in place at least until the commencement of the residential project so as to provide continuous and updated data for the foundation design of each individual building.

PROJECT IMPLEMENTATION

Based on the above described design the project implementation is as follows:

Phase A (Already under construction)
a. construction of the two main breakwaters for the protection of the marina and the facilitation of the construction of the artificial islands in the leeside.
   The northern breakwater, actually an extension of the existing lee breakwater of the commercial harbour is constructed by prefabricated concrete blocks and at its lee side a quaywall with depths of -5.00 m will be used for the reception of cruisers and large motorboats or sailing boats. The south breakwater is built of quarry stones and constitutes the future seaward limit of island D. All structures are founded on sand and gravel layers of variable thickness.

b. The existing lee breakwater of the commercial harbour is substantially widened and relatively shallow quaywalls are constructed along it. The northern part will incorporate a 6.60 m wide travel-lift with an operational length of 22 m capable of accomodating boats of up to 25 m length. Next to the travel lift an 8 m wide ramp is provided, capable of servicing all the usual sizes of boats to be accomodated in the marina. The remaining berths with depths of -2.00 m will be mainly used by small fishing boats at least until the marina housing project begins. The designers welcomed the

fishing activity (in a semi-amateur scale) as
it provides the necessary traditional character
and colour to this area while counterbalancing
any negative impressions from the boat repair
yard next to the travel lift area.
The quaywalls allocated to fishing activities
are built with superstructure in different
levels. This combined with the careful
selection of materials, the construction of a
retaining wall and the adoption of planting,
will comfortably separate the activities of the
commercial harbour from those of the marina.

Phase B (to be tendered shortly)

c. Marina basins' dredging.

d. Completion of dredging for both the basins
   and the quaywall foundation trenches.
   Construction of the pilot embankment and
   execution of the on-site geotechnical
   investigations.

e. Placement of a first sand and gravel layer in
   the foundation trenches of the quaywalls , as
   well as on the seabed area over which the
   islands will be constructed. This layer will
   not exceed the 1 meter thickness (see Fig. 2).

f. Evaluation of the geotechnical survey results
   and final determination of preloading schedule.

g. Embeddment of vertical plastic drains by means
   of pile hammer on pontoon in the areas to be
   preloaded (islands B, C, D and area E). (Fig.2)
   Completion of quaywalls in area A.

h. Completion of under and abovewater filling
   with non cohesive well graded material.
   The part of the filling above water will be
   levelled and compacted in layers of maximum
   thickness of 50 cm.

i. For preloading purposes the height of the fill
   will be raised to level +3.00 m as described
   in the previous Chapter (Fig. 2).
   Simultaneously the settlement marker plates and
   piezometers will be installed and the full
   monitoring programme will commence.

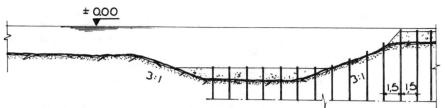

Dredging – 1st foundation layer – Embeddment of drains.

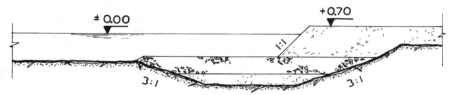

Completion of foundation layers and filling.

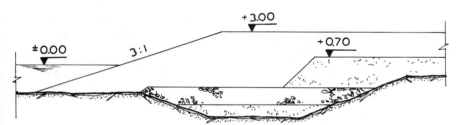

Placement of preloading mound in one or two phases.

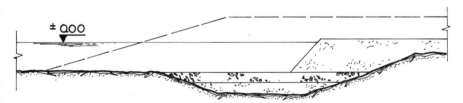

Removal of preloading after 12 months.

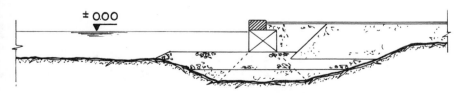

Construction of quaywalls and embankments.

FIG. 2_ CONSTRUCTION IMPLEMENTATION

j. After completion of the settlements, the fill
   in excess (for preloading) will be removed
   (Fig. 2) and the construction and placement of
   the concrete blocks will commence.
   Quaywalls around islands B, C and D range in
   depth between -2.00 and -3.50 m accomodating
   boats up to 15 m long (Fig. 3).
   At the same time the construction in dry of the
   bridges connecting the islands between them and
   with the shore will advance, together with the
   lining of the slopes with graded stones, the
   paving of the superstructure's upper level with
   paving blocks, the installation of the
   auxilliary berthing appurtenance (strip
   fenders, bollards, berthing rings etc) and the
   construction and placement of specially
   designed floating piers in order to fully
   exploit the available berthing areas.

k. Finally the architectural works will in this
   phase mainly be restricted to landscaping the
   marina entrance and the bridges connecting the
   islands with the shore.
   For these works every effort has been made to
   fit in the local vernacular Zante architecture.

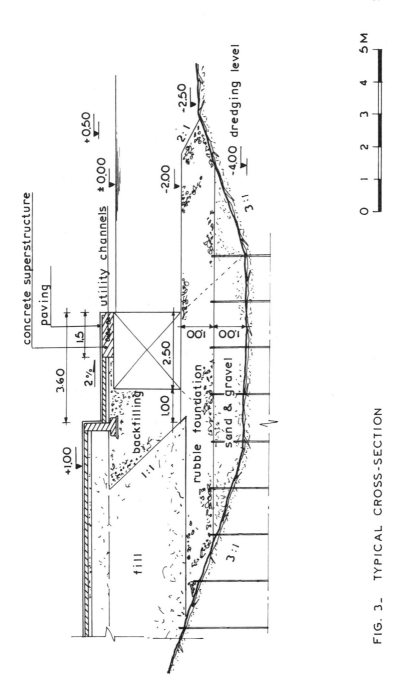

concrete superstructure
paving
utility channels
+0,50
±0,00
+1,00
3,60
2%
1,5
2,50
1,00
backfilling
1:1
fill
rubble foundation
sand & gravel
-2,00
-2,50
2:1
1,00
1,00
3:1
3:1
-4,00 dredging level

0 1 2 3 4 5 M

FIG. 3.-   TYPICAL CROSS-SECTION

On the Design and Development of Qatif Marina in the South of Israel

E. Mechrez, (*) M. Ratner (**)

(*) Coastal and Marine Engineering, P.O. Box 217, Givat Ella 10503, Israel

(**) 123 Yeffe - Nof st., Mount Carmel, Haifa 34454 Israel

Abstract

The paper describes the development of Qatif Marina in Israel.

Location:
The marina is located on the Mediterranean coast , 50 km south of Ashdod - Israel main harbour. The area , which is close to the northern coasts of Sinai is characterized by large amounts of sand movements (originated in the Nile delta) . The beach width is limited by a "kurkar " (sandstone) ridge which is undergoing natural erosion.
The marina location was chosen at the point where a stilling basin for seawater intake for the Mediterranean - Dead Sea canal is planned. This canal is intended to convey Mediteranean seawater to a point above the Dead Sea and to produce Hydroelectric power.

Marina main objectives :
1. To serve as a southern fishing harbour in Israel
2. To develop tourism and to provide shelter for yachts sailing through the Suez Canal to Europe. These yachts will take the course along the eastern Mediterranean coasts instead of crossing directly to Cyprus and Turkey.
3. To create a protected basin from waves for maritime agriculture .
4. To serve as a stilling basin for future development of the Mediterranean -Dead Sea canal.

The marina layout and architecture (described in this paper) should be able to give appropriate answers for all the above objectives at the same time.

Engineering Design :
The engineering design of the marina is based on two breakwaters (main and lee breakwaters) . The main breakwater is a shallow water breakwater (maximum wave height is limited by depth) and is a rubble mound type . The armour layer is made of 9-12 ton rocks at a 1:2.5 slope. The inner slope is 1:1.5 and made of 2-5 ton rocks. Maximum crest elevation is at +6.0 meters.

Beach erosion at the northern side of the marina is expected since the marina construction will block temporarily the south to north sediment transport. Previous experience (50 km to the north , Israel Electric Company cooling basin) has shown that sediment bypass is created within a few years.

The Mediterranean - Dead Sea hydroelectric project

Israel's power generation depends totally on fossil fuels , virtually all imported . At the same time , Israel has long been at the international forefront in utilizing solar energy . Nevertheless , large scale solar energy plants based on the solar pond concept are still at the pilot plant stage.
At present , Israel has no hydroelectric power plants , although plans have recently been completed for a 100 MW hydroelectric power station on the Jordan river . The Jordan being a small river (30 cu.m/sec) , the only possible way of producing hydroelectric power in Israel on a large scale is by making use of the difference in elevations between the Mediterranean Sea and the Dead Sea , which lies some 400m below sea level.
The main objectives of the Mediterranean - Dead Sea hydroelectric project are to enable highly flexible handling of peak loads , to provide emergency generation capacity , and to lessen Israel's dependence on imported fuels.

A unique feature of the proposed scheme is that its power-generating potential is not constrained by river flow fluctuations because the source of its water is the Mediterranean Sea. The annual and long term potential is limited only by the intake capacity of the Dead Sea.
The Dead Sea is a terminal lake . Starting from the 1930s and more pronounced from the 1950s the water level has gradually declined, mainly as a result of diversions of water from the Jordan River . To raise the Dead Sea level from 406m to the proposed 390.5 m BMSL, about 35,000 MCM of Mediterranean Sea water will be conveyed to the Dead Sea over a several -year period . When the target level is attained , conveyance from sea to sea will be equal to the rate of evaporation from the Dead Sea estimated to average 1550 MCM/year. With an installed capacity of 800 MW the Dead Sea power station is designed to supply peak power for up to 49 hours per week during the filling period and 37 hours per week at steady state.

Since the late 19th century , when the concept of an inter-seas conduit was first proposed for hydroelectrical power generation , several "alignments" have been investigated. The proposed alignment starts at the Mediterranean sea shore near Qatif , about 50 km south of Ashdod and curves south of Be'er - Sheva to a point about 10 km south of Massada on the Dead Sea shore (fig. 1).

The sea intake design and the marina concept

Among the main project facilities there is an intake basin on the Mediterranean Sea shore . Two types of intake were considered by Tahal (1983) at the design stage:
a. An intake basin , created by two breakwaters . The opening between the breakwaters is designed to ensure entry of seawater at velocities lower than the threshhold velocity at which sand grains could be transported .
b. A submarine conduit , the inlet of which is designed to ensure smooth sand-free water entrance and placed at sufficient depth not to pose a hazard to small marine craft.

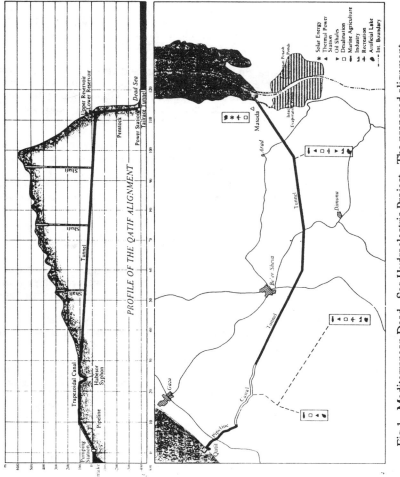

Fig.1 - Mediterranean Dead - Sea Hydroelectric Project - The proposed alignment

The basin concept was found to be far more advantageous than the submarine conduit and was therefore adopted as the preferred alternative.
It was this decision that has led the Administration of Shipping and Ports in the Ministry of Transport ,   to check in more detail the possibility to use the Mediterranean-Dead Sea stilling basin as a basis for a marina development in the area.
The location of the Katif area , in the south of Israel and very close to the Egyptian border makes it very valuable as a shelter for yachts that enter the Mediterranean Sea through the Suez Canal and sail to Turkey and Greece along the eastern Mediterranean coast. The sandy beaches, clear sea waters and desert-like atmosphere makes the area very attractive for tourism development.

Another use of a sheltered basin in the area is the possibility to develop a local fishing harbour . No such harbour exists up to the port of Ashdod and even there the fishing boats are a nuisance to the Ashdod port authorities.
It was decided to design a marina in that part of the country which will provide an answer to the development requirements of the local tourism and fishing activity. This marina will be based on the Mediterranean -Dead Sea stilling basin .

The site and the marina general layout and architecture

This particular coast development project , called "Qatif Coast " extends from an existing beach hotel (Dekelim Beach Hotel) in the south west to a large tourism site in the north east . The marina structure is one of the tourism components designed along the coast.
One of the most important values of the Juish community in the area is the observance of the Shabbat. During this day, no activity or work will take place in the marina area.

The special features of the Qatif zone is the use of high ground water for irrigation . This method produces a semi tropical vegetation pattern. The intensive use of land and the narrow beach width which is limited by a "kurkar" (sandstone) ridge, raised the necessity to create reclamation areas which become extremely important.

The design of the reclaimed land derives from the varying and sometimes contradicting functional requirement . The reclaimed area is divided by the water intake to the Mediterranean - Dead sea project, into two sections serving different purposes:
To the north of the intake point - a fishing area is created next to the wharfs and jetties designed for fishing vessels. An area of about 3 acres is secured for the infra-structure, maintenance back up facilities and fish market for the whole vicinity.

In the Southern section , tourist establishments (time sharing condominiums , beach and resort hotels) of maximum two storey buildings are planned. An area adjacent to the main breakwater in the spray affected belt , will be designed for dry docking and maintenance facilities for yachts. Because of the curving effect of the beach line, a magnificent view to the south and the north will add to the designed conglomerate of buildings , gardens and piazzas in the tourist area. The existing coast road will be widened into a panoramic beach avenue .

The total area of the marina is 40 acres from which about 15 acres are reclaimed land. The northern fishing quays can serve 75 fishing boats . The yachts quays can serve about 290 yachts of different sizes. The yachts will be moored with bows towards the approaching wave crests , diffracting from the main breakwater head. This will reduce their roll movements and improve the mooring conditions.

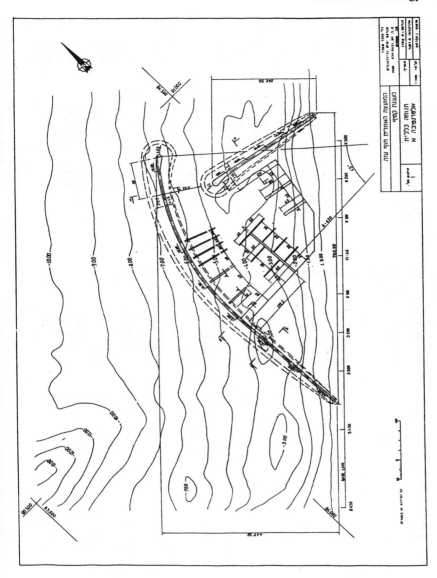

Fig.2 - General layout of Qatif Marina

The main breakwater head is located at 6.5 m water depth , 450 m from the shore base line. The entrance width is 60m at 5.5m water depth. The agitation conditions inside the marina are expected to be good since the entrance is directed towards north north east while the main wave direction during storms is west or west south west.

The central marina quay, 100m long, will serve for the Mediterranean - Dead Sea intake channel. The flow velocities inside the marina due to the pumping activity should not exceed 10 cm/sec at the nearest quays in order not to disturb the manouverability of the different vessels. The water withdrawal into the marina area will increase the deposition of sand inside it and more dredging will be neccessary in order to maintan proper depths.
Further development of the project is possible towards the north. A relatively small investment is needed in order to create a new protected basin at the northern side. This kind of development is a second option for the construction of the Mediterranean -Dead Sea hydroelectric project stilling basin (fig. 3).

Wave climatology of the Southeastern Mediterranean

A very thorough description of the Southeastern Mediterranean wave climate was presented by Victor Goldsmith and Stan Sofer (1983) based on almost all available data.
The wave climate of the southeastern Mediterranean consists of three wave seasons. The winter months (December-March) contain the highest significant wave heights. The winter wave climate is due to a non-uniform combination of storms and between -storm calm or low height waves. The maximum wave conditions during the winter occur when a large low pressure system lingers in the Aegean Sea.

The summer months (June through September) have an intermediate wave climate with most waves between 0.5 and 1.0 meters. The lowest monthly mean significant wave height occur in the months of May, October and the first half of November.

The limited duration of storm waves and the absence of a "swell -wave environment" common on ocean coasts are the main factors in the development of the southeastern Mediterranean wave climate.
The largest significant wave height observed in the 30 years of ship wave data was 10.5 m . In the 16 years of Ashdod observations it was 8.5m .

Rosen and Kit (1981) used the 16 years of Ashdod observations to evaluate statistically the long - term joint and marginal cumulative probabilities of the deep water significant heights and extreme wave heights. Their computation for return periods of extreme deep water significant wave heights are shown in the following table :

Significant wave height (m)	Return Period (years)
3.7	1
5.4	2
6.2	5
6.7	10
7.5	25
8.2	50
8.7	100

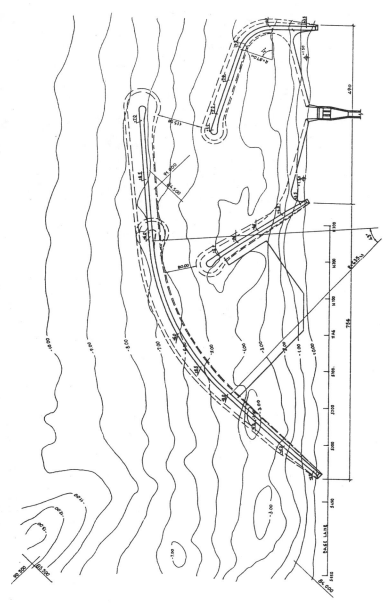

Fig.3 - Future Development of the Mediterranean - Dead Sea Stilling Basin

R

The wave direction is dominated by waves from azimuths 260-280 at nearly all times and at all wave heights. An exception occurs during the low wave seasons (May and October) when northwest waves dominate along the southern Israeli coast and in the offshore. With increasing wave heights the west - southwest direction (az=240-260) becomes more important .

The most significant aspects of the wave climate in the southeastern Mediterranean is the waves generated by storm front passage in the winter months. The passage of such storm is described in Goldsmith et al. (1982). The significant wave height increased from 1.5 to 5.2 m in just 12 h , and decreased from 3.3 to 1.7 m in 12 h . Similarly , the significant wave period increased from 8.1 to 11.1 sec and decreased from 12.5 to 8.1 sec also over 12-h time intervals. This spectral example illustrates the main aspects of storm waves in the southeastern Mediterranean : a rapid change from low wave heights and short periods to much larger values followed by a rapid decrease , a peak wave height which precedes the peak spectral period by 12 to 24h or occurs concomitantly , a wide band of wave spectra with the presence of much wave energy in the shorter period waves.

Tides and tidal currents

Due to the combined action of tide and winds , the water level may fluctuate between +0.6m and -0.3m , i.e. , a maximum amplitude of 0.9m , to which the tide contributes about 0.5m. The low amplitude of the tides result in slow rates of water level variations , usually less than 6cm/hr. The tidal currents are slow too , usually less than 6 cm/sec.

Bathymetry

During October and November 1981 , the seabed was mapped in the area . In the depth range of 5m to 14m (between 250 and 1,250m offshore) the depth contours are almost parallel , except for an irregularity 500m offshore , where a depression a few hundred meters wide extends in a west - northwest direction.

Much attention should be paid to these kinds of depressions in the process of a shallow water breakwater design. In such breakwaters the design wave height that reach the breakwater before breaking is limited by depth. The armour layer stone size is designed accordingly. When the water depth in front of the breakwater increases because of sedimentological proccesses such as the movement or deepening of a close by depression , sever damage might be caused to the structure.

Breakwaters design

The common practice for Marina structures along the Israeli coast is to use rubble mound breakwaters. This kind of breakwater were found stable with low amount of damage during the years. There are only two exceptions found in Israel , one is the use of tetrapods in the Ashdod Port breakwater and the other is the use of dollosse in the construction of Herzelia Marina which is just being built. The experience with tetrapods was, that a large amount of damage, due to broken tetrapods legs, was caused.

The present design of the marina breakwaters is based on a rubble mound type breakwater which is presented in fig. 4. The inner and outer slopes of the main breakwater are 1:1.5 and 1:2.5 respectively. The crest elevation at the head section is at +6.0m which ensure small amount of overtopping during winter storms. The armour layer stone size is 9 to 12 tons at the sea side and 2 to 5 tons at the marina side. It is possible to achieve a very low percentage of damage during overtopping at the marina side slope by an arrangement of the stones to create a relatively " smooth" slope such that the energy to pull out a single element is high. Another important factor of the low overtopping conditions is the small amount of sand which will settle in the marina basin during storms.

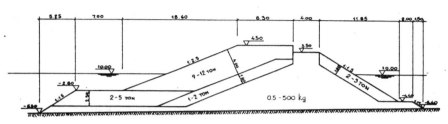

Fig. 4 - Qatif Marina Main Breakwater - Trunk Section

Coastal changes expected due to the marina construction

Large amount of sediment transport exists in the area , mainly to the north, from Sinai beaches and the Nile Delta along the eastern Mediterranean and up to the sand trap in Haifa bay. The marina basin which will penetrate into the sea to a distance of several hundred meters , may stop the longshore sand transport , thereby causing erosion to the northern beaches and coastal cliffs.

Previous experience with other coastal structures in the area (Ashkelon power station coolong basin) has shown that during the first period after construction the northern coast undergoes significant erosion during winter storms and exposed rocky beaches reveal but this erosion disappears after a while and the coastline comes back to its initial position.

Therefore it was decided not to plan any detached breakwaters or sea walls along the northern coast in order to protect the beach. It appears that after a year or two , a sand bypass is created such that the coastal structure do not block the longshore sand transport for long. Measurements and theories show that large sand quantities move under the wave action along the shore at greater water depths than 5 or 6 meters and thus the sediment transport to the north is enabled.

Laboratory tests have shown that northern and southern groins perpendicular to the shore reduce the amount of sediments which enter the basin (but do not serve as good beach protectors). Therefore it was decided to design such groins which will divert the strong longshore currents from the marina entrance.

Acknowledgement

We would like to thank the Hof Gaza Regional Council and the Ministry of Transport / The Administration of Shipping and Ports who supported this work and took part in the development process.

References

1. Victor Goldsmith and Stan Sofer , Wave Climatology of the Southeastern Mediterranean , Israel Journal of Earth-Sciences Vol. 32 1983 pp. 1-51

2. Abraham Golik and Victor Goldsmith , Coastal Changes in the Katif Gaza and Ashkelon Areas , Report no. H-9/85 , Israel Oceanographic & Limnological Research ltd. 1985.

3. Tahal Consulting Engineers ltd. , Mediterranean - Dead Sea Hydroelectric Project , Outline Design file No. 1 - Sea Intake , 1983.

4. Rosen D. and E. Kit. 1981 . Evaluation of the wave characteristics at the Mediterranean coast of Israel . Isr. J. Earth-Sci. 30: 120-134.

5. Goldsmith , V., D. Bowman and K. Kiley . 1982   .Sequential stage development of crescentic bars: southeastern Mediterranean J. Sediment . Petrol . 52 : 233-249.

Lessons from Hurricane Hugo - Marina Planning, Design and Operations in Hurricane Zones

J.G. Taylor

P.E., Inc., P.O. Box 1082, Mt. Pleasant, South Carolina 29464 U.S.A.

ABSTRACT

After causing substantial damage in Puerto Rico and the Virgin Islands, Hurricane Hugo, a Category IV (Saffir/Simpson Hurricane Scale) hurricane slammed into the coast of South Carolina at Charleston on Friday morning, September 22, 1989 - one hour before high tide. The storm surge created by Hugo peaked at an elevation of about 20 feet above sea level at Bulls Bay, approximately 35 miles northeast of Charleston. A storm surge elevation of 11 - 12 feet was experienced at North Myrtle Beach - over 100 miles northeast from the path of the eye of Hugo.

Hurricanes are not new to South Carolina, but many new and unseasoned boaters have recently located in South Carolina. During that same period, many new marinas, operated by individuals not experienced in hurricanes, were constructed. Immediately after the storm (within 72 hours) the author, a professional engineer (P.E.) who lives in the area hit by Hugo, in conjunction with the International Marina Institute (IMI) began an intensive study of over 20 marina facilities[1] to evaluate operational and design performance of marinas and coastal structures. The author has continued the investigations during a two year rebuilding program.

These studies dispel some of the myths about hurricanes and assumptions now being used in marina planning and design. New criteria for locating, evaluating, designing and operating marinas is needed. This paper presents new considerations and approaches to marina planning, design and operations in hurricane/typhoon zones.

INTRODUCTION

For the eastern coast of the United States, the normal hurricane season extends from June through November. This period also coincides with the peak tourism season along the coast. Charleston, a 300 year old city in South Carolina (USA), has a long history of recorded hurricane tragedies, however, many of the citizens (and boaters) have recently (past

20 to 30 years) moved to the coastal area and are not fully knowledgeable about hurricane dangers and how to react when a storm is coming. The last hurricane to cause substantial damage to South Carolina was Hurricane David in 1979. The last major hurricane to hit South Carolina was Hurricane Hazel in 1954.

The hurricane zone hazard for buildings in South Carolina has been confronted in the building industry with the institution of building codes, insurance requirements, and more recently a Beachfront Management Act that is administered by the South Carolina Coastal Council. However, even in the building industry, there are problems as stated in "The Risk of Hurricane Wind Damage to Buildings in South Carolina", by P.R. Sparks, undated[2]. "The problem appears not to be a failure to appreciate the severity of the storm, but our failure to properly design for it." Based on the damage to marina and coastal facilities from Hurricane Hugo, the marina consulting industry in South Carolina may not even have recognized the severity of hurricanes, much less have properly designed for them.

Before addressing the planning, design and operations aspects of marinas and the effects of hurricanes, we should dispel some of the myths and sometimes oversimplified assumptions regarding hurricanes, marinas and boaters. Following are some conclusions from Hurricane Hugo:

1.   There are no "Hurricane Holes" - quit looking for them. There are locations where it is more desirable to be during a storm, but nothing guarantees that the storm or an attendant tornado will or will not hit that particular location. Damages to boats and boating facilities can come from a complex set of circumstances - wind, current, tidal surge, heavy rains, wind driven debris, floating debris and other boats. Usually a single location can mitigate for only several of these and no one knows which of these will control.

2.   Unless the marina (1) has substantial haul-out facilities and (2) begins boat haul-out activities well in advance of the storm, complete evacuation of the boats in the marina is not possible. Boat owners are often not available for evacuation. Marina management is often indecisive about evacuation orders, timing, and commitment of funds. Partial evacuation of boats may be possible and could be a design alternative. No evacuation of boats could also be a design alternative. Either of these design alternatives requires a marina operational commitment during the design stage of the facility to be effective when the storm arrives.

3.   Staying aboard the boat during the hurricane is the least desirable alternative. Even experienced fishermen and yachtsmen can lose their boats. In Hugo, some lost their boats and their lives .

4.   Hurricanes are unpredictable. While recent advances in science have increased the predictability of the path of the storm and flooding, the intensity (wind speeds, advancing speed, pressure, etc.) are still erratic. Hugo jumped from a Category II (96 - 110 MPH) to a Category IV (131 - 155 MPH) hurricane. It is commonly believed that hurricanes

lose their power after coming ashore. Hugo did not lose much of its power and delivered hurricane force winds to Charlotte, North Carolina which is 200 miles inland. Hugo was a "one in fifty year storm" in Charleston and a "one in two hundred and fifty year storm" in Charlotte. Even with the increased reliability of predictions of hurricane movement, the maximum probability of storm arrival are:

Forecast Period Prior to Arrival	Maximum Probability
72 Hours	10 %
48 Hours	13 % - 18 %
36 Hours	20 % - 25 %
24 Hours	35 % - 45 %
12 Hours	50 % - 70 %

To reduce hurricane considerations to quantifiable planning and design parameters, the historical records should be analyzed to evaluate the following:

Study Period Years (Annual Probability)	Storm Severity In Marina Location Winds	Storm Surge
1 (100 %)	?	?
5 (20 %)	?	?
10 (10 %)	?	?
25 (04 %)	?	?
50 (02 %)	?	?
100 (01 %)	?	?

Next, take the projected design life of the marina (say 25 years) and multiply by a suitable factor of safety (say 2) to obtain a reasonable approximation of the storm event to be planned and designed for (25 years x 2 = 50 year storm event). Then, check to see the last occurrence of the 50 year storm. If the occurrence of the last 50 year storm event exceeds the design life of the marina 25 years (say 40 years) then check against the next level (say 100 years). Once a comfortable level of storm occurrence versus expected marina life is achieved, then the planning process can begin.

Planning the Marina Project in Hurricane Zones

Since marinas are market specific, use specific, environment specific and site specific, the choice of marina sites is complex. The demands on marinas for performance of all of these factors means that there are few excellent potential marina sites remaining. The consideration, given the following hurricane planning factors, is dependent on the results from previous discussion of hurricane occurrence and severity combined with the owners intended use of the marina. A secondary consideration is the effect of a hurricane disaster on vital public facilities (navigation, transportation and recovery, lifelines such as electricity, medical assistance and rescue efforts).

Marina Planning Considerations In Hurricane Zones

1. Site Selection - When protection from the elements was the prime concern, sheltered sites were the only acceptable marina locations. The multiple use marina now demands that marinas be located near urban areas, sailing lanes, excellent fishing spots, and/or other factors. At a minimum, marina sites should have as much sheltering from hills, mountains, or adjacent buildings as possible. Even more important is the free access and egress to the marina at all times and during all tide cycles. Evacuation of some boats prior to Hurricane Hugo was delayed because the access waterway was blocked by bridges that could not be raised because of the simultaneous evacuation of land facilities via automobiles over the bridges. Another prominent consideration in choosing a marina site is the potential for conflicts with adjacent properties and uses. Recreational boating must yield to more critically needed facilities during emergencies. Charleston has a major naval and defense facility. Evacuation and security of sensitive and/or vital government and defense facilities will take precedence over evacuation of a private marina. A final site consideration is the sensitivity of the marina facility to hazards created by a hurricane on the site itself. For example, boats anchored in a marina located in the mouth of a river near mountains could be safe from the winds and tidal surge of the hurricane, but could be in serious danger from swollen rivers or by land slides created by the hurricane on the region.

2. Boat Evacuation Potential - If boat evacuation via land (trailers) is to be an option, there should be adequate roads and/or storage space at the marina site. Conflicts did occur as boat ramps became crowed with boaters hauling their boats before Hugo hit. Very few marinas hauled the boats and kept them at the marinas, however, many boaters did navigate up the rivers and streams prior to the storm's arrival.

3. Boating Climate - Many areas have boaters that are very independent and expect little service, others - especially in areas that have recently been urbanized - have boaters who expect a high level of service. It's important to develop a marina user profile so the marina can plan and design to service the customers that will actually be using the marina. During hurricane crisis, many boaters really do not know how to prepare or respond. The marina must provide facilities and management to accommodate them. Sailboats, because they move slower and have little ability to move themselves in high winds, can pose special problems during a hurricane. Sailboats need to be located in sheltered marinas when possible.

4. Drystack Potential - Although there were several serious failures of drystack facilities in the path of Hugo (see Figure 1), there were some very good successes (see Figure 2). Drystack buildings probably offer one of the best protections against hurricanes for boats up to forty feet in length. These buildings can easily be designed for hurricane winds if they are located above the storm surge line. Drystack facilities offer an excellent opportunity to protect a large number of boats in a relatively small area.

Figure 1 - Two dry stacks were completely destroyed in Hurricane Hugo. (Photo Courtesy of Jon Guerry Taylor, P.E., Inc.)

Figure 2 - Some dry stack facilities sustained damage to the external structures with no damage to the racks or boats. (Photo courtesy of Jon Guerry Taylor, P.E., Inc.)

Marina Design Considerations in Hurricane Zones

As shown in Figure 3, the three primary considerations for marinas in hurricane zones: 1) marina operations during the hurricane, 2) the marina facility design and construction and 3) marina hurricane insurance.

Total Marina Hurricane Insurance Protection	=	Marina Hurricane Operations (Preparation, Evacuation and Recovery)	+	Marina Facility Design And Construction	+	Marina Hurricane Insurance

Figure 3
THE MARINA HURRICANE PROTECTION EQUATION

The first step in marina design should be the development of the marina management policy and attitude toward the hurricane threat (evacuation of boats, closure of the marina, disposition of the staff, responsibilities of management and the staff, etc.). Once this is established, the marina engineer can begin to quantify design parameters (boat loads, wind loads, wave forces, storm surge elevations, etc.). Prior to completion of design, the marina engineer should inform the owner of the engineered design limits and the hurricane operations' assumptions (wind, wave, etc.), so he can secure necessary insurance to reduce or cover hurricane risk. Following is a brief discussion of the three design considerations:

Designing Marinas in Hurricane Zones

Before initiating analysis of various facets of the marina design, the importance of the hurricane design for the marina project should be determined. For example, if the marina will contain berths for very expensive boats and there is little likelihood that the boats can be evacuated, then the hurricane design considerations will be more important. Another example, if the marina will contain smaller boats that can easily be evacuated via boat ramps or boat lifts with easy access, via good roads, then assuming the marina has a good evacuation plan, the hurricane design considerations will be less important. This is mostly a subjective evaluation process, however it is probably the last chance the marina designer has to combine the site specific and market specific factors into an overall design approach.

1. Marina Hurricane Operations Plan.

   a. The owner/operator must set policy and procedures that are realistic about the marina's capability for evacuation of boats. During the time prior to arrival of Hurricane Hugo, some marinas took on additional boats or hauled boats the day the storm hit. The marina evacuation plans must be especially mindful of potential conflicts with landside hurricane evacuation occurring on roads, bridges and boat ramps.

b. If the entire marina cannot be evacuated, the owner/operator and designer should consider designing a portion of the marina to hold the boats that remain for the storm.

c. The marina designer must be realistic about the type of boats and boaters that will be in the marina. If the boaters are absentee owners, it's difficult to get them to come into the marina in a timely manner, so the staff will probably have to handle the hurricane preparation duties for that boat. If the boats will be sailboats, they will be more difficult to evacuate and thus will require a longer notification for evacuation.

d. The marina designer must make the design criteria (size of boats, storm characteristics, etc.) known to the marina owner/operator so he can develop a companion Hurricane Action Plan to compliment the design. For example -if the marina is designed for a 75 mph hurricane (Category I) and there is an imminent danger of a 115 mph hurricane (Category III) coming ashore, he would need to consider evacuating the marina instead of leaving the boats inside.

In summary, the more involved the marina owner/operator is in the hurricane design process, the better able he will be to operate the marina consistent with the design assumptions.

2. Marina Facility Design in Hurricane Zones.

Final design of a marina is complex, especially if hurricanes are the primary design criteria. Wind is normally considered the primary design factor for hurricanes, however, when Hugo came ashore at high tide, far more damage to marina facilities was done by storm surge. Design guidelines and codes can be helpful but they cannot cover all circumstances that will occur. Usually codes are minimums and they do not relieve the marina designer of his responsibility to - as accurately as possible - estimate the proper design condition. Following are some of the more important design considerations that the marina designer should evaluate in the design of marina facilities:

a. If possible, it's desirable to have design personnel and marina management personnel who have actual experience in hurricane situations.

b. Be realistic in the application of the combination of forces (winds, currents, tides, water levels, etc.) that are likely to occur during the hurricane.

c. Design the marinas for failure of the least critical items first. On fixed piers, decking and handrails should fail before stringers, stringers should fail before piles and pile caps. On floating docks, the piles or anchoring devices should fail last. Pile failures were the major cause of floating dock failure in Hurricane Hugo. Many piles were too short and the docks just floated over the top when the storm surge arrived. (see Figure 4)

d. Make ancillary structures and facilities such as gangways, power pedestals, dock boxes, signs and utilities removable.

e. Consider new materials for marine construction very carefully. Do not rely completely on manufacturers' literature for technical information. Investigate the material and the manufacturers' performance in previous hurricanes. Manufacturers' service and repairability of docks and structures after a hurricane should be a design consideration.

f. The design firm should keep a copy of marina design documents, permits, shop drawings, manufacturers' literature and any other relevant documents that may be useful in the repair or replacement of damaged marina facilities.

Figure 4 - Storm surge and uplift forces caused fixed pier damage and damage to boats when the docks floated over the piles.
(Photo courtesy of Jon Guerry Taylor, P.E., Inc.)

3. Marina Insurance in Hurricane Zones.

Hurricane Hugo found some marina facilities with no insurance, others had insurance on the facilities and still others maintained insurance on the facilities and business interruption insurance. As previously discussed, whatever is not covered by design and supported by marina operation (Hurricane Action Plan) should be covered by insurance or it will be at risk. The insurance evaluation is clouded by the types of insurances that are available and the type of risk that they cover. The total hurricane insurance evaluation consists of the following: 1) risk of damage to boats, 2) risk of damage to the marina facility and 3) risk of damage to the marina business. Following is a brief overview of these insurance options:

A. Hurricane damage to boats: This insurance should be paid for by the boater. If the boater is to leave the marina, then he should be the beneficiary of damage reparation, however, if the boat is to remain in the marina, then the marina operator should receive some consideration for damage reparation.

B. Hurricane damage to the marina: The best insurance for the marina facility is "all risk" insurance. This insurance covers damage by wind, water (flooding) and other factors such as boats sinking, etc. If the marina insurance covers only wind damage, then flood damage will not be covered. This is especially critical when the insurance adjusters come to the project immediately after the hurricane. Since the combination of forces during a hurricane are complex, it's difficult to ascertain whether the damage came from wind, flooding or other factors. It's also very difficult to design for only one hurricane feature (wind or flooding) without designing for all of them. In other words, if you insure and design for only one factor (wind), the other factor (flooding) will be at risk and damages from flooding will not be covered.

C. Hurricane damage to the business: The marina business suffers from the hurricane because it loses its revenue source (slip rentals). Many insurance companies offer business interruption insurance. This insurance covers loss of income during the hurricane and the recovery period (reconstruction). Business interruption insurance does require verification of income by historical financial record.

Marina Operation In Hurricane Zones

Panic and indecision by marina management and operators is one of the major problems of marinas located in hurricane zones. There is also panic by boaters, especially those who have never experienced a hurricane. In Hugo, even experienced boaters had problems as they navigated their boats up the rivers and streams in an attempt to escape the storm. Many boats were lost in this effort. When the hurricane is approaching, it's too late to think about design, it's time to act. The Hurricane Action Plan can be carried out best by experienced management operating with a written action plan. In the absence of "actual hurricane experience", it is effective to have management that is trained and practiced in the implementation of the Hurricane Action Plan. Following are some considerations of marina operation in hurricane zones:

Considerations for marina operations in hurricane zones:

1. The marina management should know the limits of the marina design. His actions, to a large extent, will be dictated by the marina Hurricane Action Plan combined with size and intensity of the approaching hurricane. The marina actions will be different if the approaching hurricane has anticipated 75 mph winds (Category 1) or 125 mph winds (Category 3).

2. The marina management should know the limits of hurricane insurance and act accordingly. If the insurance covers "all risk", then the marina managers' actions will be different than if the insurance covers only wind damage or flood damage. You cannot obtain additional insurance when a hurricane threat is imminent. The marina policy relative to boater insurance should be communicated to the boater on original occupancy of the boat slip.

3. The marina should have a written Hurricane Action Plan. It should reflect the marina policy for hurricane protection in the marina and give guidance to boaters and staff on activities for preparation, evacuation and recovery. The Hurricane Action Plan should be updated periodically. It should be communicated to the marina staff and the marina staff should receive training and practice in all phases. Boater responsibility should be clearly defined and communicated to the boaters on original occupancy of the slip with reminders at the beginning of each hurricane season. Following are some of the important functions of a marina Hurricane Action Plan.

   a. Phase I - Hurricane Preparation in the Marina

      1. A Hurricane Action Coordinator for the marina should be designated. This individual would be responsible for monitoring hurricane paths and coordinating the activities of the marina staff when a hurricane is threatening. The Hurricane Action Coordinator would also be responsible for training the staff and obtaining necessary supplies.

      2. Once the hurricane threat is determined to be real and when the probability of a hurricane strike exceeds 10%, usually 48 hours before ETA (Estimated time of Arrival), then the Hurricane Action Coordinator should assume responsibility for placing the plan into action. If the marina is to use a "before storm, 24-hour locked harbor policy", then boaters must remove their boats or prepare them for the hurricane within the next 24 hours. The marina staff should also take care of their personal (family) preparation during this period.

      3. Once the hurricane threat exceeds 20% - usually 36 hours before the hurricane's ETA, the tie down and removal of anything that can be secured should begin. A final check of supplies and

preparations for possible evacuation and a recheck of communications channels to the outside must be done.

4. When the hurricane threat exceeds 35%, usually 24 hours before ETA, boat activities should cease, tie downs should be completed, utilities checked and secured, fuel tanks secured and important papers such as insurance, permits, financial records, deeds, marina plans, and leases, computer records, and any other important documents should be prepared for evacuation. Serious evaluation of the evacuate or ride-it-out decision should be made. This is not the time to take on new boats or to haul boats. It's important for marina management to realize that the goal now is saving lives and minimizing damage. This is the period when supplies such as fuel, batteries, candles, plywood, tape and cash become scarce. Contractors and suppliers should be contacted in the event that they will be needed for repairs.

5. When the probability for a hurricane strike exceeds 50%, usually 12 hours before ETA, the final decision to stay or evacuate must be made. Within four to six hours, the winds will start to increase and it will be too late to evacuate. If the marina is to be occupied during the storm, periodic checks of the docks, boats and other critical life lines (electrical, emergency generators, water, phones, radios) should be made. All people should leave (evacuate) the marina except the staff designated to stay. No one should be allowed to stay on the boats.

6. If the decision at the 24 hour preparatory level (Paragraph 4 above) is to prepare for evacuation, the Hurricane Action Plan at that time enters a new phase - evacuation of the marina.

b. Phase 2 - Evacuation of the Marina

1. If the storm is a very intense one and will test the marina's design, the decision to evacuate should be made as soon as possible - no later than 24 hours before the hurricane's ETA (see Paragraph 4 and 6 above). The actual evacuation should begin no later than 12 hours before the hurricane's ETA and should be completed at least 8 hours before the storm's ETA. This time frame is very site specific and highly dependent on land evacuation routes. Vehicles should be fueled and checked for operation so they can be dependably moved. Important papers and equipment (computers, etc.) should be prepared and loaded.

2. Since damages will probably occur, it's important to have a list of reliable contractors and suppliers for repairs. Once the decision to evacuate is made, the future planning starts to consider needs for recovery. Emergency generators, flashlights, cash, candles, radios, tools, film, plywood, lumber, rolls of plastic, nails, tape, portable stoves with fuel, first aid kits, blankets, a portable tape recorder, etc. will all be needed. The evacuation plan should contain a check list of these and other critical items.

3. The decision to prepare for evacuation must be made in a timely manner (at least 24 hours before the hurricane's ETA). (see Paragraph 4 and 6 above) The actual decision to evacuate can be revised if the storm changes direction or loses intensity, but the decision to prepare for evacuation cannot be delayed. When the evacuation order is given, you must leave whether you are ready or not. In many cases, the decision to evacuate may not be yours, some governments can order all facilities to be evacuated.

4. It's critical to establish an off-site command/communications center that will allow marina management to communicate with the staff to coordinate the return and recovery operations. All staff should know the location of other staff members during.

5. A last minute check of the marina prior to evacuation should be conducted. If in daylight, a video or series of pictures will possibly be useful in making insurance claims or defending future legal claims.

c. Phase 3 - Return and Recovery from the Hurricane

1. The anxiety during a hurricane strike is great. Even if families are safe, the concern about damages, losses, insurance coverage, etc., are there. Reviewing the return and recovery plan during this period helps to pass the time and prepare for the next steps.

2. The all clear for return will usually be given over the radio. Try to take as many supplies from the outside as possible. Food, ice, fuel, film, building supplies and tools will be very difficult to obtain, so take as many of these supplies as possible back with you. Also, take a good supply of cash for miscellaneous expenses. Every attempt should be made to get back to the marina before the boaters. Have an outside contact that can be reached when you return to the marina. The Hurricane Action Plan Coordinator should contact all staff members to coordinate the return and recovery.

3. On return, damage assessment is the first order of business. Access to the marina for the first twenty-four hours after return should, if possible, be restricted to marina staff, designated suppliers, contractors and service personnel only. If possible, set up a command post to control recovery and salvage operations.

4. Assign someone to make damage assessment of boats and the facilities. Take pictures or videos of the damage. This same individual can coordinate reports to boaters.

5. Security can be a problem as returning boaters and looters come into the environs. Police are usually very busy with public problems. The vicinity may be under martial law and it may be difficult to return. It's best to provide marina staff with identification papers, pictures and authorization to enter. A marina staff member who can recognize boaters can be very helpful. Do not

let independent contractors or individuals around the docks and/or boats. The newspapers and television reporters should not be allowed into damaged marina facilities. All salvage operations should be coordinated through marina management personnel.

6. Marina management personnel should escort and remain with insurance adjusters when they arrive at the marina. This is very serious business and deserves top management's attention. An assessment of the damage should be in hand, as some of the insurance adjusters know little about marinas. Be cautious about quick settlements as initial damage assessments are often not complete or absolutely accurate.

7. Marina management should accompany all government and/or environmental agency personnel on their evaluation of the marina. While their initial intentions may appear to be harmless and even helpful, their reports are usually accessible to the public and can be useful to insurance adjusters or potential complainants about the marina operations. Ask for a copy of any pictures that they make and a copy of the report that they make. If you do not agree with their assessment, write a letter and ask that it be reconsidered.

SUMMARY

Hurricane Hugo proved that it is time for renewed direction in planning, design and operation in hurricane zones of marinas. No longer are marinas a design feature within themselves but now must be viewed as portals to and from the water, recreation amenities to both private developments and the public, business enterprises with deep roots in the community and the local economy, and a stimulus to tourism, investment and employment generation. These positive features must be balanced with the realities of marina operations that can generate traffic (land and water), environmental impact and if not run properly, a nuisance to the neighborhood and the marina industry. Marina planning and design must now include marina operations in setting the criteria for marinas.

Growing international use of marinas indicates that marinas planning, design and operation should now be viewed from a new perspective with concerns for the boater or user. The boater needs to know the level of service he can expect for his boat, the level of protection he can expect during a hurricane and what other services and facilities are available. If he is an investor (condo marina, etc.), he needs to know the above plus the expected life of the marina. If a proposed system to consider boaters' concerns were instituted, it would make marina planners, designers and developers look past the construction contract to the operators, the end users, the investors, the market, the adjacent property owners and the community. If this is done, environmental and building permitting, acquisition of investment capital, acquisition of insurance and zoning changes will be easier and more amenable.

It could be suggested that a user rating of marinas be developed.

The planning design and operation could then be dictated by that rating. Following are the major categories that could be used:

Class I Marina   A marina for local traffic operated to support boating and other water related activity in the local area.

Class II Marina   A marina for regional and interstate traffic operated to support boating and water related activity in the local area and region.

Class III Marina   A marina capable of providing service for international traffic; operated to support boating and water related activities for the local area, the region interstate traffic and international traffic.

There could be sub categories (a, b, c, etc.) for each classification to denote different features (floating slips, fixed piers, 24-hour operation, electrical service, etc.). Once the decision on which class of marina is to be developed, the planning and design could be addressed positively to support that level of service, project life, etc. This system could be adopted internationally. Perhaps, then the boat manufacturers could agree to institute international standards for electrical service, pump outs, etc. for all new boats.

REFERENCES

1.   Taylor, Jon Guerry, P.E., P.L.S. and Dodson, Paul. Lessons from Hurricane Hugo - Marina Design. Seagrant Research Conference, Clearwater Florida, 1989.

2.   Sparks, P.R., The Risk of Hurricane Wind Damage to Buildings in South Carolina, 1990.

SECTION 10: BREAKWATERS

Approximation of Wave Forces on Arrays of Submerged Circular Breakwaters

A.G. Abul-azm

Irrigation and Hydraulics Department, Faculty of Engineering, Cairo University, Giza, Egypt

ABSTRACT

An approximate method is presented to estimate the linear hydrodynamic forces on arrays of stationary, truncated, circular breakwaters in water of arbitrary uniform depth. The breakwaters are considered to be completely submerged and resting on the sea bed. Interference effects are estimated through the modified plane wave approach, an asymptotic technique which is essentially a large spacing approximation. Numerical results are presented for a row of two to four breakwaters, which illustrate the influence of the various wave and structural parameters on the hydrodynamic loads on the breakwater array.

INTRODUCTION

In recent years the hydrodynamic interaction between the various members of large multi-component offshore and marine structures subjected to waves is of considerable practical importance. One important application is the planning and design of breakwater arrays used in the protection of Marinas, pleasure and fishing ports. For such a design, a decision has to be made on the size and spacing between each component of the breakwater used for the protection and also an estimation of the hydrodynamic wave forces experienced by each unit.

Waves incident on large obstacles are diffracted and scattered, as a result the wave field is modified substantially. Also the presence of neighboring bodies may significantly alter the fluid field and lead to hydrodynamic loads on individual components which differ considerably from the loading they each would experience in isolation.

Several methods have been developed to study the multiple scattering and wave forces on cylinder arrays. Research was mainly directed towards bottom-mounted, surface-piercing and floating, semi-immersed stationary cylindrical structures in waves. The interference between submerged, truncated circular breakwaters subjected to wave action is less well understood.

A general overview of several methods was given by Kagemoto and Yue [4]. They also presented an 'Exact' method for calculating the hydrodynamic effects on each member of a structural array using the diffraction characteristics of individual members only, and by including the evanescent waves in their formulation. Iskandrani and Liu [3] presented an alternative semi-analytical solution, but less efficient computationally, to examine the wave diffraction by fixed solid bodies. However, a computationally efficient method that has been successfully used lately termed the modified plane-wave approximation. This method has been developed by McIver and Evans [6] to study the hydrodynamic interactions between arrays of bottom-mounted, surface-piercing cylinders. They showed good agreement with the potentially more accurate method of Spring and Monkmeyer [12].

The modified plane wave method is essentially a large spacing approximation. It involves replacing the cylindrical diverging wave scattered (and/or radiated) from one body in the vicinity of another by a plane wave together with a nonplanar first correction term, and neglecting the local evanescent wave components of the wave field when calculating interference effects. This method was applied by McIver [5] to study the scattering and radiation of waves by two floating circular docks, also by Williams and Demirbilek [7], and Williams and Abul-Azm [8] for the scattering and radiation of waves by arrays of multi floating cylinders. Results for the diffraction loadings, added mass and radiation damping coefficients of each dock in an array showed good agreement with those obtained by Matsui and Tamaki [11] using a source distribution technique.

In the present paper, the modified plane-wave technique is utilized to calculate the hydrodynamic forces due to the interaction between different members of an array of submerged circular breakwaters of equal radius. The breakwater array is located in a single row parallel to the shore line, and the incident wave angle is arbitrary, the fluid is assumed to be inviscid, incompressible undergoing small-amplitude irrotational motion. Numerical results are presented for a row of two, three and four breakwater units, which illustrate the influence of the various wave and structural parameters on the hydrodynamic loadings.

THEORETICAL FORMULATION

The geometry of the problem is shown in figure (1). A group of N circular submerged breakwaters of equal radius a are situated in water of uniform depth d. The height of the breakwater is denoted by h. Cartesian coordinates will be employed with the x and y axes in the horizontal plane with the z-axis pointing vertically upwards from an origin on the sea-bed. The center of unit j, at point (x_j, y_j), j = 1,..., N is taken as the origin for a local polar coordinate system where θ_j is measured anticlockwise from the positive x-axis. The center of the breakwater k has polar coordinate (R_{jk}, α_{jk}) relative to that of breakwater j and j, k = 1,...N. $\alpha_{jk} = 0$ for $x_k-x_j > 0$ and equals to π for $x_k-x_j < 0$. The theoretical formulation of the wave-structure interaction problem is based on the assumption of a homogeneous, ideal, incompressible, inviscid fluid undergoing irrotational motion. Subject to these restrictions and assumptions the motion of fluid may be described in terms of a velocity potential $\phi(r_j,\theta_j,z,t)$ where t denotes time. The fluid velocity vector is then given by $\underline{q} = \nabla\phi$, then the following boundary-value problem may be obtained, see, e.g. Sarpkaya and Isaacson [1].

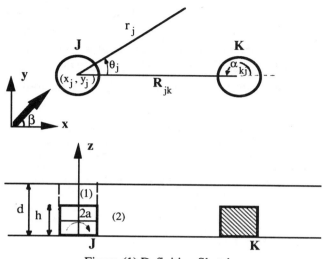

Figure (1) Definition Sketch

$$\nabla^2\phi = 0, \qquad \text{in the region of flow} \tag{1a}$$

$$\frac{\partial\phi}{\partial z} = 0, \qquad \text{on } z = 0 \text{ for } a \le r_j \le \infty, \tag{1b}$$

$$\frac{\partial\phi}{\partial z} = 0, \qquad \text{on } z = h \text{ for } 0 \le r_j \le a, \tag{1c}$$

$$\frac{\partial\phi}{\partial r_j} = 0 \qquad \text{on } r_j = a \text{ for } 0 \le z \le h, \tag{1d}$$

$$\frac{\partial^2\phi}{\partial t^2} + g\frac{\partial\phi}{\partial z} = 0, \qquad \text{on } z = d \text{ for } r_j \ge 0, \tag{1e}$$

where g is the acceleration of gravity.

In subsequent sections the total velocity potential ϕ will be decomposed into incident and scattered components

$$\phi = \phi_I + \phi_s, \tag{2}$$

where ϕ_I is the potential due to a straight-crested wave propagating at an angle β to the x-direction as shown in figure (1). Referred to an origin at unit j, ϕ_s is the scattered potential which is the principal unknown of the problem. The incident potential on body j is given by

$$\phi_{Ij} = \frac{\omega H}{Re} \frac{\cosh k_o z}{2k_o \sinh k_o d} I_j \sum_{n=-\infty}^{\infty} i^{n+1} J_n(k_o r_j) e^{in(\theta_j - \beta)} e^{-i\omega t}, \tag{3}$$

where J_n denotes the Bessel function of the first kind of order n, Re denotes the real part of a complex expression, I_j is a phase factor defined by $I_j = \exp\{ik_o(x_j \cos\beta + y_j \sin\beta)\}$, the wavenumber k_o and angular frequency ω are related through the dispersion relation, $\omega^2 = gk_o \tanh k_o d$, and H is the wave height.

There is a further condition to be satisfied by ϕ_{sj}, the radiation condition as $r_j \rightarrow \infty$. For time dependence of the type $e^{-i\omega t}$, this takes the form

$$\lim_{r_j \to \infty} \sqrt{r_j} \left\{ \frac{\partial \Phi_{sj}}{\partial r_j} - ik_o \Phi_{sj} \right\} = 0, \tag{4}$$

in which $\phi_{sj} = Re[\Phi_{sj} e^{-i\omega t}]$. Equation (4) ensures that ϕ_{sj} behaves as an outgoing wave at large distances from the breakwater. The pressure P at any point in the fluid may be obtained by knowing the potential, through the linearized Bernoulli equation namely

$$P = -\rho \frac{\partial \phi}{\partial t}, \tag{5}$$

in which ρ is the fluid density.

SOLUTION FOR THE VELOCITY POTENTIALS

The eigenfunction expansion solution to the problem of wave diffraction by each cylinderical breakwater in an array presented in this section, is essentially that of Issacson [10] for an isolated submerged circular cylinder. The fluid domain for each unit j is divided into two regions, figure (1), an interior region (1) above the breakwater and an exterior region (2) external to the structure and extending to infinity in the horizontal plane. The potential in the outer region is decomposed into incident and scattered components and so

$$\phi_j = \phi_{Ij} + \phi_{sj1}, \qquad \text{for } r_j \geq a, 0 \leq z \leq d, \tag{6a}$$

$$\phi_j = \phi_{sj2}, \qquad \text{for } 0 \leq r_j \leq a, h \leq z \leq d. \tag{6b}$$

Then, the wave-structure interaction problem for the breakwater array will be solved by assuming that the spacing R_{jk} between the breakwaters is much larger than the incident wave length, i.e. $k_o R_{jk} \gg 1$. Under this assumption a diverging wave emanating from one cylinder is replaced at another by an equivalent plane wave, together with a non-planar correction term, so that the overall effect of such radiated waves can be estimated, (McIver and Evans, [6]). If the breakwater array is subjected to a train of regular waves incident from a direction β, and the incident velocity potential on cylinder j is defined as in equation (3), then a suitable scattered potential due to the diffraction of the incident wave from outside the cylinder group will have the following form

$$\phi_{Sj} = \frac{\omega H}{Re} \frac{1}{2k_o} I_j \sum_{n=-\infty}^{\infty} \Delta^{(n)}(r_j, z) e^{in(\theta_j - \beta)} e^{-i\omega t}, \tag{7}$$

where in the inner region (1)

$$\Delta_{s1}^{(n)}(r_j,z) = A_o^{(n)} \frac{J_n(\beta_o r_j)}{J_n(\beta_o a)} Q_o(z) + \sum_{p=1}^{\infty} A_p^{(n)} \frac{I_n(\beta_p r_j)}{I_n(\beta_p a)} Q_p(z), \tag{8}$$

this form satisfies equations (1a), (1c) and (1e), in which I_n denotes the modified Bessel function of the first kind of order n, β_o is defined by $\omega^2 = g\beta_o \tanh \beta_o(d-h)$, and β_p, $p = 1, 2,...$, are the positive real roots of $\omega^2 + g\beta_p \tan \beta_p (d-h) = 0$. The eigenfunctions $Q_p(z)$, $p = 0, 1, 2, ...$, are

$$Q_p(z) = \begin{cases} \sqrt{2} \cosh \beta_o(z-h) / [\ 1+ \sinh 2\beta_o(d-h)/2\beta_o(d-h)]^{1/2} & p = 0, \\\\ \sqrt{2} \cos \beta_p(z-h) / [\ 1+ \sin 2\beta_p(d-h)/2\beta_p(d-h)]^{1/2} & p \geq 1. \end{cases} \tag{9}$$

and in the exterior region (2)

$$\Delta_{s2}^{(n)}(r_j, z) = \left\{ -\varepsilon_n i^{n+1} \frac{\cosh k_o z}{\sinh k_o d} \frac{J_n'(k_o a)}{H_n(k_o a)} + B_o^{(n)} \frac{Z_o(z)}{H_n'(k_o a)} \right\} H_n(k_o r_j)$$

$$+ \sum_{q=1}^{\infty} B_q^{(n)} \frac{K_n(k_q r)}{K_n'(k_q a)} Z_q(z), \tag{10}$$

in which ε_n is Neumann's number, $\varepsilon_o = 1$, $\varepsilon_n = 2$ for $n \geq 1$, H_n is the Hankel function of the first kind and K_n is the modified Bessel function of the second kind, both of order n, and primes denote differentiation with respect to argument. In equation (10) the k_q, $q = 1, 2, 3, ...$ are the positive real roots of $\omega^2 + gk_q \tan k_q d = 0$ and the eigenfunctions $Z_q(z)$ are

$$Z_q(z) = \begin{cases} \sqrt{2} \cosh k_o z / [1 + \sinh 2k_o d/2k_o d]^{1/2} & q = 0, \\\\ \sqrt{2} \cos k_q z / [1 + \sinh 2k_q d/2k_q d]^{1/2} & q \geq 1. \end{cases} \tag{11}$$

To determine the local complex potential coefficients $A_p^{(n)}$ and $B_q^{(n)}$, n, q, $p = 0, 1,...$, for breakwater j, continuity of mass flux and pressure across the fluid interface between the two regions will be utilized. These matching conditions implies the following,

$$\phi_{sj1} = \phi_{Ij} + \phi_{sj2}, \tag{12a}$$

$$\frac{\partial \phi_{sj1}}{\partial r_j} = \frac{\partial \phi_{Ij}}{\partial r_j} + \frac{\partial \phi_{sj2}}{\partial r_j}, \tag{12b}$$

which are valid on $r_j = a$, $j = 1,..N$, for $h \leq z \leq d$. Applying the matching conditions (12) and utilizing equations (3), (7), (8) and (10), lead to the following equations linking the coefficients $A_p^{(n)}$ and $B_q^{(n)}$

$$A_p^{(n)} + \sum_{q=0}^{\infty} F_{pq}^{(n)} B_q^{(n)} = R_p^{(n)} \qquad n, p = 0, 1, 2,.., \tag{13a}$$

and

$$B_q^{(n)} = \sum_{p=0}^{\infty} G_{qp}^{(n)} A_p^{(n)} \qquad n, q = 0, 1, 2, .., \tag{13b}$$

Expressions for the coefficients $F_{pq}^{(n)}$, $G_{qp}^{(n)}$ and $R_p^{(n)}$ are found in Abul-Azm and Williams [13]. Equations (13a) and (13b) now constitute two infinite simultaneous matrix equations for the potential coefficients, these equations may be truncated after a finite number of terms and the coefficients obtained by standard matrix solving techniques.

By invoking the Bessel function addition theorem, Abramoitz and Stegun [9], the scattered potential in the exterior region (2), may be written in terms of the local polar coordinate system with origin at breakwater k, namely

$$\phi_{Sj} = \frac{\omega H}{Re} \frac{\cosh k_o z}{2k_o \sinh k_o d} I_j \sum_{n=-\infty}^{\infty} i^{n+1} D_n (-1)^{n\gamma_{jk}} e^{in\beta} \sum_{m=\infty}^{\infty} i^m J_m(k_o r_k)$$
$$H_{n+m}(k_o R_{jk}) (-1)^{m\gamma_{jk}} e^{-i(m\theta_k + \omega t)}, \tag{14}$$

in which $\gamma_{kj} = 0$ for $x_k - x_j > 0$ and $\gamma_{kj} = 1$ for $x_k - x_j < 0$, and

$$D_n = -\frac{J_n(k_o a)}{H_n'(k_o a)} - i^{1-n} B_o^{(n)} \frac{N_o^{-1/2}}{H_n'(k_o a)} \sinh k_o d. \tag{15}$$

Utilizing the asymptotic form of the Hankel functions, Watson [2], the scattered wave emanating from body j may be written approximately in the vicinity of breakwater k by

$$\phi_{Sj} = \frac{\omega H}{Re} \frac{\cosh k_o z}{2k_o \sinh k_o d} I_j \{ P_{kj}(\beta) \sum_{m=\infty}^{\infty} i^{m+1} J_m(k_o r_k) (-1)^{m\gamma_{jk}} e^{-im\theta_k}$$
$$+ Q_{kj}(r_k, \theta_k, \beta) \} e^{-i\omega t}, \tag{16}$$

in which

$$P_{kj}(\beta) = \sum_{n=0}^{\infty} \varepsilon_n i^n D_n H_n(k_o R_{jk}) (-1)^{n\gamma_{jk}} \cos n\beta, \tag{17}$$

is the plane wave factor, and

$$Q_{kj}(r_k, \theta_k, \beta) = \frac{i}{2k_o R_{jk}} \{ \sum_{m=0}^{\infty} \varepsilon_m i^{m+1} J_m(k_o r_k) (-1)^{m\gamma_{jk}}$$
$$[m^2 P_{kj}(\beta) \cos m\theta_k + 2m T_{kj}(\beta) \sin m\theta_k] \} \tag{18}$$

is a non-planar correction term, and

$$T_{kj}(\beta) = \sum_{n=1}^{\infty} - \varepsilon_n \, i^n \, D_n \, H_n(k_o R_{jk}) \, (-1)^{n\gamma_{jk}} \sin n\beta. \tag{19}$$

When contributing to a plane wave component the asymptotic form of Hankel function H_n may be used (Abramoitz and Stegun [9]), and taken to $O[(k_o R_{jk})^{-3/2}]$, however when appearing in the correction term it is necessary to take H_n to $O[(k_o R_{jk})^{-1/2}]$ only. Thus, the total incident potential for breakwater j may be written to $O[(k_o R_{jk})^{-3/2}]$ as

$$\phi_{Ij} \stackrel{=}{\overline{Re}} \frac{i\omega H}{2k_o} \frac{\cosh k_o z}{\sinh k_o d} \{ I_j \, e^{ik_o r_j \cos(\theta_j - \beta)} + \sum_{k=1, k\neq j}^{N} [E_{jk} \, e^{ik_o r_j \cos(\theta_j - \alpha_{kj})}$$
$$- i \, I_k Q_{jk}(r_j, \theta_j, \beta)] \}, \tag{20}$$

in which the first term corresponds to the potential of the incident wave from outside an array, referred to an origin at unit j; the second term is the plane wave approximation to the total wave field scattered by unit k evaluated in the vicinity of breakwater j, and the third term is the non-planar first correction to the wave field scattered by unit k and is essentially the correction to the scattering of the incident wave from outside the array. The complex amplitude E_{jk} appearing in equation (20) consists of the plane wave approximation to the scattering of the incident wave from outside the group by breakwater k, evaluated in the vicinity of breakwater j and the plane wave component of the back-scattered wave from unit k described above. Neglecting the scattering of correction terms, it can be shown that

$$E_{jk} = I_k \, P_{jk}(\beta) + \sum_{q=1, q\neq k}^{N} E_{kq} \, P_{jk}(\alpha_{qk}) + O[(k_o R_{jk})^{-2}] \quad , \, j,k=1..N, \, j\neq k. \tag{21}$$

Equation (21) constitutes $N(N-1)$ simultaneous equations for the complex plane-wave amplitudes E_{jk} which may be solved numerically by standard matrix inversion techniques. The total potential for waves incident on breakwater j may now be written in the following form

$$\phi_{Ij} \stackrel{=}{\overline{Re}} \frac{\omega H}{2k_o} \frac{\cosh k_o z}{\sinh k_o d} \sum_{n=0}^{\infty} \varepsilon_n \, i^{n+1} \, J_n(k_o r_j) \, f_n(\theta_j) \, e^{-i\omega t}, \tag{22}$$

in which

$$f_n(\theta_j) = I_j \cos n(\theta_j - \beta) + \sum_{k=1, k\neq j}^{N} (-1)^{n\gamma_{kj}} [E_{jk} + n^2 a_{jk} \cos n\theta_j + n b_{jk} \sin n\theta_j] \tag{23a}$$

and

$$a_{jk} = i I_k P_{jk}(\beta)/(2k_o R_{jk}), \tag{23b}$$

$$b_{jk} = i I_k T_{jk}(\beta)/(k_o R_{jk}). \tag{23c}$$

Utilizing the solution for the isolated breakwater problem with the incident potential adjusted according to equation (22), the total potential in the vicinity of each cylinder j , j = 1, 2,..., N may be written to $O[(k_o R_{jk})^{-3/2}]$ as

$$\phi_j \; \bar{\underset{Re}{}} \; \frac{\omega H}{2k_o} \sum_{n=0}^{\infty} \Delta_{s1}^{(n)}(r_j,z)\, f_n(\theta_j)\, e^{-i\omega t}, \tag{24a}$$

which is valid in the inner region (1), and

$$\phi_j \; \bar{\underset{Re}{}} \; \frac{\omega H}{2k_o} \sum_{n=0}^{\infty} \{\varepsilon_n\, i^{n+1} \frac{\cosh k_o z}{\sinh k_o d}\, J_n(k_o r_j) + \Delta_{s2}^{(n)}(r_j, z)\}\, f_n(\theta_j)\, e^{-i\omega t}, \tag{24b}$$

in the exterior region (2).

DIFFRACTION LOADINGS

Once the total velocity potential has been known for each breakwater in the array, the hydrodynamic loading components may now be computed by utilizing the hydrodynamic pressure (5). Although the first two terms in equation (24), i.e. n = 0, 1, will contribute to the loading components shown later, but still a larger truncation for the index n need to be considered, as for the computation of the plane-wave amplitude and the non-planar correction terms in equations (21) and (19), respectively. The total diffraction force in the x-direction on each breakwater j is given by

$$F_{xj} = -a\rho \int_0^h \int_0^{2\pi} \frac{\partial \phi_j}{\partial t}\, (a,\theta_j,z,t)\, \cos(\pi - \theta_j)\, d\theta_j\, dz \tag{25}$$

Substituting (24b) into (25) and carrying out the integration yields

$$F_{xj}/F' = \mathcal{F}\, \{I_j \cos\beta + \sum_{k=1, k\neq j}^{N} (-1)^{\gamma_{kj}}[E_{jk} + a_{jk}]\}. \tag{26}$$

where $\mathcal{F}$ is the horizontal force experienced by a body in isolation, an explicit expression may be found in Abul-Azm and Williams [13], $F = \rho \pi a^2 g H/2$, and E_{jk} is the plane-wave amplitudes as defined in (21), and a_{jk} is defined in (23b).

Similarly, the force components in the y and z-directions on unit j are

$$F_{yj} = a \int_0^{2\pi} \int_0^h [P]_{r_j=a} \sin(\pi + \theta_j)\, dz\, d\theta_j , \tag{27a}$$

$$F_{zj} = -\int_0^{2\pi} \int_0^a [P]_{z=h}\, r_j\, dr_j d\theta_j . \tag{27b}$$

Utilizing equation (5) for the hydrodynamic pressure, and substituting (24b) into (27a) for the hydrodynamic force in the y-direction, and (24a) into (27b) for the vertical force, and carrying out the integration yields

$$F_{yj}/F' = \mathcal{F}\, \{I_j \sin\beta + \sum_{k=1, k\neq j}^{N} (-1)^{\gamma_{kj}} b_{jk}\}, \tag{28a}$$

for the force in the y-direction, and the force in the z-direction becomes

$$F_{zj}/F^{'} = Z\{I_j + \sum_{k=1, k\neq j}^{N} E_{jk}\}.$$ (28b)

where Z is the vertical force experienced by a body in isolation, an explicit expression may be found in Abul-Azm and Williams [13].

The wave-induced pitching and rolling moments, M_y and M_x, are taken about $z = 0$, and is positive in the direction indicated in figure (1). The total rolling moment experienced by breakwater j may be written

$$M_{xj} = \int_0^{2\pi} \left\{ \int_0^a [P]_{z=h}\, r_j^2\, dr_j - \int_0^h [P]_{r_j=a}\, z\, dz \right\} \sin\theta_j\, d\theta_j.$$ (29)

Explicit expressions for the rolling moment may be written as

$$M_{xj}/aF^{'} = \mathcal{M}\{I_j \sin\beta + \sum_{k=1, k\neq j}^{N} (-1)^{\gamma_{kj}} b_{jk}\},$$ (30)

Similarly, an expression for the total pitching moment on breakwater j may be written as

$$M_{yj}/aF^{'} = \mathcal{M}\{I_j \cos\beta + \sum_{k=1, k\neq j}^{N} (-1)^{\gamma_{kj}}[E_{jk} + a_{jk}]\}.$$ (31)

in which $\mathcal{M}$ is the moment experienced by a body in isolation, an explicit expression may be found in Abul-Azm and Williams [13].

NUMERICAL RESULTS AND DISCUSSION

Numerical results will be presented for a number of parameters to illustrate the effect of hydrodynamic interactions on the wave loads. In all the examples considered a value of $d/a = 3.0$ was considered. In obtaining the results presented herein, the infinite summations in equations (13a) and (13b) for the potential coefficients were truncated after 20 terms for each mode $n = 0$ and 1. The infinite series in the non-planar correction terms in equations (17) and (19) were truncated after eleven terms. Increasing the number of terms did not change the computed results by more than 0.05%, indicating that the numerical convergence had essentially been achieved. A consistent notation is adopted throughout the figures, (_____) for isolated breakwater, (O) for breakwater 1, (x) for the breakwater 2, (+) for breakwater 3 and (Δ) for breakwater 4.

Figures (2.a) to (2.f) present the amplitude of the dimensionless horizontal, vertical forces and pitching moment about z=0 for a pair of breakwaters for different $k_0 a$ values. Results are presented for $R/a = 5.0$, $\beta = 0°$ and different h/d values. A value of $h/d = 1.0$ presents the limiting case of a bottom-mounted, surface-piercing, vertical cylinder, primarily for comparison. Significant increase in the horizontal force and moment on breakwater 1 as shown in figure (2.a) for $h/d=1.0$. The sheltered unit 2 experiences less forces and moments as shown in figure (2.d). As the breakwater height to depth ratio, h/d, decreases, forces and moment decrease, and the interference effects become negligible, particularly for the upstream breakwater 1, as shown in figures (2.a), (2.b) and (2.c). Unlike the surface-piercing case, figures (2.d), (2.e) and (2.f) show that the sheltered breakwater 2 experiences an increase in the hydrodynamic loadings over which it would experience in isolation. This increase is about 12% in the horizontal force, 15% in the vertical force and 11% in the moment between $k_o a = 0.5$ and 1.5 for h/d =0.75.

The effect of the spacing ratio R/a on the hydrodynamic horizontal force for the individual units is presented in figure (3). The values considered in this example are, $k_o a=1.5$, h/d=0.5 and $\beta=0°$. The horizontal forces is nondimensionalized by the corresponding force experienced by an isolated body. Force on the upstream breakwater 1 shows to oscillate about unity and decays as R/a increases, while force on the sheltered breakwater 2 decays monotonically with R/a.

Figures (4.a) and (4.b) show the effect of the incident wave angle β on the vertical and lateral forces, respectively. In this example, $k_o a=1.5$, h/d=0.5 and R/a =5. Figures (5) and Figures (6) show the results for a group of three and four breakwaters, respectively. Results are presented for R/a = 5.0, $\beta = 0°$ and h/d=0.5. Generally, interference effects has shown a little influence on the hydrodynamic loadings for $k_o a > 1.0$.

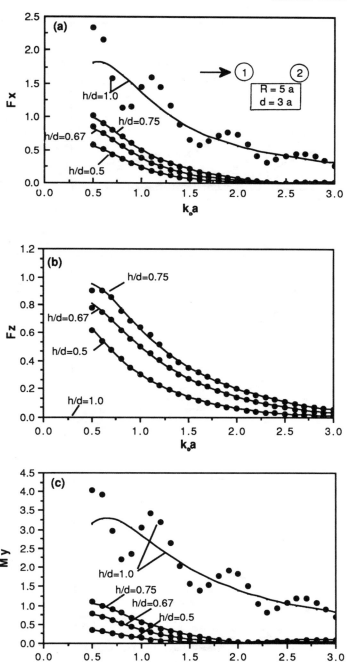

Figure (2) Nondimensional amplitude of (a) force in x-direction, (b) force in z-direction, (c) Pitching moment, on Breakwater 1, as a function of $k_o a$.

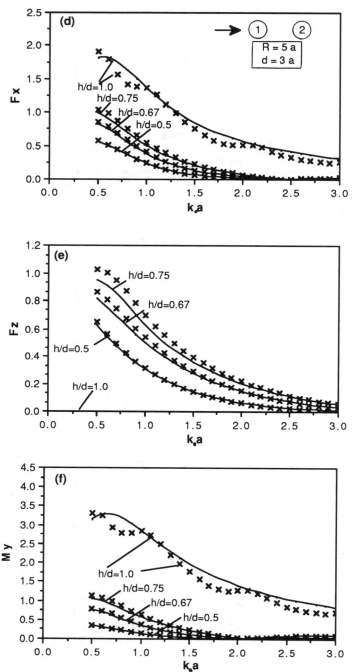

Figure (2) Continued, (d) force in x-direction, (e) force in z-direction and (f) pitching moment, on Breakwater 2. For legend see text.

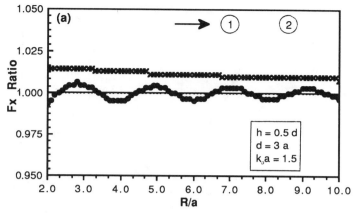

Figure (3) Effect of spacing ratio R/a on the force ratio in the x-direction, for two breakwaters array. For legend see text.

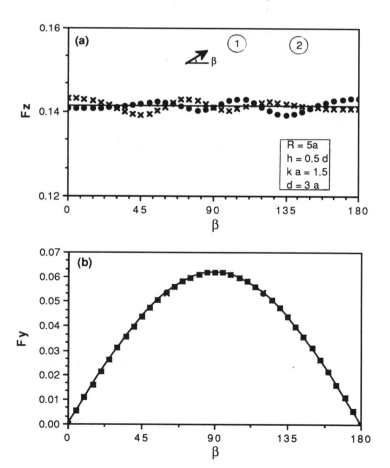

Figure (4) Effect of incident wave angle on the nondimensional force amplitude, (a) in the z-direction, (b) in the y-direction, for two breakwaters array.

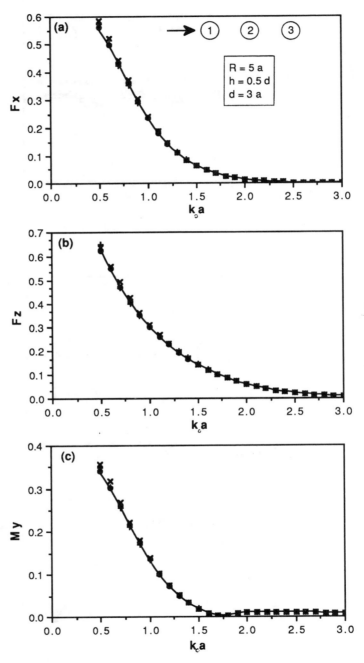

Figure (5) Nondimensional amplitude of (a) force in x-direction,   (b) force in z-direction, (c) pitching moment, for a three breakwaters array, as a function of $k_c a$. For legend see text.

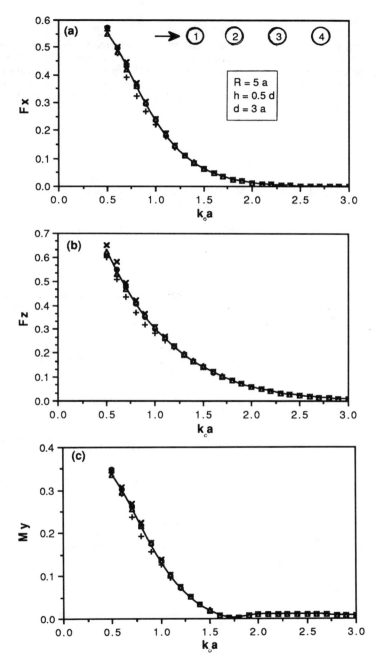

Figure (6)   Nondimensional amplitude of (a) force in   x-direction, (b) force in z-direction,   and (c) pitching moment, for a four breakwaters array as a function of $k_o a$.

CONCLUSIONS

An approximate method has been utilized to determine the hydrodynamic loading on an array of submerged circular breakwaters. Numerical results are presented for arrays of breakwaters which illustrate the influence of the various wave and structural parameters on the forces and moments. It has been found that interference effects may increase the hydrodynamic forces and moments particularly for smaller values of $k_o a$ and larger values of h/d.

REFERENCES

BOOKS

1. Sarpkaya, T. and Isaacson, M. Mechanics of Wave Forces on Offshore Structures. New York: Van Nostrand Reinhold, 1981.

2. Watson, G. N. A Treatise on the Theory of Bessel Functions. 2nd Edition, Cambridge University Press, 1944.

Paper in a journal

3. Iskandarani, M. and Liu, P. L.-F. Multiple Scattering Of Surface Water-Waves And Wave Forces On Cylinder Arrays, Applied Ocean Research, Vol.10, pp. 170-179, 1988.

4. Kagemoto, H. and Yue, D. K. P. Interactions Among Multiple Three-Dimensional Bodies in Water Waves: An Exact Algebraic Method. Fluid Mechanics 166, pp. 186-209, 1986.

5. McIver, P. Wave Forces on Arrays of Floating Bodies, Engineering Mathematics, Vol.18, pp. 273-285, 1984.

6. McIver, P. and Evans, D. V. Approximation of Wave Forces on Cylinder Arrays, Applied Ocean Research, Vol.6, pp. 101-107, 1984.

7. Williams, A. N. and Demirbilek, Z. Hydrodynamic Interactions in Floating Cylinder Arrays, Part I - Wave Scattering, Ocean Engineering, Vol.15 , pp. 549-584, 1988.

8. Williams, A. N. and Abul-Azm, A. G. Hydrodynamic Interactions in Floating Cylinder Arrays, Part II - Wave Radiation, Ocean Engineering, Vol.16 , pp. 217-264, 1989.

Chapter in a book

9. Abramowitz, M. and Stegun, I. A. Handbook Of Mathematical Functions. Chapter 9, Bessel Functions of Integer Order, (Ed. Abramowits, M. and Stegun, I. A.), pp. 358-433, Dover Publications, Mineola, New York, 1972.

Paper in a Conference Proceedings

10. Isaacson, M. Wave Forces on Compound Cylinders, pp. 518-530, Proceedings Civil Engineering in the Oceans IV, 1979, San Francisco, California, U.S.A , 1979.

11. Matsui, T. and Tamaki, T. Hydrodynamic Interactions Between Groups of Vertically Axisymmetric Bodies Floating in Waves, pp. 817-836, Proceedings International Symposium on Hydrodynamics in Ocean Engineering, 1981, Trondheim, Norway, 1981.

12. Spring, B. H. and Monkmeyer, P. L. Interaction Of Plane Waves with Vertical Cylinders, pp. 1828-1847, Proceedings 14th Conference on Coastal Engineering, 1974, Copenhagen, Denmark, 1974.

Technical reports

13. Abul-Azm, A. G. and Williams, A. N. Second-order Wave Loading on Truncated Circular Cylinders in Water of Finite Depth, Research Report No. UHCE 87-1, Department of Civil Engineering, University of Houston, Houston, Texas, U.S.A, 1987.

Reef Breakwater Design for Lake Michigan Marina

J.C. Cox

Ocean Technology Department, CH2M Hill, 777 108th Avenue Ne. Bellevue, Washington 98004, U.S.A.

Abstract

An innovative breakwater concept consisting of a conventional breakwater and a submerged-reef breakwater operating in tandem has been designed for the southern end of Lake Michigan. The tandem breakwater system was found to cost $1 million less to build than a single structure designed to meet the same operating criteria. Because of the depth-limiting behavior of the reef, the tandem design possesses a lower design risk for extreme events. This paper provides a summary of the design development and physical model testing results for the concept.

Introduction

The demand for protected moorage for recreational craft in the Chicago metropolitan area is high, but there are few, if any, natural harborages. To meet the demand, small craft harbors must be created along the exposed shoreline. Currently under construction at Hammond, Indiana, a neighboring community of Chicago, is the third-largest marina to be built in the United States. It also is the largest dedicated small craft harbor in the country to be constructed at an exposed site.

The marina is located on the southern tip of Lake Michigan. It is exposed to a fetch of several hundred miles over deep water, so that waves are fully developed when they reach the site. In addition, the lake level can fluctuate 1 to 2 feet annually,[2] with the annual mean varying up to 4 feet in roughly 7-year cycles. Storm surges at the southern end of the lake can add 2 to 3 feet to the local water level.

---

[1]Senior Coastal Engineer, Ocean Technology Department, CH2M HILL, 777 108th Avenue NE, Bellevue, Washington 98004, (206) 453-5000.

[2]To convert feet to meters, multiply feet by 0.305.

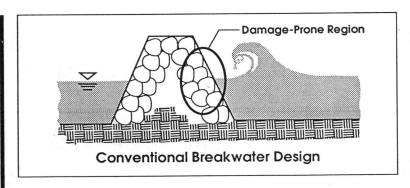

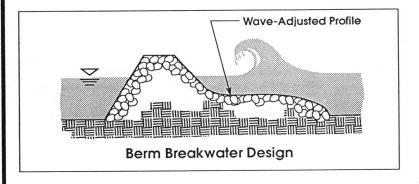

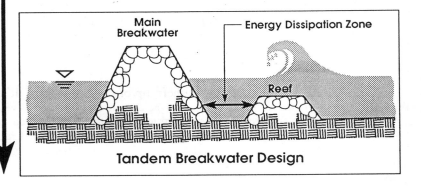

Figure 1
Tandem Breakwater Design Evolution

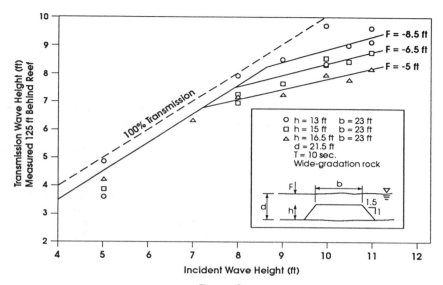

Figure 2
Effect of Reef Submergence on Wave Transmission

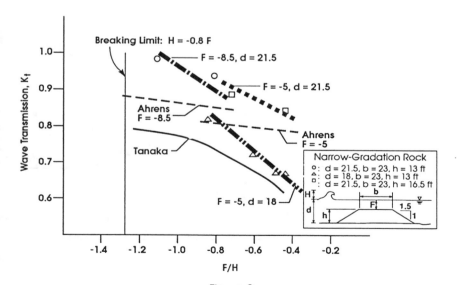

Figure 3
Comparison of Measured and Theoretical Reef Submergence Effects

The data assembled from these tests are compared in Figure 3 with the existing transmission theories. Note the generally different trends in the data versus the theory. Tanaka's (1976) theory appears to underestimate transmission until the relative freeboard becomes small; however, the trend is correct. Ahrens' (1987) theory seems to bisect the data, making it more conservative at small freeboards, but, like Tanaka, still less conservative at deeper relative freeboards. Because the general objective of this study was to create the minimum structure necessary to achieve the desired transmission, the theories were found to be inadequate for design and would need to be refined further to better describe behavior at deeper freeboards before they would become useful.

Crest Width. The effect of the reef crest width was also explored with the physical model. Tanaka (1976) suggested that increases in crest width at small ratios of crest width to wavelength would cause a substantial reduction in wave transmission, although, as the crests become wider, the rate of increased efficiency would decline. However, the results of these tests, presented in Figure 4, reveal no such improvement, even though crest widths were trebled between the widest and the narrowest. Comparing Figures 2 and 4, it is clear that the greatest efficiency, for the same section volume, is achieved by increasing height rather than width.

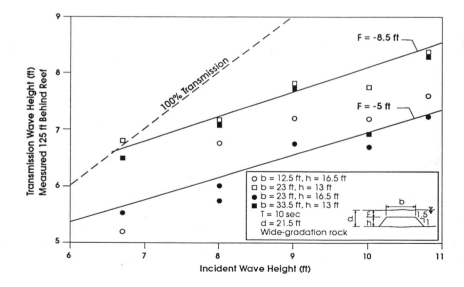

Figure 4
Effect of Crest Width on Wave Transmission

Reef Porosity. Plotted in Figure 5 is a comparison of wave transmission past two model reefs constructed from stones of different size gradation, but with essentially the same mean stone diameter (d_{50}). Figure 6 defines the size gradations (wide and narrow) used for the model reefs. The data suggest that porosity of the stone matrix, even for the deepest submergence, plays a significant role in the transmission efficiency. Decreasing the porosity by widening the stone gradation appears to lower transmission by 10 to 15 percent. As an example, sand bars and coral reefs make the best attenuators because of their more solid nature.

Breakwater Spacing. Artificial reef structures tend to be relatively narrow crested and have widths that are typically only a small percentage of the wave length. The breaking process is therefore interrupted as the overtopping wave re-enters deeper water on the lee side. Battjes and Janssen (1978), in observing this phenomenon on natural bars, suggest that the attenuation process of breaking waves over bars might extend as much as 40 wave heights beyond the initial breakpoint, reaching a stable transmitted height of approximately 50 percent of the incident wave height. This is generally consistent with the findings of Horikawa and Kuo (1966), who present an exponential rate of decay in wave height from the initial onset of breaking, which occurs at roughly 80 percent of the water depth over a shelf, to an asymptotic stable height limit of roughly 35 percent of the water depth (50 percent reduction in wave height).

The data set presented by Horikawa and Kuo included only wave heights large enough to just break at the shelf edge. However, by assuming the shape of the attenuation curve to be the same for waves initially much higher than the break-point threshold, a nondimensional attenuation versus breaking distance curve was developed. This curve, shown in Figure 7, was used to guide the initial model tests and design. Comparison of this curve with the actual data does show a similar trend, but with the curve projecting more and earlier attenuation. This would be expected because the wet stilling basin associated with the reef would be expected to be somewhat less efficient in attenuating the wave energy than a rock bench.

Reef Stability. Reef stability was predicted first by using Ahrens' (1987) theory for graded stone structures. The reef section was then physically tested for a broad range of submergences ranging from slightly emergent to deeply submerged. The tests indicated that the structure remained stable under all design waves and water levels. Further, Ahrens' stability theory was verified.

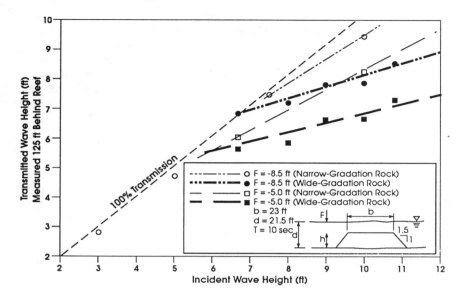

Figure 5
Effect of Reef Porosity on Wave Transmission

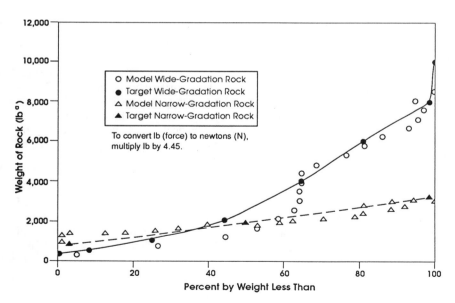

Figure 6
Definition of Model Reef Rock Gradations

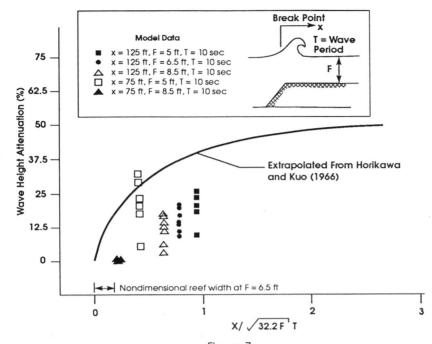

Figure 7
Nondimensional Wave Attenuation Distance

Summary and Conclusions

The final tandem breakwater design used for the Hammond break-water is shown in Figure 8. For scale, the equivalent performing "conventional" breakwater is overlaid. The engineer's estimated cost difference between these two designs was near $1 million. The reef concept was readily accepted by environmental interest as a new fish habitat area. The U.S. Coast Guard does not consider the reef as a navigation hazard, provided it has adequate signage. Fishing piers extending outward from the main break-water and straddling the reef further divert boaters away from the reef and provide fishermen direct access to the reef.

A tandem breakwater might not always provide the least cost alternative, because section costs are strongly related to material costs and water depths. However, the tandem breakwater does offer significantly reduced design risk because high-return-period events are also reduced to a more manageable level, thus reducing the chance of structure failure in storm events outside the design envelope.

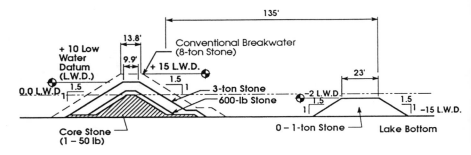

Figure 8
Tandem Breakwater Concept Schematic Cross Section

APPENDIX 1. REFERENCES

1.   Ahrens, J.,   Characteristics of Reef Breakwaters,   USAE CERC
      TR 87-17, 1987.

2.   Battjes, J., and J. Janssen, "Energy Loss and Setup Due to Breaking of
      Random Waves," *Proceeding of Coastal Engineering Conference*, 1978.

3.   Danel, P., "Wave Theory Applied to Shore and Harbor Structures,"
      Centennial Convention Paper No. 54, ASCE/ASME, 1952.

4.   Horikawa, K., and O. Kuo, "A Study of Wave Transformation Inside
      Surf Zone," *Proceeding of Coastal Engineering Conference*, 1966.

5.   Tanaka, N., "Effects of Submerged Rubble-Mound Breakwater on
      Wave Attenuation and Shoreline Stabilization," *Proceeding of Japanese
      Coastal Engineering Conference*, 1976.

Traditional Concrete vs Sand Concrete for Breakwater Armouring

M. Denéchère

Department of Port and Coastal Engineering, SOGREAH, 6 rue de Lorraine, 38130 Echirolles, Grenoble, France

ABSTRACT

This paper discusses the recent renewed interest in sand concrete in France as a replacement for the more traditional types of concrete. Equivalent performance values can be obtained by the use of admixtures. An industrial application of sand concrete is then described, involving the manufacture of ACCROPODE® armour blocks for a sea defence project on the French Atlantic coast. From the comparative results obtained and the technical and economic assessment, the use of sand concrete for ACCROPODE® blocks would seem to be an economically and technically viable possibility.

INTRODUCTION

Sand concrete is not really a new technique. Since 1950, sand concrete has been used on a large scale. Some of the oldest structures, such as the Port Said Lighthouse in Egypt, were even erected last century.

Nowadays, there has been a renewed interest in sand concrete because, in certain cases, it can offer superior technical characteristics and can provide substantial savings compared to traditional concrete.

The optimum use of local material while minimising the transport of heavy materials has always been a much sought-after requirement.

However, to promote the use of sand concrete has required not just a research programme to be conducted in the laborarory, but also real-life experiments such as those performed by the French maritime association called Sablocrete (Bordeaux, France).

The purpose of this programme was to investigate the potential fields of application of this particular type of concrete.

The three-year Sablocrete programme was open to all interested organisations and included specific projects in various fields like buildings, roadways, bridges and marine structures. One of these experiments was carried out in 1989 on a new marina project located on the Atlantic coast of France. It consisted in placing a limited number of ACCROPODE® armour units made of sand concrete.

WHAT EXACTLY IS SAND CONCRETE?

Generally, traditional concrete consists of fine and large aggregates and requires betweeen 250 and 400 kg of cement per cubic metre for proper compactness. Should only fine aggregates be mixed with cement, then a cement content ranging from 450 to 700 kg is necessary (mortars).

Sand concrete mixes are determined on a different basis:

- fine content and wetting agent to produce optimum compactness with the help of filler materials (generally limestone),

- standard cement content (less than 400 kg/m³) to obtain sufficient strength.

Thus, it is possible to avoid the high cost of cement and the drawbacks related to the use of mortars, as sand concrete contains the same proportion of cement as traditional concrete.

Chauvin[1] states that sand concrete is particularly interesting in terms of the following advantageous characteristics:

- higher workability,

- lack of segregation,

- attractive appearance,

- economy

Coastal regions where sand is plentiful are particularly well-suited to the use of sand concrete. In these areas, it can offer a low-cost substitute to traditional concrete.

On the other hand, its use to make precast armour units may be hindered by the following drawbacks:

- lower modulus of elasticity (about 20% less),

- lower concrete density (10 to 15% less),

- durability, which remains to be evaluated.

As far as mechanical strength is concerned, optimum performance depends on obtaining the correct proportion of cement and filler. However, it has already been confirmed that a similar level of strength can be achieved compared to traditional concrete.

CASE STUDY: NEW MARINA AT SAINT-DENIS-D'OLERON

Project Description

As part of the Sablocrete research programme[2], it was decided to test sand concrete during the construction of a new marina located north of Oleron island (see Figure 1).

It was known in advance that, for marine works, concrete density is a very important consideration. However, because of the abundant sources of local sand, especially from the project itself, this experimental choice was made purely for economic reasons. In other words, in view of the sandy nature of the coastline, priority was given to the economic benefits while knowing the technical drawbacks.

The ACCROPODE® armour block

In order to respect the environment, the "Commission des Sites" stipulated that the breakwaters for the 600-boat marina should be armoured with artificial blocks.

Indeed, the natural diorite rock usually available in this area was considered too dark in colour to blend in properly with this project.

Figure 1: Project location, north of Oleron island

Of the different types of block available, the ACCROPODE® block (see Figure 2) was selected to protect the new harbour, involving about 6000 units of 0.8 and 1.5 cubic metres (2 and 3.5 tons).

Casting of ACCROPODE® blocks with sand concrete

At first, the project technical specifications originally called for a traditional concrete having the following ingredients:

- CPJ45 RPM cement: 310 kg/m³

- 0/2 mm sand: 530 kg/m³

- 2/6 mm aggregates: 265 kg/m³

- 10/31.5 mm aggregates: 1200 kg/m³

• Water, plasticiser and air entrainment additives.

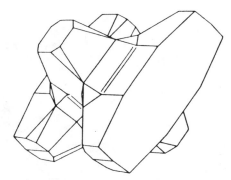

Figure 2:
The ACCROPODE® block

In February 1989, the experiment was conducted on part of the daily production: 12 units of 3.5 tons out of the 50 made daily were made with sand concrete.

After checking the properties of various possible sources of sand from the site, trial tests on concrete specimens were performed and, from the results obtained, the following sand concrete mix was adopted:

• CPJ45 RPM cement: 350 kg/m³

• Beach sand from site: 1550 kg/m³

• Calcareous filler (< 80 μm) : 200 kg/m³

• Water, plasticiser and air entrainment additives.

From this Sablocrete pilot scheme[3], it was noted that mixing time had to be twice as long compared to traditional concrete owing to the greater attention that had to be given in incorporating the materials in the mixer, although almost no vibration was required (see fig. 3). After curing, however, the general aspect of the block outer surface was excellent.

Figure 3: Casting of 1.5 m^3 ACCROPODE® blocks with sand concrete

Selection of test area

The experimental blocks were placed on the west groyne (see Figures 4 and 5). Considering their lower unit weight, the sand concrete blocks needed to be adequately scattered within the armour to keep its stability as homogeneous as possible (see Figure 6 below).

By adopting this scheme, it was possible to obtain good interlocking between all the blocks while making sure that the resulting average concrete density was superior to the design value (see Fig. 6).

Figure 4: Plan view of test experiment site

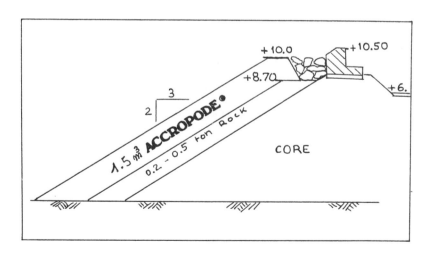

**Figure 5: Cross-section of the breakwater
where the experiment was carried out**

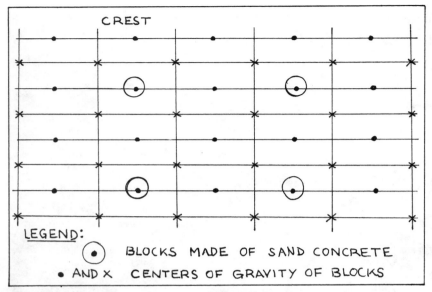

Figure 6: Plan view showing the position of block centres of gravity

TECHNICAL ASSESSMENT

The mean values obtained with concrete test cylinders were:

- compressive strength after 3 days = 20 MPa
- compressive strength after 28 days = 33 MPa
- flexural strength after 28 days = 4.5 MPa
- tensile strength after 28 days = 2.5 MPa

The results in SOGREAH's report[4] showed that the concrete strength values met the design standards. In this experimental case, they were slightly below the values obtained with traditional concrete.

However, it is believed that adjustment of the concrete mix, with even better proportions of cement and fines, would probably lead to characteristics greater than those obtained with traditional concrete.

Concrete density and hydraulic stability

Concrete density was measured on cylinders made of sand concrete and was found to be 2.2 t/m^3.

The unit weight of the blocks is a function of the ratio $\dfrac{d}{(d/d_0-1)^3}$ in Hudson's formula.

If a density value of d = 2.2 t/m³ is considered for sand concrete instead of d = 2.4 t/m³ in the case of traditional concrete, the weight of armour units needs to be increased by 15% in order to obtain the same degree of stability under wave attack. This means 15% more concrete, but a corresponding 24% reduction in the number of blocks to be cast and placed.

For example:

Type of concrete	ACCROPODE® unit volume	Armour thickness	Filter layer unit weight	Thickness	Number of blocks required to cover 100 m2 of slope
(1)Traditional	2 m3	1.6 m	min. 0.3 t max. 0.7 t	1.3 m	41.5
(2) Sand concrete	3 m3	1.85 m	min. 0.5 t max. 1.0 t	1.5 m	31.5

Cross-sectional comparison

① ②

2 m³ ACCROPODE® blocks made of traditional concrete	3 m³ ACCROPODE® blocks made of sand concrete
0.3 to 0.7 t rock (filter layer)	
	0.5 to 1.0 t rock (filter layer)
Core material	
	Core material

Quantity differences in core material and rock are minor. Less filter rock is required with the traditional concrete solution. However, the sand concrete solution requires a little less core material.

Durablity

Once they were in place on the breakwater, it was necessary to monitor the behaviour of the sand concrete ACCROPODE® blocks (see figure 7). Additional tests were carried out in the laboratory to check the behaviour of the concrete with time by freezing/ unfreezing and saline mist tests. To date, it has not been possible to detect any significant increase in fragility of the sand concrete blocks.

Figure 7: Completed breakwater including sand concrete blocks

ECONOMIC ASSESSMENT

For this particular case, the savings obtained by using sand concrete were estimated to be 15% compared to the traditional concrete solution. This was possible only because sand was readily available on site at no cost. In fact, the incorporation of filler material resulted in a substantial saving in cement.

CONCLUSIONS

This experiment shows there are definite benefits to be gained in using material from sites on condition that it is readily available in quantity and quality and, most important of all, free of charge.

When sand and aggregate materials are available, both solutions (traditional concrete and sand concrete) are equivalent in cost. However, if sand only is available, then the sand concrete solution presents a definite advantage.

The major problem in using sand concrete for making armour units remains its low density. Several possibilities exist, however, to solve this problem, such as the use of an efficient water admixture or finding fillers of higher specific gravity.

Nevertheless, it has been proved economical to make and place sand concrete ACCROPODE® blocks.

In any case, regardless of the type of concrete specified, the ACCROPODE® block offers not only a safe solution for protective armour facings, but also an approximate 50% saving in concrete compared to most other conventional types of artificial unit.

REFERENCES

1. J.J. Chauvin, Les Bétons de Sable, Bulletin Liaison Labo Ponts et Chaussées 157, Sept-Oct 1988, Ref. 3336.

2. Report: Expérimentation de blocs ACCROPODE® en béton de sable, Projet National Sablocrete, 1989.

3. Centre d'Etudes Techniques de l'Equipement du Sud-Ouest, Action Pilote de Développement, Opération ACCROPODE®, May 1989.

4. Création d'un port de plaisance à St Denis d'Oléron - ACCROPODE® technique - SOGREAH Report No. 1, January 1989.

oOo

SECTION 11: PONTOON SERVICES AND MARINA MAINTENANCE

Water and Electrical Supply for Marinas: An Italian Survey on Actual Consumptions Towards Updated Design Guidelines

L. Franco (*), R. Marconi (**), C.A. Marconi (***)

(*) Politecnico di Milano, Piazza Leonardo da Vinci 32, Milano, Italy

(**) Acquatecno S.r.l., Via della Camilluccia 35, Roma, Italy

(***) Consortile Ponza e Ventotene r.l., Via Cesalpino 1, Roma, Italy

ABSTRACT

A field survey has been carried out in a number of typical Italian marinas in order to investigate on the actual characteristics of the water and electricity consumptions, and on the performance of the related supply systems. The results of the analysis represent a useful guidance for the marina designer due to the present lack of criteria and to the rapid evolution of yachtsmen requirements.

1. INTRODUCTION

The supply of water and electricity at the berth are basic requirements for marina users. Therefore a careful design of the hydraulic and electric supply system is needed, taking into account both the particular (each boat) and general uses. The evaluation of these consumptions is still very uncertain for the designer, given the large variability in time and the special needs of the yachtsmen. Moreover very little guidance is given on this subject in the literature or technical regulations.

This paper aims to fill this gap and make a step forward more reliable and updated design criteria for water and electrical supply in marinas by means of a field analysis of recent consumption rates (some even differentiated by type of use) in a few important Italian craft harbours. The research survey includes an evaluation of the technical characteristics and performances of the existing distribution networks.

2. PRESENT CHARACTERISTICS OF THE WATER SUPPLY SYSTEM

Usually the design of marina aqueducts is still based on simple oral traditions assuming a peak daily consumption of 500 litres per boat.

However the "water resource" is among the most important utilities in a marina and its deficiency is badly tolerated by the users. This is confirmed by the results of a recent interesting survey carried out by the Chamber of Commerce of Forli: the interview of some 300 boaters in Northern Italian marinas showed that the water is the most used utility (70% of all Italian users, increasing to 90% for boats above 7 m l.o.a.), followed by auto-parkings, boat repair yard and electricity (50%).

Moreover the survey showed that the foreign yachtsmen (mainly Germans, Austrian and English) are less "water-dependant" (50% preference), thus indicating different habits of users from various nationalities (also the American boaters make less use of water in marinas).

The present uncertainties in the design of a marina hydraulic system are related to the lack of knowledge of the variable rates of different types of consumption and their fluctuations in time. The main water uses can be classified as follows:
a) Drinking and sanitary (for boat freshwater tanks and for the marina service blocks)
b) Industrial (boat washing and gardening)
c) Fire-fighting
d) Commercial
e) Losses in the network

a) Drinking and sanitary water
Typically around 95% of boats have tanks to be filled with drinking water, with an average capacity of 300-400 litres (8-14 m boats). However the sanitary water use inside marinas is mainly concentrated in public service blocks, since flushing aboard is forbidden. Usually one hygienic unit (sink-w.c.- shower) is addressed to 15-25 berthed boats with an equal share between gentlemen and ladies; but the survey undertaken in Nettuno marina just showed that a 60% (men) - 40% (ladies) distribution is better balanced. Facilities for handicapped people should also be provided. This freshwater consumption is very variable with time: high peaks typically occur in summer and at weekends (especially if a regatta or special event takes place).

b) Industrial water

This water use category may produce the largest consumption rates, mainly because of boat washing, which is concentrated in the same limited boat use periods. It involves the waste of 200-400 l of freshwater (or seawater) at each washing operation, even with detersives, carried out once or twice a day when the boat owner is aboard. This habit is typical of the Italian yachtsmen, maybe due to the lack of summer rainfall and to the lavish and wasteful Italian character.

For this reason in most marinas boat washing is strictly regulated in peak use periods, although without much success. Therefore different costly solutions have been applied, such as separate supply system for industrial and drinking water, pay toll distribution (Punta Ala Marina) or push-button switch supply (Cala Galera Marina). However in some Italian marinas and also in other Mediterranean countries (France, Greece, Turkey) the supply is often limited to only a few hours a day. This critical problem should then be carefully tackled at the design stage.

Additional not-drinking water is required for irrigation of the green areas: the consumptions vary a lot for each marina. The water quality needs anyway to be controlled to avoid gardens drying up.

c) Fire-fighting plant

The characteristics of the fire-fighting network change with the national regulations. The pipes typically deliver water with a pressure of 2-3 atm (1 atm = 101325 Pa) at hydrants spaced 45 to 70 m with the standard movable rubber pipe (10-20 m long) located in the vicinity for emergency.

Four different systems are typically in use in Italy:
1) Unitary line for all uses directly linked to the hydrants
2) Separate line fed by sea water through electric pumps with own energy supply.
3) Unitary line supplying freshwater up to a limited pressure (eg. 5 atm) or saltwater for higher pressures in emergency conditions. This solution has the disadvantage requiring a full aseptic wash of the lines after each use with sea water.
4) Separate line fed by not-drinking freshwater. If sea water is not used, a 50-100 m3 filled tank is necessary for fire-fighting emergency only. Significant freshwater consumption are to be considered during the periodic fire-fighting tests, needed for training personnel and checking

T

equipment.

d) Commercial water
Further use of water not directly related to nautical
needs is made by the various shops and restaurants
(usually up to 20-40% of total marina uses)
particularly by the laundries, bars and ice cream
shops. Some commercial units are closed in the winter
period. Water is also consumed in the boat repair
yard, especially for hull cleaning operations with
pressurised jets.

e) Water losses
Marina hydraulic systems are subjected to various
types of losses similarly to any other system.
Galvanized steel pipes are vulnerable to corrosion in
saltwater environments; moreover the damage location
is difficult to detect, because the spilling water
quickly reaches the harbour basin without causing
apparent damage to marina structures. PVC pipes are
sensitive to freeze and thaw cycles and require many
joints vulnerable to pressure changes. High density
polyethylene pipes also have sensitive joints but
their number is much less. An annual check of the
plant watertightness is normally carried out, also to
guarantee the water potableness, which can be assessed
with periodic water quality control.

     Beside the structural losses, further wastes are
caused by imperfect closure of boat user's outlets or
by mere vandalism. Accidents of this kind occur almost
every day in marinas, and may produce a relevant water
consumption, which can amount up to 25-30% of the
total supply. The losses due only to boaters
inattention are more regular, reaching a 2-4% of the
total water volume.

3. PRESENT CHARACTERISTICS OF THE ELECTRICAL SUPPLY
   SYSTEM

Present design criteria largely vary from country to
country according to boat characteristics and
yachtsmen habits. Total electrical consumptions
exhibit wide scatter, whereas individual uses of
similar-size boats are rather uniform, due to the
extended standardization. Some well known "structural"
differences remain, such as the frequency of 50 Hz in
Europe and 60 Hz in USA.

     Typically the design rated electric power per
boat, excluding commercial users, is around 1 kW,
mainly consumed during the peak periods (summer
weekends). The main electrical consumptions in a
marina can be distinguished as: nautical, industrial,

commercial   and   lighting,   each   with   its   own
characteristics and fluctuations.

a) Nautical uses
Power is usually supplied at each berth to   outside
sockets with breaker switches in order to feed boat
batteries,   lighting,   fridge,   air-conditioning,
heaters, maintenance, tools and any other electrical
gear carried aboard. Consumption rates increase more
than exponentially with boat size and comfort level:
craft larger than 15 m l.o.a. may absorb more than 15
kW, while small craft (less than 8 m) usually use less
than 3 kW. The available capacity can be 32 or 64 A
for the large craft and 8 or 16 A for the smaller
boats. In Italy the voltage is generally 380 V three-
phase, 4 wires for large yachts and 220 V single-
phase, 2 wires for the small ones.

b) Industrial uses
Typically the major consumption is located in the boat
repair yard and workshops for the use of electrical
tools and equipment and even launching machines. A
total power of 40 to 100 kW is normally supplied with
a   voltage of 380 V. Further constant   current
absorption may be related to the use of fixed dredging
plants or water flushing and oxygenation facilities
(against pollution and stagnation). The pumps for
enhancing water circulation may absorb 15 kW for
periods of 3 to 18 hours/day and the oxygenators can
use 5-7 kW each. A booster plant for sand bypassing,
combined with a suction dredger, can absorb 30 to 100
kW depending on the discharge distance. Other general
"industrial"   uses   are   those   of   the   marina
administration complex and sanitary service blocks.
The equipment installed in the headquarters (air
conditioning, computers, radio station)   are usually
supplied at 220 V single-phase with consumption rates
rarely exceeding 20-25 kW. In the sanitary units the
boilers, hair-dryers and electric razors require on
average a supply of 12-15 kW for each unit, the rates
varying with the number of facilities (showers, etc.).

c) Commercial uses
It   is   generally recommended to   separate   the
electricity supply network for the shopping centre
from that for the berths, due to the different
consumption rates and the risks of claims   on
individual metering. The electrical needs of each shop
typically amount to 3 kW (as in any urban shop) but
some commercial activities involving the use   of
electric machines, air-conditioning and computers can
require a larger power (10-15 kW) sometimes even at
380 V three-phase voltage.

d) Lighting
Dock and yard lighting in marinas should strike a
compromise between the need of ensuring safe vehicular
and pedestrian circulation and that of minimizing
disturbance to sleeping yachtsmen and to the
visibility of navigational lights. Therefore piers are
illuminated by low level pedestal lights (40-60 W
incandescent lamp at 7-8 m intervals), while dock and
pedestrian walkways have small High Pressure Sodium
Lamp posts 4 m high at 20-25 m spacings with a power
of 75-150 W. Lighting represents an important part of
the electrical consumption of marinas since it may
amount to 15% to 40% of the total use.

4. A SURVEY OF SOME ITALIAN MARINAS

A specific survey has been carried out by interviewing
numerous marina managers in order to obtain updated
information on the actual hydraulic and electric
consumptions (with particular reference to the
specific nautical uses) and on the variable
characteristics of both supply systems. Difficulties
arose in gaining reliable and homogeneous data,
especially from the small single-owned marinas of
Northern Adriatic Sea. Acceptable data have been
received from six important marinas on the Italian
west coast, namely Loano, Rapallo and Lavagna in
Liguria, Punta Ala and Cala Galera in Tuscany and
Nettuno in Lazio (see location map in fig. 1).

The total annual recorded consumptions are
summarized in tab. I and II. The monthly variation of
water consumption is shown in fig. 2, while the annual
average of specific daily use is given in fig. 3.
Similarly the monthly distribution and total annual
average of specific daily electrical consumption are
illustrated in fig. 4 and 5. Finally a graph showing
the contribution of the only industrial water uses is
given in fig. 6. A brief description of the supply
systems of each surveyed marina is following.

Loano Marina
It is a municipal harbour built in the early 70's with
470 wet berths (80 of which deserved to fishing boats)
and 150 dry berths. In summer it is fully occupied,
while in January 1990 a 60% berth use was recorded.

The municipal Water Authority only supplies water
to the boat users, the general service blocks, offices
and a small maintenance area. Two independent
galvanized steel networks provide drinking water also
for fire-fighting. Pier distribution takes place at
pedestals every two berths. Extensive pipe corrosion
and related water losses have recently called for a

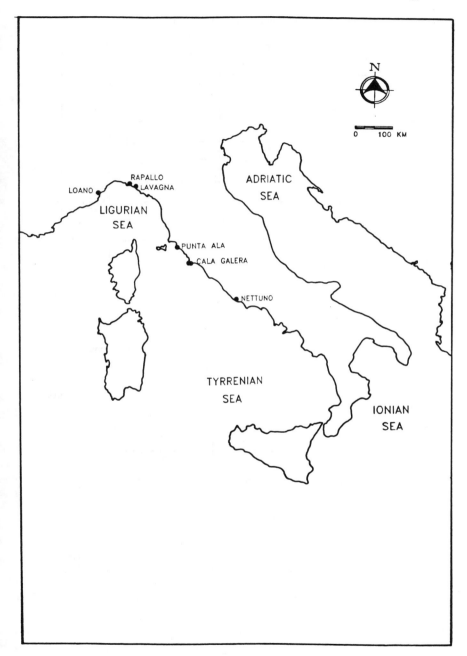

Fig. 1: Location map of some Italian marinas

Table 1: Annual water consumption in six Italian marinas ($10^3 m^3$)

Marinas (n. berths)	LOANO (620)	RAPALLO (380)	LAVAGNA (1440)	P. ALA (800)	C.GALERA (700)	NETTUNO (800)
1982	–	–	–	32.5	42.7	–
1983	–	–	–	42.5	47.2	–
1984	–	–	–	69.0	59.7	–
1985	–	–	–	42.5	67.0	–
1986	–	–	–	32.0	42.8	–
1987	54.1	* 22.8	* 105.8	35.0	40.2	–
1988	* 50.2	* 22.9	* 156.9	40.0	44.1	+ 37.8
1989	* 47.6	–	* 231.0	* 43.0	47.9	+ 47.0
1990	–	–	–	* 36.0	51.0	* 63.3
1991	–	–	–	* 39.0	–	–
Typical specific daily average (1/day boat)	219	167	317	137	189	220

* Reference years for the definition of specific consumptions
+ Reference years for the definition of monthly consumptions

Table 2: Annual electrical consumption in four Italian marinas ($10^3 kWh$)

Marinas (n. berths)	RAPALLO (380)	P. ALA (800)	C. GALERA (700)	NETTUNO (800)
1987	–	–	* 840.9	–
1988	–	–	* 897.9	–
1989	* 741.8	–	* 872.4	341.6
1990	–	* 660.0	* 971.4	* 420.0
1991	–	* 640.0	–	–
Specific daily average (kWh/day boat)	5.35	2.3	3.6	1.3

* Reference years for the definition of specific consumption

rehabilitation of the whole system: in fact the rates
metered   in   the   last   3   years (with   full   berth
occupation) show a reduction of water consumption due
to reduced losses. Data for 1990 are not reported in
tab. I, since an extreme draught forced to limit the
water supply to just 2 hours a day from July until
October.

As far as electric power supply is concerned, the
transformation from high to low voltage takes place
inside the marina, and it is then distributed at 220 V
for all uses, except for the service blocks, the large
craft   (> 25 m) and the fishing   vessels   (ice
production) at 380 V. The electric energy is supplied
at all berths through outside sockets on the same
pedestals of the water outlets. The system is now
being completely renovated to satisfy the CEI
regulations. Electric consumption is paid as a
lumpsum, simply related to the berth size: therefore
no detailed reliable data are available. However it
was found that 40% of the total consumption is due to
street lighting (around 5.000 hours/year) and 60% to
boat and general service uses.

Rapallo Marina
This private marina receives up to 380 boats all year
round. The municipal aqueduct supplies water also to a
shopping centre with 11 commercial services, two bars,
a yachting club and a summer restaurant. The drinking
water distribution system includes two 300 m3 tanks,
which ensure a regular supply during the critical
summer peaks and a galvanized steel pipe network
subjected to periodic repair to reduce water losses.
Instead the fire-fighting plant consists of "GEBERIT"
plastic pipes fed with seawater by two pumps (one
spare diesel motorized for emergency). Monthly fresh
water consumption rates are shown in fig. 2 and
reached a maximum around 4.400 m3.

With regards to the electrical consumption
monthly data (including commercial uses) could only be
obtained for 1989, as shown in tab. II and fig. 4:
peak rates are observed during winter holidays.

Lavagna Marina
Lavagna (completed in 1977) is the largest private
marina in Italy receiving up to 1440 yachts. The
municipal Water Authority supply system also serves
part of the nearby town. A unitary high density
polyethylene line network normally supplies drinking
freshwater, unless the pressure drops below 7 atm to
start up the pumps using sea water for fire-fighting
emergencies and periodic tests. The pipes then need a
long and costly sterilization procedure, before

refilling with freshwater. Outlets along piers are available at each 4 berths for boats up to 10 m; at each 2 berths for boat lenghts between 12 and 26 m; at each berth for longer boats. The aqueduct also supplies water to some 50 shops, a restaurant and a swimming pool open all year round, although it was not possible to deduct the water volumes only used for the pool. This additional use also explains the larger consumption rates at Lavagna, compared with those of other marinas. Total annual water consumption (metered every 3 months after 1987) are listed in tab. I. No data could be collected on the electrical uses and supply system.

Punta Ala Marina

This private craft harbour can receive up to 890 boats (90 for transit) during the summer (peak occupancy in August), whereas in winter no more than 650 boats are at berth. Separate lines provide drinking water and industrial freshwater from inland wells (for irrigation, washing and fire-fighting). Both networks are made with galvanized steel pipes, prone to corrosion and related water losses. One outlet serves 3 boats with lengths of 6 to 12 m, and the supply is metered. The water consumption rates are generally smaller than those of the other marinas due to the exclusion of the shopping center and to the recent introduction of a coin payment system with the aim to reduce water use.

As far as an electric energy supply system is concerned the survey at Punta Ala marina has shown that the voltage is usually 220 V, apart from boat berths greater than 12.5 m and the general services at 380 V. The recorded consumptions do not include the commercial uses. Total annual data are available only for 1990 and 1991 (tab. II), but it appears that the rates of previous years were much the same. It was found that public lighting absorbed 15% and 10% of total consumption in 1990 and 1991 respectively, the reduction being attributed to the use of new energy-saving lamps.

Cala Galera Marina

Constructed in 1975 this 780-berth (80 for transit boats) private marina is very busy in summer. Drinking and industrial waters are supplied by two distinct networks, partly made with PVC pipes, partly with galvanized steel; the latter are being replaced by PVC to reduce water losses. Water shortages are frequent and a push-button switch device was introduced to limit consumption. Monthly consumption rates are recorded since 1976 for both distinct water uses. Annual total volumes listed in tab. I outline the

Fig. 2: Monthly distribution of total water consumption in three
Italian marinas

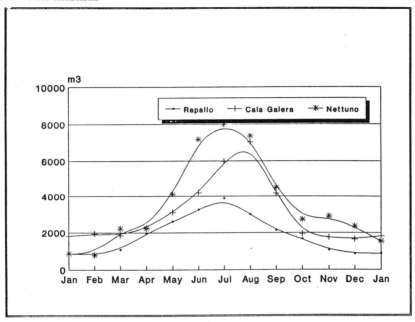

Fig. 3: Annual average of specific daily water consumption in six
Italian marinas

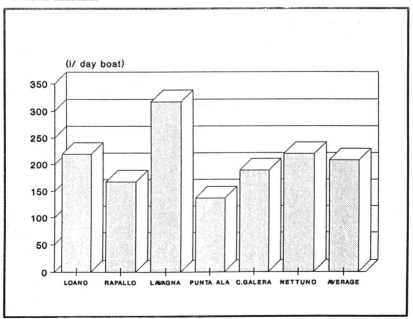

reduced values recorded after the rehabilitation of the supply lines. Daily rates could also be obtained for the peak-use months of July and August 1989: the maximum metered daily water volumes were 350 m3 and 580 m3 respectively, i.e. specific consumptions of some 450 and 750 l/boat day.

With regards to the electrical supply the plant is similar to that of other marinas with low voltage transformation taking place inside the harbour area. Voltage is 380 V three-phase for boat berths longer than 10.5 m and 220 V for the smaller ones. Monthly records of each type of electric use are available since the time of construction: peak rates typically occur in September (fig. 4) when the yachts come back to their berth after the summer cruise. The "nautical" consumptions are metered at the pier head for boat berths shorter than 15.0 m, and at each berth for the larger craft. The annual use rates have been derived for the last years (1987-90), when all berths had been sold, as shown in tab. II. The data also include the commercial consumption. An interesting analysis of the only nautical electric consumption, subdivided for boat categories, has also been carried out for 1990 with the results given in tab. III. It is interesting to observe that the recent increase of electrical equipment aboard (especially on motor-boats) makes the present supply inadequate: therefore the yachtsmen are often using simultaneously the electric outlets of adjacent berths!

Nettuno Marina
The new private marina at Nettuno near Rome was opened in 1986 and "filled up" in 1990. It can receive 800 yachts and 60 fishing boats. A unitary water supply system, including fire-fighting, serves the whole marina together with the shopping center. A 50 m3 tank is devoted to fire-fighting emergencies and is fed by one electric or one spare motorized pump. The lines are made with high density polyethylene. The increasing consumption rates shown in tab. I are related to the progressive occupancy of the available berths. The annual data have been derived for the last year (1990), when no monthly rates are available. Therefore in fig. 2 the monthly data are related to 1988-89 when no full berth occupancy was yet achieved. The supply outlets serve two boats at berths between 8.5 m and 16.5 m and each berth for boat categories up to 22 m l.o.a. The recorded monthly distribution in 1989 (with August peak) is shown in fig. 2.

Regarding the electric plant, three transformation cabins (a main central one and two secondary peripheric ones) are located inside the

marina with a nominal total power of 200 kW, which so
far has not been fully used. The boat repair area and
the shopping centre have independent supply networks.
General consumption rates, excluding the shopping
centre and the boat yard, are given in tab. II for
1989-90. At the berths the waterproofing breaker
switches are placed on the same pedestals carrying the
water outlets.

5. DESIGN GUIDELINES FOR THE WATER SUPPLY SYSTEM OF A MARINA

The results of this survey allow the derivation of
some useful guidelines for the design of similar
marina hydraulic plants, particularly in Mediterranean
conditions.

With regards to all pipe material the best
solution seems to be high density polyethylene, since
it is not subjected to the corrosion deterioration
typical of galvanized steel (with 20-30% water losses
after 2-3 years!) and it is supplied in long flexible
rolls without the frequent vulnerable joints needed by
PVC lines.

For the distribution network it is recommended to
have separate lines for drinking and "industrial"
(washing, irrigation and fire-fighting) waters.

The increasing shortage of water resources in
the crowded coastal tourist centres demands suitable
devices for limitation of the boat washing use.

Berth supply outlets are best located in low
pedestals along the piers shared by 1 to 4 boats
according to size, which is acceptable given the small
contemporaneity factor. Adjacent pedestals may be used
in peak periods for multiple simultaneous use.

Tanks are usually not necessary except for the
fire-fighting plant. The best solutions for the fire-
fighting system are either an independent line fed
with sea water by two pumps (one electric, one spare
diesel activated), or combined with the industrial
water network. In the latter case a 50-100 m3 tank is
necessary to compensate possible pressure failures.

As far as design water discharge is concerned the
results of this survey show that, despite the
influence of variable and often not quantifiable local
factors, the average total annual consumption per boat
berth is 60-70 m3, including all general drinking
uses, even for the commercial area. If the latter is
not existent the discharge reduces to 50-55 m3/year.

Fig. 4: Monthly distribution of specific daily electrical consumption in three Italian marinas

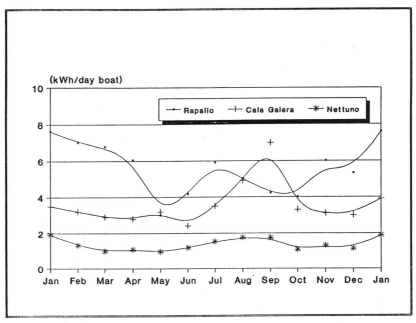

Fig. 5: Annual average of specific daily electrical consumption in four Italian marinas

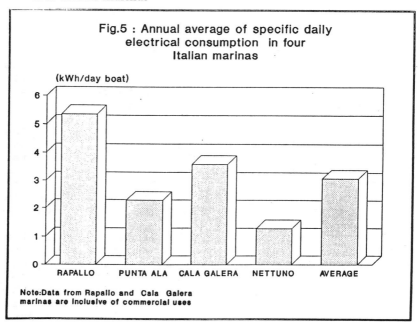

On a daily basis these average values correspond to nearly 200 l/boat and 150 l/boat respectively. A maximum daily peak of 750 l/boat was recorded in August. These values seem to be larger than those recorded in other countries, where the boat washing habit is not so common.

The graphs drawn for different marinas show similar monthly distributions of water consumption rates in the last years, with maximum use recorded in July (Rapallo) or August (Cala Galera) due to different regional holiday habits. The summer rates are typically 4 times larger than the winter ones.

6. DESIGN GUIDELINES FOR THE ELECTRICAL SYSTEM OF A MARINA

From the results of our survey useful design guidelines can also be derived for the marina electric plant. As far as materials are concerned, the aggressive environment recommends the use of high quality cables, fireproof and rodent-bite resistant. The number of cable joints should be minimized since they are the weakest points of the system.

All outside power outlets should be protected from humidity with a standard above IP66. All steel material needed for the installation of electrical equipment ought to be stainless to avoid corrosion. However, the complexity of a relatively widespread electric network demands not only a great care and conservatism in the design, but also a continuous maintenance work by the marina management staff.

As far as the distribution plan is concerned, the high power absorption rates suggest a direct connection of the marina electric system to the national high voltage network, and to locate the low voltage transformation inside the marina for further distribution to the nautical, industrial and commercial users. The transformer substation should possibly be in a central position to minimize drop voltage and losses. It is also recommended to subdivide the network in sections to be made independent in case of localized damage, in order to limit the inconvenience to just a few berths. Due to the presence of loads such as motors and fluorescent lamps, with consequent low power factor ($\cos\phi < 0.9$), suitable power factor correction equipment should be provided in the transformer substation.

Earthing has to be provided according to standards, typically by connecting all electric equipment and metal mass to a continuous ring copper

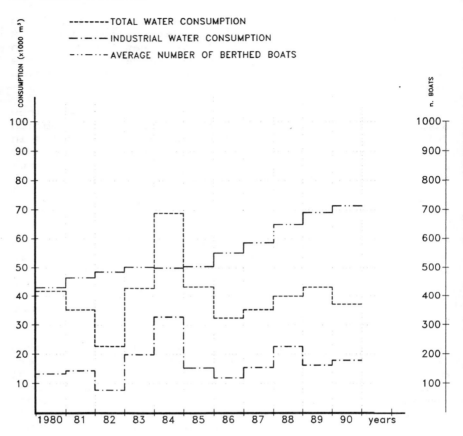

Fig. 6: Evolution of water consumption in Punta Ala Marina

cable of adequate cross section connected to underground earth pit.

Still the main design problem remains the prediction of the highly variable consumption rates. The data collected during this survey show that peak use rates, especially in northern marinas, occur during the Christmas period when boats are at berth (being used as holiday homes) with simultaneous use of electric heating. In summer instead the high electricity demand of air-conditioning systems produces another peak especially in the southern areas, although the contemporaneity use factor is typical smaller.

Interesting results are summarized in tab. IV, where boat users have been classified in five categories having a nominal power increasing with size and related amount of electric gear aboard. The actual average consumption is obtained after consideration of two factors, namely the "boat use factor" (Fu) which accounts for the simultaneity of use of different equipment on each craft and the "contemporaneity coefficient" (Fc) which gives the percentage of boats in the marina making simultaneous use of electric power. The former factor Fu may be lower for large craft having various equipment and uses, but it can be assumed with a nearly constant value of 0.3. The second factor Fc strongly depends on the actual berth occupancy and operational presence aboard: it is not then a function of the boats size but of the number of berths, being smaller for large marinas. An average value of 0.2 was found to be adequate for design purposes. The actual average electric consumption is finally obtained by multiplication of the nominal power for the product of the two factors (0.3x0.2 = 0.06, total use factor) leading to the results shown in tab. IV where the available capacity are also given.

As far as the industrial uses and public lighting are concerned the surveyed data show large scatter, due to the high variability of electrical equipment installed in each marina and repair yard and to the different size of buildings and harbour areas to be illuminated. The design criteria however do not differ from those of urban areas.

7. CONCLUSION

The results of the survey described in this paper are useful to fill a gap of the technical literature regarding the water and electric consumptions in marinas. General design guidelines for both supply

Table 3: Average electric consumption for different boat classes at Cala Galera marina in 1990

Boat Berth size (m)	Class	Users (n°)	Nominal power (kW)	Annual average consumption per berth (kWh)
up to 8.6	1	4	5	97
" 10.75	2	4	5	150
" 12.50	3	2	15	388
" 15.00	4	2	15	450
" 17.50	5	1	18	776
" 20.00	6	1	18	1.436
" 25.00	7	1	18	2.440
above 25.00	8	1	50	*

* Data not reliable due to limited number of vessels

Table 4: Recommended values for electrical power and capacity for different boat berths sizes

Boat class	Nominal power (kW)	Fu	Fc	Actual average used power (kW)	Available current (A)
1 l.o.a. < 8 m	3	0.3	0.2	0.18	16
2 l.o.a. 8-12 m	7	0.3	0.2	0.42	32
3 l.o.a. 12-16 m	10	0.3	0.2	0.60	32
4 l.o.a. 16-22 m	15	0.3	0.2	0.90	60
5 l.o.a. > 22 m	20	0.3	0.2	> 1.20	60-120

systems have been derived for middle size marinas in typical Mediterranean conditions. The high fluctuation of uses is found to be influenced by the marina geographical location and climate, by the yachtsmen habits (such as that of frequent boat washing) and even by the distance of the harbour from metropolitan areas.

It is hoped to extend the survey in the future, with stimulation of the control and the publicity of data by the marina operators. An accurate knowledge of the variable water and electric uses is infact of great importance because of the present rapid development of marinas and the high peak summer requirements in tourist coastal zones.

REFERENCE
Franco L. and Marconi R., Marina Design and Construction. Chapter 6, Marina Developments, Ed. Computational Mechanics, Southampton, 1992.

ACKNOWLEDGEMENTS

The Authors (who equally contributed to this paper) wish to thank the managers of the marinas of Loano, Lavagna, Rapallo, Cala Galera and Nettuno for the kind provision of data throughout this survey. Useful discussions were also undertaken with the engineers of SISTEMA WALCON (Italian branch) and CEI-CASILLO S.p.A. The help of ing. Enrico d'Argenzio and Anna Maria Socini is gratefully acknowledged.

Dredging as an Influence on the Redevelopment of Port Edgar, South Queensferry, Edinburgh

A.S. Couper (*), M. Wakelin (**)

(*) Landscape Development Unit, Department of Planning, Lothian Regional Council, Castlebrae Business Centre, Peffer Place, Edinburgh, EH16 4BB U.K.

(**) Port Edgar Marina and Sailing School, Lothian Regional Council, South Queensferry, Edinburgh, EH30 9SQ U.K.

ABSTRACT

The Port Edgar Draft Provisional Parliamentary Order proposes reclamation works that will remove the worst accumulations of silt within the harbour and with the east breakwater extension, will make the future management of the marina less costly, once development is complete. Until that is achieved the Regional Council must maintain the marina in a condition that ensures full use at low tide for all users but that is constrained by a lack of capital and a trading loss for the Centre. Clearly interim dredging measures must be built in to the redevelopment programme.

Plough dredging has been tried to minimise the cost to the Council but with mixed results. What plough dredging focused attention upon is the desirability of a purpose built craft that could operate in a marina or a small harbour environment and effectively deal with interim and long term dredging whilst working beside or around a pontoon system.

INTRODUCTION

The Regional Council, in taking a decision on 13th March 1990 to proceed with the redevelopment of Port Edgar as a marina village, by inviting developers to submit tenders in line with a brief for the development, has embarked upon a programme aimed at achieving this objective. The first steps in this programme of consulting the public on the draft development brief are complete. The second and third steps of preparing a draft Parliamentary Provisional Order eg. Welsh and Dyson Bell Martin [6] for the engineering works in tidal waters and an outline planning application have also been done.   Although outline planning permission is expected soon from the local planning authority, the City of Edinburgh District Council, a petition against the Draft Parliamentary Order has been submitted to The Scottish Office.

DEVELOPMENT TIMESCALE DETERMINED BY PARLIAMENTARY ORDER

Consultation on the draft development brief, eg. Department of Planning [1], was widespread and all affected parties had an opportunity of either commenting or hearing of the proposals at local meetings.   All comments made were considered and many lead to modification and refinement of the design concept before to submission of the draft Parliamentary Order.   The end product of that process was an eminently more achievable development yet sensitive to local concerns.   By prior consultation, it was hoped that the approval of the Draft Order would be smooth and that it would be confirmed by Parliament within a year, by the end of 1991.   By submitting the outline planning application in parallel with the Draft Order, it was hoped that marketing and selection of a developer could be undertaken in Spring/Summer 1992 with development commencing early in 1993. The petition lodged by the local Community Council against the draft Order has seriously affected this timescale and unless it is withdrawn there is a strong prospect of a Parliamentary Inquiry into the Order, which would delay the development programme even further.   It would also delay when dredging of the harbour can take place.

DEVELOPMENT BRIEF REQUIREMENT

The feasibility study, eg. Landscape Development Unit [2], Halcrow [3], Wimpey [4] and L&R Leisure Group [5], undertaken in connection with the redevelopment project indicated clearly that, apart from the ongoing maintenance dredging requirement, it would be necessary also to remove all clay and soft alluvial materials from those areas of the harbour which it is intended to reclaim.   The Council considered that, the developer having to mobilise dredging plant and hopper craft in order to dispose of the very substantial volumes of this material, approximately 586,000 cu.m., the developer should also be required to undertake 'arrears' of maintenance dredging throughout the remainder of the operational harbour and thereby establish post-dredge soundings of at least -2.00 m C.D.

The imperative for placing the burden of removing this accumulated siltation on the developer lies in the fact that Port Edgar operates at a substantial annual trading deficit due partly to the financing costs of the original purchase of the site and development of the marina facilities, and partly to the cost of operating the Council's sailing school and boat hire operation, which is run at least partially as a public service and is heavily used and much appreciated in an area not otherwise well provided with boating facilities.

The annual operating deficit in recent years has been in excess of £100,000 and this despite the fact that very little significant maintenance dredging has been undertaken since 1987. Monitoring of siltation and the Admiralty Dredgemaster's records from the days when the Royal Navy carried out a regular programme of maintenance dredging in Port Edgar indicate that the harbour as a whole receives approximately 44,000 cu.m. natural siltation annually, of which about 22,000 cu.m. settles in the approximately 3 hectares of principal operations, that is beneath the marina pontoon system and in the harbour mouth. It can be seen that an ongoing programme of routine maintenance dredging in these parts of the harbour alone would represent an annual commitment of some £40,000 to £50,000 simply to maintain the status quo, without tackling the accumulated siltation which now threatens the operational viability of the harbour.

INTERIM DREDGING NECESSARY

Since the Regional Council purchased Port Edgar in 1978, two fairly small campaigns of maintenance dredging have been carried out. In 1983 some 13,000 cu.m. of material was deposited at the Beamer Rock (close outside Port Edgar) by the grab-hopper dredger 'Breckland' and in 1987 12,495 cu.m. was dumped on the Oxcars spoil ground, some 5 nautical miles downstream, by the small grab-hopper 'Jean Ingelow'. In neither case was any material taken from within the area of the marina pontoon system, the dredgers working on shallow areas around the margins of the pontoon layout.

Within the pontoon system, depths have shoaled to the extent that in places there is less than 0.5m at M.L.W.S. While yachts tend to make their own beds in the deep, soft mud and therefore have not tended to lay over dangerously, siltation ongoing at a general reate of almost 300mm per year, means that the marina is living on 'borrowed time'. Movement of boats to and from their berths is hampered for up to 2 hours either side low water on extreme spring tides, and this is particularly acutely felt at Port Edgar because low springs tend to occur at mid morning on weekends in July and August, when of course the facility is most heavily used. Those who keep their boats in marinas with lock access perhaps restricted to 2 hours either side high water

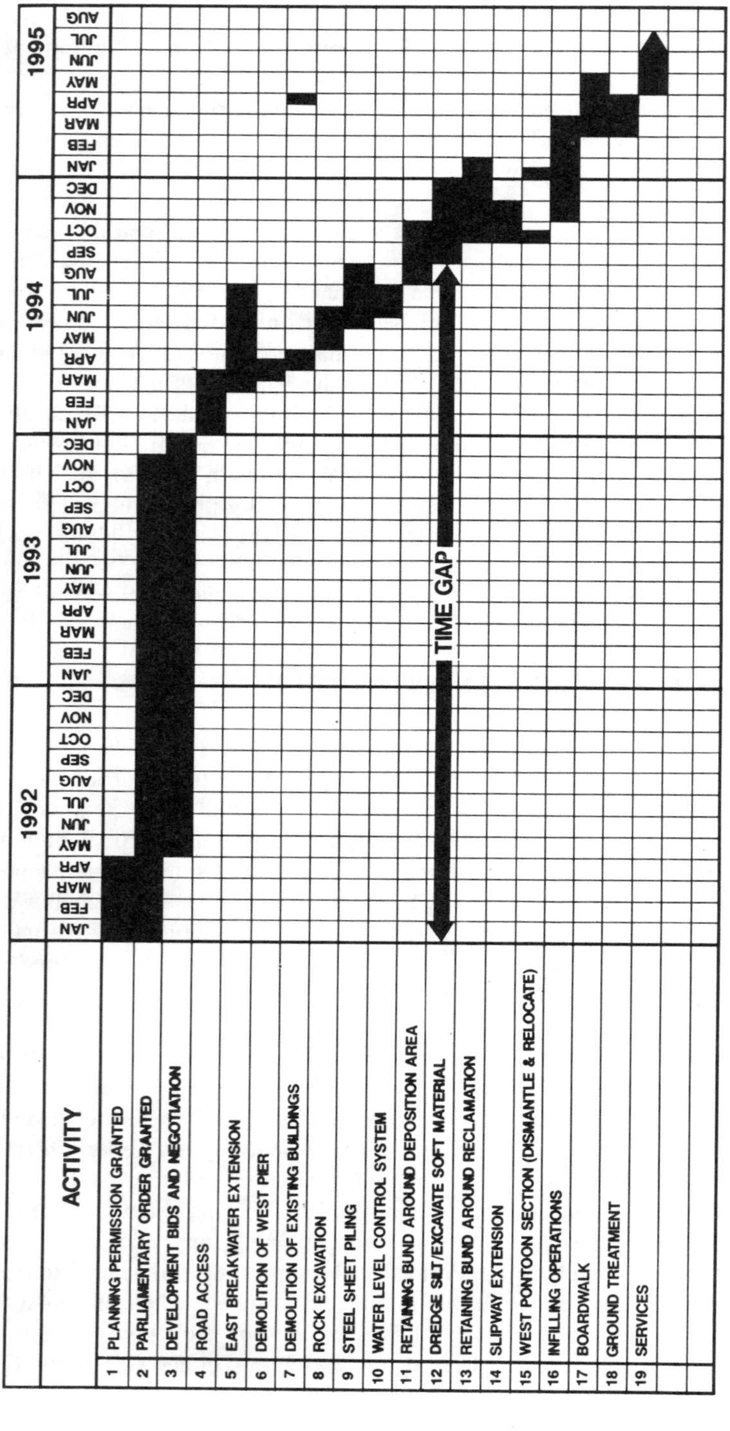

Fig 1 DEVELOPMENT PROGRAMME FOR RECLAMATION WORKS AT PORT EDGAR

whatever the tide, may not feel that our problems are too severe, but Port Edgar is the only true marina on the east coast of Scotland and is promoted as the all-tides access harbour it has always been. Absence of other attractive and amenable destination harbours within the immediate area means that much boating activity at Port Edgar is of the 'day sailing' type for which restricted tidal access is a particular problem.

Until the Parliamentary Order process is complete, Outline Planning Consent obtained and a developer selected and appointed to start work on the redevelopment scheme, the Council is obviously unable to achieve its original aim of having the marina dredged as part of the developer's consideration for his stake in the project. Assuming a best possible case, that is that objections to the Parliamentary Order will be withdrawn, Outline Planning Consent obtained, and there will be a strengthening in the housing market in the short term, it will still be at least two and a half years before reclamation works can commence and with them the needed maintenance dredging [1]. Recognising that the marina could lose trade in the interim if siltation is allowed to carry on unchecked, the Council has allocated £125,000 in the Capital Plan for 1991-94 for dredging within the harbour. Assuming the Council adopts this item in its Capital Budget for 1992-93, the earliest that dredging is likely to take place is after the summer sailing season of 1992.

PLOUGH DREDGING EXPERIMENTS

In 1990, using a small underspend in the Centre's revenue budget, an experiment using a Plough Dredger was tried. The river bed immediately beyond Port Edgar's breakwater ends falls away steeply to deep water and the intention was to use a bottom plough to pull material out from the marina into this deeper water, from whence it would be dispersed by the fast running tidal streams in the main river [2]. A big advantage of the bottom plough dredging method is that it appears to be regarded by the powers that be as an extension of the bottom levelling process from which evolved, and accordingly does not require the dredging and sea disposal licences necessary for other dredging methods, which can be extremely time consuming and difficult to acquire.

Dredging Method

For those who are unfamiliar with the method, Plough Dredging consists of a sizeable steel box plough with open front and cutting edge being pulled across the harbour bed by a powerful workboat. As it is

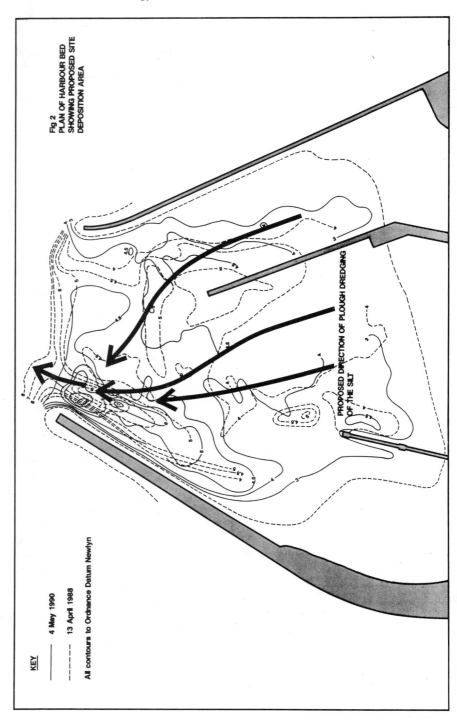

Fig 2
PLAN OF HARBOUR BED
SHOWING PROPOSED SITE
DEPOSITION AREA

PROPOSED DIRECTION OF PLOUGH DREDGING
OF THE SILT

KEY

——— 4 May 1990

– – – – 13 April 1988

All contours to Ordnance Datum Newlyn

dragged along, the plough fills with material until it overflows. The plough is hung on a lifted wire rigged to an 'A Frame' or gantry on the stern of the work boat and this wire is kept at a predetermined length so that, when the bed shelves away to deeper water, the plough leaves the bottom and the material within it drops out through the underside of the plough, which is partly open [3].

As each towing cycle removes only one 'ploughful' of material it follows that the vessel's cycling time must be short if reasonably affordable production rates are to be achieved.

First Experiment

In the case of the first experiment in Plough Dredging at Port Edgar, the mistake was made of approaching a marine contractor without previous experience of this specific dredging process. The contractor fabricated a plough by modifying a pre-existing and unrelated item of equipment based, it is assumed on the usual sort of longshore espionage of hanging round other people's yards in the gloaming armed with a fag packet and a stub of pencil. The plough was supplied hung on the back of a small ship handling tug about 70' long and drawing nearly 9' of water. It was hung from another 'borrowed' piece of equipment being a small fixed derrick with hydraulic winch to raise and lower the plough, this being driven by the tug's main engine. Sadly the hydraulic pump and winch were a complete mis-match with the result that at working revs. the winch would operate only at a snail's pace. To raise and lower the plough, the tug had to be taken out of gear and the engine raced. The towing wires, led round turning blocks on the vessels rail toward were controlled by a quite inadequate hand winch and were consequently virtually incapable of adjustment in use. The angle and length of the towing wires is critical to the success of the operation and must be capable of adjustment to suit differing depths and materials.

Not surprisingly this rig did not work. The tug (single screw) was far too unhandy to manoeuvre safely in the close confines of the marina, not least because of her draft of 9', which in turn severely limited the hours during which she could work. Handling a ship like this, the skipper dares not to get close to the marina pontoons or any boats lying on them: one touch would have done thousands of pounds of damage. The combined effects of her unmanoeuvrability and her pathetically slow winch speeds meant that her cycling time was running at about 24 minutes, of which the plough was actually on the ground and 'working' for only about 5 minutes. Even had the plough been digging in and shifting material over the approximately 350 metre distance to the deeper water it was clear that this was likely to be the

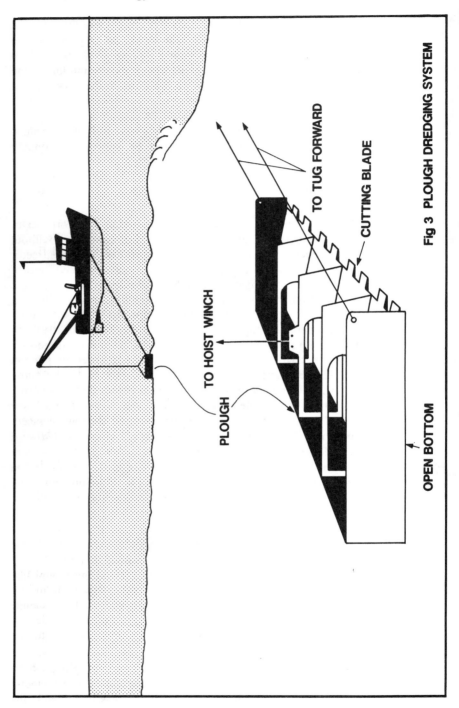

Fig 3 PLOUGH DREDGING SYSTEM

TO TUG FORWARD

CUTTING BLADE

TO HOIST WINCH

PLOUGH

OPEN BOTTOM

dearest mud ever shifted. In fact however the plough was not working at all: when dropped into the mud with the vessel stopped it was possible to see (on the following low water) that it was digging in under its own considerable weight, leaving a neat rectangular bite in the mud. However as soon as sufficient power was applied to the big tug to start the plough ahead, it was clear that the plough's leading edge was coming straight up out of the mud and the whole thing was skittering away over the surface achieving precisely nothing. Even at dead slow ahead the tug had too much way and momentum to allow the plough to bite. When lifted clear of the surface for the return run (something proper ploughing rigs do not need to do) the plough was coming up beautifully washed and totally innocent of any trace of the mud that was supposed to be dragged out off the harbour.

After two days of this and various excuses, promises and attempted obfuscations from the contractor this experiment was knocked firmly on the head.

A Second Attempt

This however is not to say that Plough Dredging does not have its uses: early in 1991 the opportunity of using Westminster Dredging's small ploughing unit F49, was taken which had been working at Rosyth Dockyard. This was an entirely different proposition. The vessel herself is a 50' twin screw standard Damen type workboat with twin 325hp diesels. As such she is vastly more manoeuvrable than the previous experimental vessel and can work close to marina pontoons with precision and much reduced risk of damage. Her draft at about 6' is obviously much more manageable as well. Her plough was not dissimilar in design to that produced by the original contractor, but it was hung from a purpose designed A frame on her stern complete with hydraulic lifting/lowering winch and independently controllable winches for her plough towing wires.

The objective of this Plough Dredging effort was much more limited than before: it was to deepen a channel over the short distance between the visitors' berths at the seaward end of the marina and the naturally deeper scour pit just inside the west breakwater end, and to deepen an area in the vicinity of the visitors' berths, so that vessels returning to the marina at low water could at least get alongside the visitors' berths even if they had to wait for sufficient rise of tide to get to their own berth.

The distance over which the plough was being asked to work in this case was much shorter - approximately 150 metres - and the working area generally unconfined.   The vessel was able to work in a circular rotation, never having to reverse to position herself, and this in turn allowing her to operate without having to raise and lower the plough at the beginning and end of each cycle: at the end of the ploughing run more power was applied to each engine causing the plough to 'kite' up in the water dropping its load. The vessel returned to the beginning of the plough run at speed where she was slowed down, the plough sinking again into the mud. The unit was achieving cycling times of approximately 5-6 minutes of which the plough was on the bottom and working for 3-4 minutes.

In this way the vessel was surprisingly effective and was able to deepen the intended channel (between two sets of transits established on the breakwater) and the visitors' berth area remarkably rapidly [4]. It was noticeable however that the rate of progress slowed down markedly as the plough filled up the end of the scour pit nearest the dredge area: as the vessel worked her way back into the dredge area and further from the scour pit she was also filling up the pit, needing to drag material even further to deeper water.

After a few days working and well within the fortnight's work budgeted the unit had deepened the channel and the visitors' berth area. She was therefore deployed to creating a channel along the west edge of the marina. This involved a total drag to deep water of about 350 metres and, because of depth restriction, it had to be tackled in two bites, with the plough pulling material during the high water period the first 180 or so metres into the area already deepened and returning during low water to rehandle the material the remaining distance out to the deep. Progress with this was much slower and although a narrow channel was established it was clear that the production rate was such that it was not a cost effective way of going on.

Observations on this Method

From these Plough Dredging experiences it is possible to derive several important principles about the method;

1.   The vessel and plough must suit the intended application in terms of size, draught and manoeuvrability. The first tug was totally inappropriate for working in the close and comparatively delicate confines of a marina. The second unit, whilst much handier and more controllable, and able to work close to the pontoons in relative safety, was still too large a 50' length to work effectively between the main walkways. While able to work at speed in circular pattern round the margins of the marina, never going

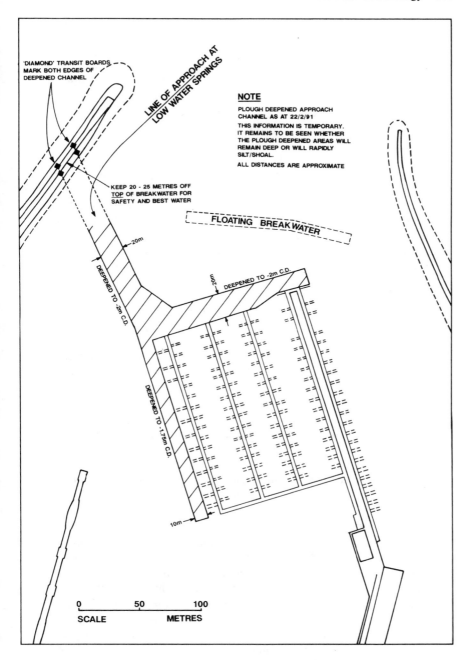

'DIAMOND' TRANSIT BOARDS
MARK BOTH EDGES OF
DEEPENED CHANNEL

LINE OF APPROACH AT
LOW WATER SPRINGS

NOTE

PLOUGH DEEPENED APPROACH
CHANNEL AS AT 22/2/91

THIS INFORMATION IS TEMPORARY.
IT REMAINS TO BE SEEN WHETHER
THE PLOUGH DEEPENED AREAS WILL
REMAIN DEEP OR WILL RAPIDLY
SILT/SHOAL.

ALL DISTANCES ARE APPROXIMATE

KEEP 20 - 25 METRES OFF
TOP OF BREAKWATER FOR
SAFETY AND BEST WATER

FLOATING BREAKWATER

20m

DEEPENED TO -2m C.D.

20m

DEEPENED TO -2m C.D.

DEEPENED TO -1.75m C.D.

10m

0   50   100

SCALE   METRES

Fig 4 SECOND PLOUGH DREDGE EXPERIMENT

astern she achieved creditable output. She would not however have been able to do this within the pontoon system itself. The need to back and fill and precisely position the plough, raising and lowering it each cycle, would have slowed her down immensely. For working within a pontoon system a smaller vessel of length perhaps 10 metres is needed, with rapid plough raising and lowering capability and the manoeuvrability afforded by twin screws. Plough winch controls should be operable from the helm position for single man operation, and the helmsman should have unobstructed vision of all parts of his boat. The second of the ploughing vessels did not meet either of these requirements.

2.   The method is cost effective only in suitable materials (not a problem at Port Edgar) and over limited distances. This implies that there must be a deep or non depth-critical area in the immediate vicinity into which to plough the material. As very few marinas will be fortunate to have these natural assets it follows that for most marinas the plough dredging method on its own will be of value only for small ad hoc projects or stop gap measures, rather than as a means of carrying out substantial volumes of maintenance dredging.

3.   The Plough Dredging /Bed Levelling method will also have application in conjunction with other types of dredging plant. Marina maintenance dredging frequently aims to remove a thin skin of material, perhaps with a bench height of less than one metre, over a wide area. If for any reason a stationary cutter/suction unit pumping to pipeline discharge is not suitable (probably because there is no available nearby spoil area), a conventional grab-hopper vessel may have to be used. The problem with such units is their inability to dredge to precise levels and leave a regular and even finish. If in-hopper measurement is used, the client may end up paying for large quantities of over-dredging in order to achieve an area generally down to the specified depth. The use of a plough dredger or bottom leveller to level up the worked area, pulling the high spots into the over dredged areas may speed the operations and may achieve a better finish. Indeed the grab-dredger may intentionally concentrate on dredging a deep pit in an area of easy access, thereby speeding its own cycle time and boosting its productivity, while a plough is used to pull material from less accessible areas into the over-dredged pit. Again, while a trailer suction dredger may not be ideal for dredging most marinas due to lack of space for manoeuvre, a plough can be used very effectively to level up the 'corrugated' effect of repeated passes by the trailer's draghead.

PURPOSE BUILT SOLUTION

Mobilising for a major cutter suction dredging exercise will undoubtedly account for the majority of the cost of dredging at Port Edgar, especially as the depth of water available for such a dredger is limited within the actual harbour (only half of the harbour is navigable at low tide.)

A grab-hopper could be used as has been in past at Port Edgar but it is not as manoeuvrable a vessel within a marina environment, tends to be more crude in its ability to execute a consistent and even dredge to predetermined levels. If their rotating grab could be mounted on a more efficient platform than a traditional powered hull, then it possibly may be an ideal solution for a smaller harbour containing a marina. However such a craft does not exist in the market place.

Design Concept

What is needed is a small, self-contained backhoe/hopper dredger capable of being operated by a small crew (2 men ideally) and of being mobilised and work-readied with minimum inconvenience and expense, eg. Wakelin [7]. It must be of a size and manoeuvrability capable of entering and working safely within marina access gates and pontoon systems with minimum operational disturbance and yet capable of carrying a meaningful and viable payload on a limited draught. It must be truly versatile in being able to:

a) load itself,
b) load other barges,
c) discharge itself by bottom dump,
d) place material ashore, and
e) stationary suction dredge.

No such craft is thought to exist at present in the United Kingdom.

The drawing shows [5] a novel but practicable solution to the problem eg. Wakelin [8]. This is a small self powered bottom dump hopper (capacity about 200 cu.m.) which carries with it, it's own 360° backhoe excavator and a spudded pontoon on which the excavator will ordinarily work, although for smaller jobs, or for shore placement work, the excavator can work on the aft deck of the hopper. The whole unit is self sufficient and can work entirely on its own, or can form the basis of a larger operation, loading other hopper craft as well as itself. It is capable of sea mobilisation with a crew of two men, although more

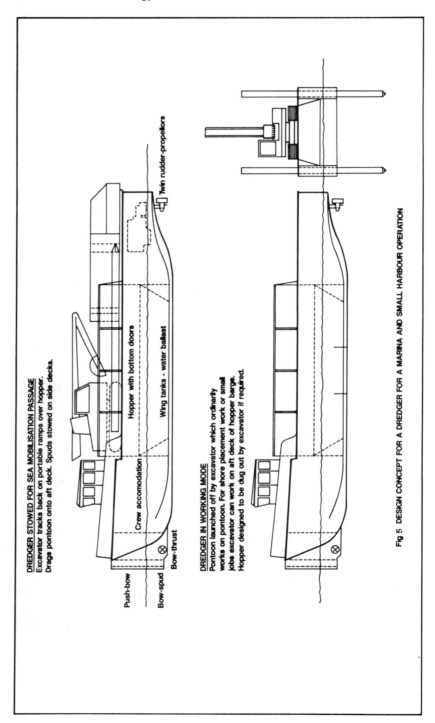

DREDGER STOWED FOR SEA MOBILISATION PASSAGE
Excavator tracks back on portable ramps over hopper.
Drags pontoon onto aft deck. Spuds stowed on side decks.

Twin rudder-propellors

Hopper with bottom doors

Wing tanks - water ballast

Push-bow

Crew accomodation

Bow-spud

Bow-thrust

DREDGER IN WORKING MODE
Pontoon launched off by excavator which ordinarily
works on pontoon. For shore placement work or small
jobs excavator can work on aft deck of hopper barge.
Hopper designed to be dug out by excavator if required.

Fig 5 DESIGN CONCEPT FOR A DREDGER FOR A MARINA AND SMALL HARBOUR OPERATION

may be required depending on the passage. Once on site, the lifting and dragging power of the excavator is harnessed to launch the pontoon and install the spuds etc., without need of shore crane assistance, and likewise on completion of contract, the excavator can reload the pontoon and unship spuds etc. Depending on operational and site conditions, the dredging operation can be accomplished by two men.

Advantages

The advantages of the 'dismountable pontoon' are that the hopper does not have to carry the dead weight of the excavator away to sea dump each trip, thereby increasing its own payload. Instead, the excavator is left at the dredge site where it can load other barges, if needed, level up the area it has just worked, and reposition itself ready for the hopper's return. Because the digging machine remains at the worksite, the delays and inaccuracies involved in repeated repositioning, which lead to missed patches and over dredging ,are eliminated.

Very significantly the 'dismountable pontoon' will allow the use of the recently introduced variable counterbalance, long-reach 'semi-dragline' type of excavator. These machines are at their best in soft digging, such as most marina silt, and have colossal reach. They have never been successfully mounted on a hopper dredger however, for the reason mainly that, owing to the great length of their main and dipper arms, they become slow and awkward if asked to place material close to themselves, as is obviously necessary if mounted on a hopper craft. With the excavator mounted on the pontoon, the hopper, which is also provided with spuds, can be stationed for loading at the appropriate distance from the pontoon to allow efficient digging. The 'variable counterbalance' aspect of the design of these machines wherein the centre of balance shifts but little, even though the working radius varies considerably, has stability advantages for pontoon operation.

The proposed vessel would be extremely versatile: with the pontoon launched off and the excavator operating on the aft deck it can be used for conservancy and riverbank maintenance work and for shore placement of materials at inaccessible locations, the hopper being designed to be dug out by the excavator. The unobstructed aft deck and tractive and lifting power of the excavator would make it a very viable diving support, or salvage vessel, or buoy tender. It could readily carry overhead access equipment for bridge or harbour wall maintenance work.

U

CONCLUSION

If marina village developments are to be built by inviting developers to submit bids for their development, then to ensure the promoter succeeds in achieving this goal it would be prudent to give full consideration not only to the influence dredging could have on the bid offered but more importantly to the effect the development timescale has or could have on the need to dredge. If full account is not taken at this predevelopment stage it will be a serious impediment to success. Furthermore, although every developer is aware that maintenance of dredged levels is important for sustaining custom, few take real account of how to resolve the matter in future, because by then it is highly likely to be an operator's rather than a developer's problem. Perhaps the design concept for the new type of dredging vessel suggested here is the solution for marina operators in the future. Port Edgar's essential interim dredging will protect its potential for redevelopment but it will not unfortunately test out this idea.

ACKNOWLEDGEMENTS

This paper represents the views of the authors which are not necessarily those of Lothian Regional Council. Gratitude is expressed to the Director of Planning for allowing publication of this paper and to Chris Bushe in the Unit for his assistance with the illustrations.

REFERENCES

Report

1.   Department of Planning, Lothian Regional Council, Development Brief, Port Edgar Marina Village, Queensferry, Edinburgh, Draft for Consultation, December 1989.

2.   Landscape Development Unit, Department of Planning, Lothian and Regional Council, Port Edgar Development Phase 11 Study, Volume 11, February 1989.

3.   Sir William Halcrow and Partners Scotland Ltd, Port Edgar Development Project, Hydraulic Study, Glasgow, August 1988.

4.   Wimpey Laboratories Limited, Proposed Redevelopment of Port Edgar, Report on Site Investigation, Lab. Ref. No S/25863, Broxburn, May 1988.

5. L & R Leisure Group, In association with CASCO, GRM Kennedy and Partners, W J Cairns and Partners, Tozer Gallacher, LRC Landscape Development Unit, PriceWaterhouse, Port Edgar Development Phase II, Volume I, Edinburgh, November 1988.

Draft Provisional Order

6. Welsh, G.F.G and Dyson Bell Martin & Co, Lothian Regional Council (Port Edgar) Draft Provisional Order, Edinburgh and London, November 1990.

Paper In A Journal

7. Wakelin, M, A New Idea for Dredging Marinas, Marina Management International, pp 15-17, August 1991.

8. Wakelin, M, Self Contained Dredging Vessel, Marina Management International, November 1991.

U*

The Use of Floating Pontoons for the Construction of Italian Marinas

F. Prinzivalli

SISTEMA WALCON S.r.l., Via Sutter, 29 - Ferrara, Italy

ABSTRACT

The aim of this article is to illustrate the use of floating pontoons in Italy; in fact, with the introduction of these products in the Italian market, marinas on a par with the highest European standards have been able to be constructed with reduced costs.

Numerous examples of marinas built using floating pontoons are described here, with the main technical features of each.

INTRODUCTION

In order to examine the use of floating pontoons in marina construction in Italy, first of all it is necessary to take into consideration some elements which have a direct influence on these structures use. The first concerns the country's geographical location: almost completely surrounded by the sea, necessity has compelled it to build structures for mooring boats. The second, subtly connected with the first, regards the boat market, which has developed both in terms of quantity and quality during the last few years: in other words, as well as increasing numerically, pleasure boats have also grown in size.

This phenomenon has caused a constant increase in the need for berths, inducing operators of this sector trade to find an alternative solution to marinas realized with fixed jetties, looking for flexible, modifiable structures able to be adapted to the changing needs of the nautical service market.

On this subject, trade operators have had to reconcile environment and landscape requirements (unable to be changed in any way) with Italian and foreign boats demand for berths.

On the basis of these requirements, it was therefore better to use modular floating structures, which enable flexible mooring systems to be built. At the same time yacht harbours should not be fixed and defined, but continually evolving structures in which the type of berths and their layout can be varied according to the yachtsmen's needs.

From the point of view of its utility, a completely floating harbour system represents the right solution from an economic and rational point of view; the ease and speed with which the floating modules can be assembled and moved in the water and the flexibility of the connections between pontoons and fingers allow berth set-up to be reorganized to suit different types of fleets and bearing in mind different market demands.

The above-mentioned ideas, already known to marina designers throughout the world, have only been put to use in Italy during the last few years.

Up until a few years ago, very limited use was being made of floating pontoons in Italy; marinas were built with permanent jetties, and the only modular floating structures known were small plastic pontoons, which were extremely unstable. This brought users and manufacturers to consider floating pontoons as unstable things to be used only where necessary while awaiting the construction of a fixed jetty.

The diffusion on the market of safe, stable jetties revolutionized ideas on the instability and precariousness of floating pontoons and actual marinas were able to be built using floating pontoons, generally combined with mooring systems using fingers.

The installations shown here illustrate how the construction of safe, flexible and modifiable mooring systems with reduced costs has been possible in Italy as well, thus offering the most up to date technological solution for fitting out a sheltered basin, already used with considerable success in numerous marinas throughout the world.

THE INTRODUCTION OF FLOATING PONTOONS

The introduction and knowledge of floating pontoons in Italy concided with the development of one of Europe's most important nautical centre: Lignano Sabbiadoro. In this area in fact, from 1983 until today, berths and marinas with a capacity of more than 5000 boats have been built.

Lignano is situated in the northern Adriatic in a lagoon area alongside the mouth of the River Tagliamento.

Its particular geographic position and ease of access for tourists arriving from countries north of the Alps (Austria and Germany) has meant that there has been a great increase in the demand for berths in this area, already famous as a seaside tourist resort.

Due to its geographic configuration, the Lignano zone has a wide tidal range and building marinas with fixed jetties would have been difficult, so marinas with floating pontoons had to be designed and built instead.

New technology and products already widely used abroad, but which Italian trade operators considered as innovations, were therefore introduced; after designers' and trade members' initial mistrust of floating jetties had been overcome, the marinas described here were able to be built.

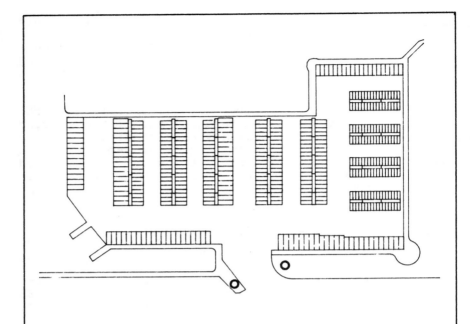

FIG. 1– MARINA UNO LAYOUT

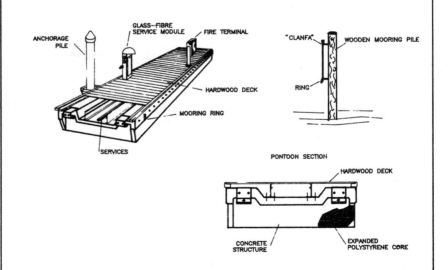

FIG. 2– CONCRETE FLOATING PONTOON WITH HARDWOOD DECK

MARINA UNO (Lignano Sabbiadoro - Udine province)

The Marina Uno yacht harbour (fig. 1) is at Lignano at the mouth of the River Tagliamento and can hold 535 boats.

In the 18,000 sq.m. dock, there are 9 pontoons built by connecting 37 floating elements, each measuring 11.5 m. x 2.5 m. (fig. 2) and built from a single suitably reinforced concrete block which floats thanks to closed cell polystyrene foam blocks inserted during manufacturing; the concrete sections also have holes and brackets for mounting deck and mooring fittings.

An "Iroko" wood deck is mounted on the concrete structure with special brackets which leave a space between the deck and the top of the concrete, along which the cables and tubes necessary for the boats services have been laid.

The floating elements weigh 11.5 tons, have 0.5 m. freeboard and (thanks to their displacement) are extremely stable. The dimensions of the steel piles used to anchor the pontoons were calculated according to the wind force which the boats are subject to. The boat moorings on the other hand are wooden piles fitted with "clanfe" with sliding rings which allow the boats to follow the rise and fall caused by the river rate of flow, always ensuring correct mooring line tension.

Tubes and cables for water, electricity and fire fighting have been laid in the special duct formed in the floating section.

Special fibreglass service modules (fig. 3) mounted on the pontoons have been designed not only as indispensable accessories for pleasure boats, but also to fit in with the marina fixtures. Each module has water and electricity plugs with power calculated on the basis of the boat consumption: they are also fitted with a low-voltage lamp positioned in such a way as to illuminate the deck round the module.

There is a fire-fighting system throughout the entire marina and special fibreglass cabinets have been mounted on the pontoons for this purpose (fig. 4).

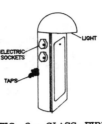

**FIG. 3– GLASS–FIBRE
SERVICE MODULE**

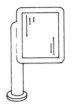

**FIG. 4– GLASS–FIBRE
FIRE TERMINAL**

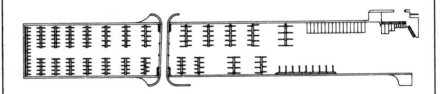

FIG. 5– DARSENA SABBIADORO LAYOUT

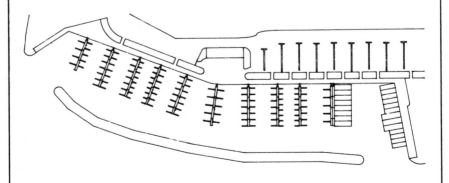

FIG. 6– MARINA PUNTA VERDE LAYOUT

DARSENA SABBIADORO - MARINA PUNTA VERDE

(Lignano Sabbiadoro - Udine province)

The Lignano Sabbiadoro dock (fig. 5) is an example of reutilization of a sheltered basin; this dock was in fact built in 1934 and during World War II used as a harbour for seaplanes. It was later used as a fishing port and lastly (in 1988) converted into a marina.

The 39,000 sq.m. dock has been fitted with floating pontoons and fingers with technical facilities enabling 500 boats to be moored.

Punta Verde (fig. 6), a 308-boat marina located in a sheltered 23,000 sq.m. basin on the left bank of the River Tagliamento, is an example of how these floating structures can fit in harmoniously with the surroundings.

The inner pontoons of both marinas were built by connecting 11.5 m. x 2.5 m. floating elements (fig. 7), each consisting of a support frame built from TA6082 aluminium alloy for marine use making it light, stable and extremely versatile and ensuring low costs in the event of any changes in the set-up of the dock pontoons which could be necessary if the fleet to be moored changes in coming years.

The frame is covered with a deck in Yellow Balau, a high quality exotic wood from Borneo which has proven to be the most suitable for marine use, as it maintains its looks and technical characteristics.

The floating element has a displacement of 2500 kg., 0.5 m. freeboard and is supported by floats manufactured in fibre reinforced concrete. The concrete shell is filled with closed cell polystyrene foam, which ensures that it floats even in the event of accidental breakage. The fact that there is no metal reinforcement eliminates the risk of rusting, guaranteeing the floating section unlimited life.

Lastly, the use of concrete for the floats ensures greater stability for the entire pontoon, since it gives a lower centre of gravity.

There are two separate ducts in the pontoon metal structure for the cables and tubes necessary

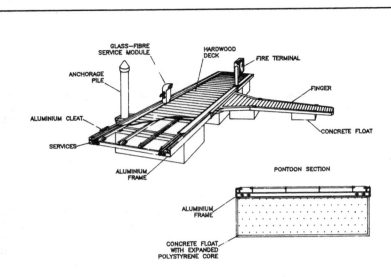

FIG. 7— ALUMINIUM FRAME FLOATING PONTOON

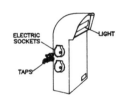

FIG. 8— GLASS—FIBRE
SERVICE MODULE

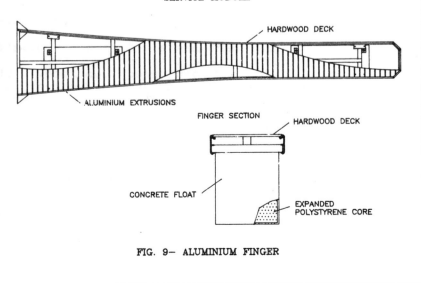

FIG. 9— ALUMINIUM FINGER

for the boat electricity and fire fighting facilities. Special fibreglass service modules (fig. 8) with all the necessary fixtures for supplying water and electricity to the boats are mounted above the ducts. Fire fighting cabinets are mounted along the pontoons in such a way as to cover the entire area occupied by the boats.

The particular shape of the pontoon frame beams also enables all the accessories to be fitted (fingers, mooring cleats, service modules, etc.) and positioned wherever they are required along the frame beams.

The dimensions of the steel piles which anchor the pontoons were calculated according to the wind force which the boats are subject to (since this is an internal marina, it was assumed that the action of the waves could be overlooked). The pontoon metal structure is connected to the piles by means of special pile guides fitted with buffers to allow the structure to move along with the rise and fall of the tide.

The boats are moored using fingers, the most up-to-date, functional solution for ensuring boats and yachtsmen safety and comfort under any tidal conditions.

The mooring fingers are connected to the pontoons to exploit the available space to the utmost and give the largest possible number of berths.

This is a trend being followed by all marinas throughout the world, making the entire structure even more functional and smart.

The 0.72 m. wide finger (fig. 9) is set at right-angles to the pontoon and forms a useful deck running along at least two thirds of the length of the boat, enabling yatchsmen to carry out landing and embarking from the side of the boat as well. Moreover, a boat moored at a finger of a floating pontoon seems to be part of it, following the rise and fall of the tide, without annoying differences in level between the pontoon and the boat.

Lastly, a floating pontoon with fingers is more stable and makes the entire set-up smart and more pleasing to the eye.

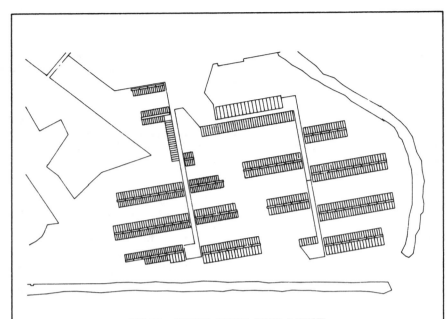

FIG.10– MARINA PUNTA FARO LAYOUT

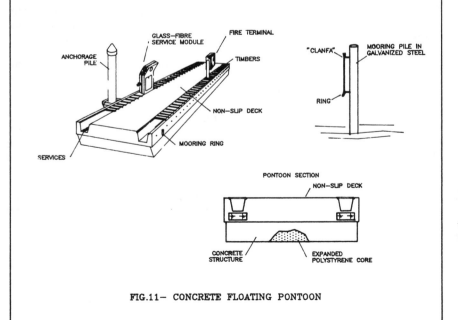

FIG.11– CONCRETE FLOATING PONTOON

MARINA   PUNTA   FARO   (Lignano   Sabbiadoro   - Udine
province)

A large marina (fig.   10) situated at the end of
the Lignano peninsula and divided into two   parts: a
dock   situated right alongside the mouth and an inner
part   with   the   jetties directly   in   front   of   the
houses.

The   dock   has a capacity of up to 1,400   berths
and was built in two stages: the first in 1981/82 and
the second in 1988/89.

The   dock   inner pontoons were built   using   101
modular   elements 10 m.   long and   3   m.   wide.   Each
section   was built from a single suitably   reinforced
concrete   block (fig.   11) which floats thanks to the
insertion   of   blocks   of   closed   cell   expanded
polystyrene during manufacturing; there are holes and
brackets   on   the   concrete   structure   for   fitting
mooring accessories.

The   upper   side   of the block (in   concrete   as
well)   forms a deck with a nonslip paint   finish.   At
the sides of the deck there are two ducts covered   in
Yellow Balau wood which hold the cables and tubes for
the boats' services. Each floating element has 0.5 m.
freeboard   and   13 tons displacement which   makes   it
very stable.

The dimensions of the steel piles used to anchor
the   pontoons   were calculated according to the   wind
force which the boats are subject to.

The boats on the other hand are moored at wooden
piles   fitted with "clanfe" with sliding rings   which
allow the boat to follow the rise and fall caused   by
the   rate   of   flow of   the   river,   always   ensuring
correct mooring line tension.

The special fibreglass service modules (fig.   12
mounted   on the pontoons were designed to fit in with
the port fixtures.

Each module has water and electricity plugs with
power   calculated   on   the   basis   of   the   boat
consumption.   They are also fitted with a low voltage
lamp   positioned   in such a way as to illuminate   the
deck round the module.

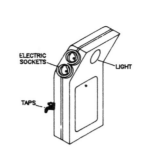

FIG.12– GLASS–FIBRE
SERVICE MODULE

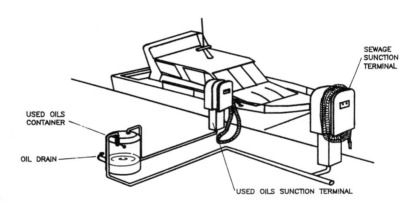

FIG.13– SEWAGE AND USED OILS SUNCTION PUMP

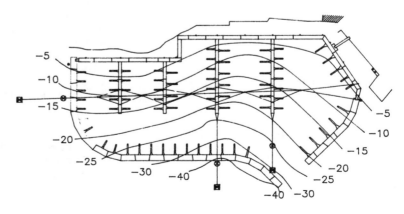

FIG.14– PORTOLABIENO LAYOUT

There is a fire fighting system throughout the entire marina and special fibreglass cabinets have been mounted on the pontoons for this purpose.

The marina also has a system which enables sewage and used oils to be removed directly from the boats (fig. 13).

THE USE OF FLOATING PONTOONS ON ITALIAN LAKES

As has already been mentioned, nautical tourism in Italy is constantly increasing. As well as involving the coast, this phenomenon also regards inland waters. In the last few years in particular, there has been an increase in the number of boats on the lakes and therefore a constantly increasing demand for berths.

At first glance, the building of a port on a lake could seem easy, since the area is sheltered, but at the same time there are environmental and orographical factors which can technically condition both the design and building of a marina; in particular, due to the rate of flow of water in and out, lakes have a wide water level range. Moreover, lakes are not sheltered areas, since, even if the fetches are limited, the characteristics of the wave motion created by the wind mean that pontoons and boats can only be moored if they are protected by a breakwater.

PORTOLABIENO Laveno Mombello (Varese province)

Portolabieno (fig. 14) was built at Laveno Mombello (Varese) in 1989 on the eastern shore of Lake Maggiore and was the first example of an Italian marina completely built using floating elements.

The orographic situation of the territory, in particular its water (over 40 m. deep just a short distance from the shore) and a range of over 4.5 metres in the water level of the lake had not allowed the construction of a traditional marina.

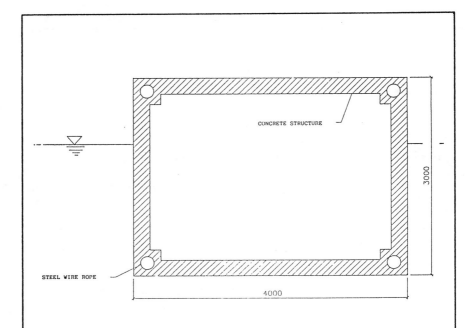

FIG.15— PORTOLABIENO FLOATING BREAKWATER SECTION

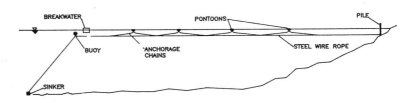

FIG.16— PORTOLABIENO FLOATING PONTOONS ANCHORAGE SYSTEM

The concept of using floating elements was thus developed, not only for the fixtures inside the dock (already done in numerous marinas), but also for constructing the outer breakwater.

In fact, the realization of the work using fixed structures would have had prohibitive costs and considerable technical difficulties, without taking into consideration the extremely serious environmental impact on the lake shore.

The solution was therefore reached by using a completely floating system able to automatically adapt to the lake level.

Portolabieno has a capacity of 160 berths for boats of up to 14 metres and it occupies approximately 14,500 sq.m.

Mooring is done using fingers, a type of berths which means that the boat, the pontoon and the mooring form a single floating body, able to follow every variation in the water level perfectly.

To protect the inshore water from the wave motion, a breakwater had to be built.

The orography of the site excluded in advance the possibility of building a breakwater of the traditional type with rock filling or with a foundation on piles, so a completely floating reinforced concrete structure was decided on.

The breakwaters are in pre-stressed reinforced concrete (fig. 15) which forms a monolithic (joint-free) element. The overall dimensions are as follows:

- Outer breakwater length 696.5 m.
- Inner breakwater length 54 m.
- Width 4 m.
- Freeboard 0.7 m.
- Draft 2 m.

An excellent reduction (80-85%) of relatively short waves (wind) was achieved with these dimensions.

The breakwater was built directly in a dock yard in single sections approximately 9.5 m. long and weighing approximately 70 tons; to standardize the sections to the utmost, a trapezoidal shape was chosen, enabling the breakwater to be positioned in a straight line or a curve. The sections were launched and then assembled using steel ties inserted in specially prepared slots on the four corners of the structures.

The sections are hollow, can be inspected using watertight hatches and have internal ducts for electricity and water supplies.

To enable the connection of all additional fittings, there are two rails set in the concrete along the inner side of the breakwater: one for mounting the wooden fender, the other for the mooring fingers, which can therefore be positioned according to the dimensions of the boats to be moored, enabling all the space to be exploited to the utmost. The basin was equipped with floating pontoons in aluminum alloy with the characteristics already mentioned. The marina also has water and electricity supplies for servicing the boats.

Innovative anchorage systems for the port structure were studied in order to ensure that structures were firmly held and at the same time allowing them to follow the rise and fall of the lake level.

In the shallower water along the shore, a series of anchor piles were driven into the lake bed: by means of a system of chains and struts, the stress sustained by the breakwater is discharged on other piles driven into the bed at a depth of 20 metres.

An anchor cable for the pontoons (fig. 16) was laid under them and moored at a depth of 6.5 m. below mean lake level.

One end of the cable is connected to two breakwater anchor piles and the other to a buoy/anchor-block system which guarantees that the cable is kept at the set depth and with the tension necessary to resist the wind force on the pontoons. The pontoons are connected to the cable by means of chains.

The heads of the two longest jetties have also been anchored by means of a buoy/anchor-block system similar to the one described above.

PORTOSCUSO - Portoscuso (Cagliari)

In the southern part of Sardinia, a 2,900 sq.m. basin in the town of Portoscuso, previously used by the local fishermen, has been converted to a marina.

After having renovated the piers, floating modular elements in aluminium alloy for marine use were installed, forming pontoons with mooring space for 300 boats (fig. 17).

The pontoons are anchored with piles and have been fitted with fingers, thus ensuring the users greater convenience.

The port has water, electricity and fire fighting-systems. A system for the removal of sewage and used oil from the boats is also planned.

To ensure users's security, an entry gate equipped with an electric lock and opened by means of a magnetic card has been mounted on each pontoon.

The care taken with details, material quality and technology used has made this marina one of Italy's best from the point of view of functionality and convenience for its users.

OTHER INSTALLATIONS

As well as the previously described ports, there is a considerable number of ports or sheltered basins round the Italian coast which were not expressly designed for yachting and are under-used or even unused.

Using floating structures, some of them could be converted into marinas able to hold a considerable number of pleasure boats.

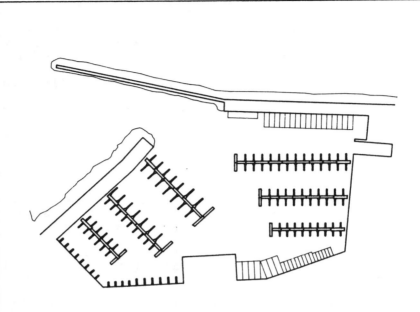

FIG.17- PORTOSCUSO LAYOUT

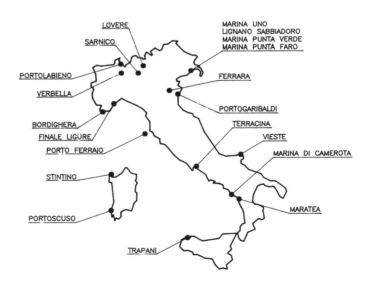

FIG.18- FLOATING PONTOONS MARINAS LOCATION

Some examples of these are: Verbella at Sesto Calende (Varese) on Lake Maggiore; Sarnico e Lovere (Bergamo) on Lake Iseo, Bordighera (Imperia) and Finale Ligure (Savona) in Liguria, Stintino (Sassari) in Sardinia, Trapani in Sicily, Portoferraio in the Island of Elba (LI) then Ferrara, Porto Garibaldi (Ferrara), Terracina (Latina), Marina di Camerota (Salerno), Maratea (Potenza), Vieste (Foggia) and numerous other smaller places (fig. 18).

In short, there are over 14,000 boats moored at floating pontoons in sheltered basins in Italy today: this has been possible thanks to reduced costs and flexibility, as well as the great ease with which these structures can be installed.

CONCLUSIONS

From what has been said up until now, it is clear that technical designers' and marina users' lack of trust in these floating structures has been overcome and that they have now become an integral part of Italian nautical life.

Marinas and yacht harbours have been described in which the use of floating pontoons has allowed not only better use of the available space without altering the environment or landscape, but sometimes even reviving otherwise unused areas, resulting in top class marinas.

So this experience, along with that gained abroad, will serve as examples for the construction of new berthing.

In Italy infact, there are still a great many harbours and sheltered basins which, using modular floating structures, could be rapidly converted into convenient, functional, flexible marinas to meet Italy's increasing, variable demand for mooring space with reduced costs.

Durban's Marina Extension - a Modular Flexible Design using Fasttrack Methods of Construction

A.J. Tollow

Department of Civil Engineering, University of Durban-Westville, Private Bag X54001, Durban 4000, South Africa

ABSTRACT

The Durban Marina on the subtropical Natal Coast required enlarging with facilities for a mixed fleet of craft. High corrosion and marine life vectors were present. The site was exposed, so a flexible design capable of withstanding storms was required. Design and construction time were limited, so a fasttrack concept was adopted. This used a minimum number of different prefabricated steel units which were easily erected. The concrete deck was cast on site before attaching the rota-moulded polyethylene floats. The units were launched before being assembled in sections. They were linked by flexible neoprene blocks. The structure is readily modified to suit the needs of the various craft. The design has been subject to 80 knot winds from the south-west and withstood the conditions well.

Key Words: Flexible, Modular Design, Fasttrack Construction,

INTRODUCTION

A small part of Durban harbour, the largest commercial port in Southern Africa, has been set aside for leisure craft moorings. Access to the open sea is gained through the deep dredged commercial entrance which was constructed with great effort (Bender[1]) during the nineteenth century. The harbour is well protected from the Indian Ocean swell and has the advantage of only a small tidal range (2.0m spring tidal range). However the large expanse of water can generate a short steep sea when the South-Westerly "buster" blows. The harbour was originally the estuary of two rivers but one has migrated north-east as a result of sand washed down from inland and from littoral drift. Mangrove swamps used to surround this sub-tropical inlet but they have been almost completely wiped out by commercial development.

Some years ago a single spine floating jetty was constructed using floating concrete monoliths to provide additional mooring facilities adjacent to the yacht clubs, one of which is located on the protecting breakwater. A need for additional facilities developed. However, these had to be located on the site of existing moorings and space was limited (Fig. 1). It was originally intended to extend the jetty using concrete monoliths, but this proved too inflexible and expensive. The area was also likely to be re-developed within the next few years by the Port Authority. The Marina extension was required to:

1)   take craft varying in type and size (from 7.5 to 20 m),

2)   withstand more exposed conditions than the existing structure,

3)   be easy to remove and re-position,

4)   be constructed in a short period of time,

5)   require little maintenance and,

6)   be *relatively* inexpensive.

THE DESIGN CONCEPT

Any marina design has to cope with local conditions such as:

1)   the direction and maximum strength of prevailing wind,

2)   the effects of current and tide,

3)   variations in water level due to tide, wind seiche, or drought, and

4)   the chosen method of locating the structures.

In this case the design also required;

i)   the extension and repositioning of the existing jetty, and

ii)   two long spine units with fingers at irregular intervals.

The basic dimensions of the structure, such as the width between the main spine units and fingers were similar to those given in Adie[2] with approximately 50 metres between main spines.

Because of the aggressive conditions encountered in an area located in the sub-tropical region and fed by the warm Indian Ocean currents the choice of materials from which to construct the extension was strictly limited. They had to be resistant to corrosion from the humid salty atmosphere as well as impervious to attack from abundant marine life. The use of timber and timber piling was not practical due to the short life expectancy (Treadwell[3]). In addition the client wished the structure to blend in with the existing units. Different forms of composite construction were considered. This made it possible to adopt the approach of a strong but flexible structure which would fulfil the prime objects of:

1) holding the craft safely and

2) providing access to and from the shore.

The sandy bottom and relatively small tidal range made the use of anchors feasible. It was therefore possible to consider the structure as a whole without the need to worry about the effect of piles forming fixed points along the length of the spine units. There was thus no need to design a stiff structure. To aid flexibility and provide limited movement between units the primary coupling would be formed from memory retaining plastic blocks pinned to each unit. To speed construction a modular system was devised with a limited number of different components. These consisted of main spine and finger units. The units would be mounted on standard floats giving as much free passage to the water as possible to permit the flotsam usually associated with commercial harbours to pass through (Fig. 2).

The modules were designed to be constructed by factory production methods. Fortuitously a large hot dip galvanising bath was available. It was therefore feasible to consider the use of steel for the main framework. However, the floats proved more difficult as glass fibre, either surrounding polystyrene or hollow was too expensive because of the skill and techniques required (Treadwell[3]) and it had shown a tendency to bacterial degradation in Durban.

The client, the Yacht Basin Mooring and Development Association (YBMDA), wished to moor a mix of craft so there was no guarantee that the fingers would occur opposite one another and a strong cruciform design was not feasible if the maximum use was to be made of the space

Figure 1. General view of the Durban Marina

Figure 2. Detail of the spine and finger units
        (International Jetty)

available. Thus a flexible approach was the best compromise but this had implications in the layout and assembly of the final structure. Although the design may appear unbalanced, the lack of strict regimentation enhances the geometric form when the flexibility is exploited during bad weather. Another advantage is that it is relatively straightforward to adapt the berthing arrangements to suit the demand as well as accommodating the maximum number of craft, which would not have been feasible if fixed piles had been used.

THE MODULAR STRUCTURE

The way to speed construction was to maximise the number of similar units and to ensure that as much work as possible was constructed under ideal conditions where the work will not be hindered by adverse tides or weather. To achieve this a modular structure was adopted. This consisted of spine units, finger units with separate entry and end units, fenders, connectors, anchors and buoyancy tanks. Locally available materials were used in the fabrication to reduce costs. The total length of the structure is 367 metres (Fig. 1). Figure. 2 shows, as an example, the new International Jetty with spine, fingers, fenders and bollards in place. The floats under the finger units can be clearly seen.

The Spine Units
The six metre long by two metre wide spine units were fabricated from 156 mm x 76 mm channel which formed an outer frame work. The legs of the channel faced outwards. By drilling at regular intervals pinning points for coupling the units and fitting accessories were provided within the main dimensions of the unit. This permitted the adjustment of the fingers to suit the demand for different sized craft by providing berths of different widths. The units were assembled in jigs. They were then dimensionally checked on the diagonals and for sag. Membrane grating was welded into the centre of the unit. As well as being a structural member it served both as a key and reinforcement for the concrete deck. Welding was by manual metal arc with random x-ray tests for quality control. After fabrication the steelwork was shot blasted and hot dip galvanised.

The deck was designed to take an area loading of 150 kg/m^2 with a maximum displacement of 200 mm and was made from 40 MPa concrete carefully vibrated into the membrane. The whole unit was considered as one, with the concrete taking some compression loading and the steel the tensile loads. The vulnerable edges of the concrete deck slab were protected by the steel channel section. The concrete was made from Portland cement with slagment (ground blast furnace slag) replacement for some of the cement and an additive of potassium chromate (the dichromate being equally acceptable) to counteract any reaction between the galvanising and the cement. The aggregate was crushed quartzite and the sand a mixture of washed and sieved beach sand and sieved river sand to give an acceptable grading.

The Finger Units
The seven metre by one metre finger units were of similar construction to the spine units. However, one metre, three metre, five metre and seven metre long joining pieces were provided to fit between the spine and the finger to accommodate the variation in craft length. In Figure 2 the junction of the splayed inboard end of the finger and the main unit is clearly visible just beyond the second float. It is these splayed units which were manufactured to the different sizes quoted.

The Connectors
The pre-drilled holes in the units, which were all of standard pitch permitted the units to be coupled together in a number of different combinations. Because of the expected relative horizontal, vertical and torsional movement of each unit, a flexible coupling which would gradually provide additional stiffness was needed. To achieve this a compression-moulded block of a predetermined shape was made of a 'memory' neoprene plastic material. Since the parallel flange channel sections were not readily available the block had to be shaped to fit the taper flange section and was held in position by galvanised pins.

An initial design was tensile tested at the University of Durban-Westville to ensure that the load distribution was satisfactory. The shape of the block was modified so that it could resist over three tonnes of axial load. It is stiff in shear so as to resist vertical movement and torsion but is more flexible in tension so as to cope with horizontal and bending movements. As an added precaution steel restraining safety straps are fitted to restrict the longitudinal deformation.

The Floats
Because the preliminary design proposed the use of non-structural resin impregnated timber walkways (Gowans[4]), the use of heavier concrete decking required a complete reappraisal of the float design. The heavier decking had raised the metacentric height and the weight of each unit so that additional buoyancy was required. The original concept of using glass reinforced cement floats to give the necessary stability was rejected. Alternatives such as coated polystyrene and closed cell polyurethane were considered. These were also rejected because polystyrene was hygroscopic and would require a very heavy duty case to withstand effects of waves and fouling, while closed cell polyurethane blocks lacked the necessary

strength. After laboratory tests it was decided to use rota-moulded polyethylene to form the floats. These could be pressurised and were pressure tested after forming. The material was flexible so that external and internal forces could be balanced to reduce skin stresses. It was also possible to adjust the flotation of the units by varying the pressure. If necessary it would be feasible to ballast the units as well, or to fill them with foam or other material. Since the units had to be standardised only two sizes of floats were manufactured. The larger, 2 m x 1 m x 0.5 m, was for the spine units and the smaller, 1 m x 1 m x 0.7 m was for the fingers. These units were to be manufactured over a relatively short timescale (see Table 1).

The flotation units required careful hydrostatic design to ensure that:

1)   there was sufficient lateral stability as the units heeled under loading,

2)   the units remained at the same relative height above water,

3)   the units could accommodate loading from both large numbers of people and from those standing at the ends of the fingers, and

4)   the movement of the fingers would be in phase with that of the craft moored alongside.

The floats were made as wide as possible. The deck units would be supported by several floats so that there was an additional safety factor as well as allowing free passage of water between the units. This let any flotsam escape readily, so debris did not build up in any berth. It also allowed a good water circulation among the craft, which was not critical in this case, but could be in more confined locations. The hydrostatic design was checked using a computerised stability programme. The spine units were shown to be stable at all times, including the assembly process. However, the much narrower finger units were shown to be only marginally stable as individual units before final attachment to the spine, as they relied for some of their stability on the connection to the spine unit.

The Fender Units
Because of the unforgiving main structure of steel and concrete adequately designed fenders were essential. Timber has some resilience if a craft should accidentally come into contact. Originally the fenders were

to be formed from rubber, neoprene or other similar shock absorbing material and were to have a "D" section. However, the essential flexibility, ease of distortion, and relatively short runs made it difficult to devise a suitable fixing to the taper flange channel. However, the short lengths could be turned to advantage if rota-moulded polyurethane was used. A different principle was finally adopted which used the air within the flexible case to absorb the energy of impact. Connections between the fender and the channel were simplified so that the rear of the fender slotted into the channel. It was held in place by pins inserted into the pre-drilled holes in the channel and pre-formed apertures in the fender. It is also relatively easy to demount the fenders for removal and maintenance.

THE CONSTRUCTION

The extension was required to be completed in a very short time necessitating the use of "fasttrack" methods of design and construction. In addition, as soon as any berths could be made available they would be occupied by craft that had been displaced from the swinging moorings which had been located in the area under development. The original schedule drawn up at the planning stage is shown in Table 1 as a bar chart. In the event computerised critical path planning was used to co-ordinate the construction and to monitor progress. The planning routine included:

1)   manufacture (at different locations),

2)   testing and basic assembly,

3)   launching,

4)   anchoring the units, and

5)   repositioning the moored yachts during final assembly.

Advantage was taken of the industrial facilities available, so that the lack of space at the final assembly and launching site was offset by the availability of undercover premises at the fabricators. However, units had to be transported by road between the fabricators, galvanisers and final assembly, where the concrete deck was cast. During the casting slump tests were performed and cubes were taken for later testing. The concrete was vibrated under very careful supervision to achieve the necessary compaction around the reinforcement. The flotation units were attached when the concrete had reached 20 MPa and launched using a specially adapted lifting frame. The units were towed to the assembly area.

Table 1   Bar Chart – for construction of Units

Function	Week 1	2	3	4	5	6	7	8	9	10	11	12
Order	#											
Fabricate Jigs		<---------->										
Mould for Floats		<---------->										
Mould for Pads		<--------------------->										
Receive Materials				#								
Construct and Test First Spine and Finger Units					<---------->							
Adjust Float Mould						<----->						
Fabricate							<-------------------->					
Galvanise							<-------------------------->					
Cast Deck								<---------------------->				
Launch								<---------------------------->				
Assemble								<---------------------------->				

However, during floating out and assembly of the fingers to the main spine, special flotation units were temporarily attached to ensure greater stability and ensure that there was no risk of a finger unit capsizing before it had been attached to the main spine.

The spine and fingers were anchored approximately in position using danforth type anchors and chains which were secured temporarily with rope ends. The spines were then aligned with the wind in one direction and then when it veered round to the other side they were realigned. Before finally shackling the chains in position on the units divers checked the setting of the anchors. The anchoring system had been designed for wind loadings on moored craft from a 50 year return period storm. This represented a maximum of 160 km/hr. As the units were brought together they were joined by the neoprene coupling units by means of pins. On completion the services (water and electricity in this case) were threaded along through the channels.

OPERATING EXPERIENCE

The flexible design received much comment to start with, especially when the main spine adopted curves during gales. However, it has withstood several severe storms. One lesson has been learnt and that is not to moor too many large craft in areas which have been designed and fitted out for small craft. However, the main spine units have proved themselves in addition to be very suitable as fingers for the larger and heavier craft provided they are adequately anchored. Another benefit of the flexible design is the damping effect it has on the waves, unlike fixed installations which do not have energy absorption capabilities. The floats have required modification. A valve has been fitted to permit the air in the floats to be topped up from time to time.

POSSIBLE IMPROVEMENTS

Because of greater movement at the sides of the main spines, threading the services along the channel will not be pursued in future designs. Despite greater inaccessibility the services, such as water and perhaps electricity would be located under the centre of the main spines. This would require a minor modification to the float profile. The use of the parallel flange channel would also simplify the design of fixtures and fittings such as fenders, couplings and safety connectors. In addition the floats would be improved by the inclusion of a pressurised membrane, or a sealed egg-crate structure to ensure that air could be held satisfactorily under all operating conditions and thus reduce maintenance.

CONCLUSIONS

The basic concept met the need to accommodate additional craft in a limited space at minimum cost, in a short construction time and in a congested location. Fasttrack methods accelerated the whole project. The design has proved itself in working conditions and has been adapted for use elsewhere.

ACKNOWLEDGEMENTS

The help of Brian Gowans of Marine Technology (the designers), Peter Chenery of Chenery and Associates (the clients consultants) and the YBMDA (the client) was much appreciated, as was the assistance of staff in the Faculty of Engineering at the University of Durban-Westville. The photographs are reproduced by permission of Peter Squire.

REFERENCES

1.   Bender, C. Who Saved Natal?, Colorgraphic, Durban, 1988.

2.   Adie, D.W., MARINAS a working guide to their design and development, Third Edition, The Architectural Press Ltd, London.

3.   Treadwell, G.T., Dearstyne, S.C., Dunham, J.W., Kvammen, K.R., Lee, C.E. and Sembler, E.L, Report on Small Craft Harbors, ASCE Manual No. 50, New York, 1969.

4.   Gowans, B.J.M., Marina design for South African Harbours and Dams, The Civil Engineer in South Africa, Vol. 32, No.2, 1990.

SECTION 12: MANAGEMENT AND MARINA INCOME

Computer Management and Control

H.N.E. Sheppard

Marina Projects (NZ) Ltd., P.O. Box 54-021, Bucklands Beach, Auckland, New Zealand

INTRODUCTION

With the massive expansion of the leisure industry, there has been an enormous market demand for anything that could broadly be labelled as a marina (the word "marina" needs to be considered in the wider context of a complete shoreside environment). With a changing world economy and a variety of other pertinent factors, the euphoria of the past is fast dwindling into the sunset and a professional approach to management is now imperative if the industry is to prosper and grow and operators are to receive a commercially acceptable return on the asset through which they have an investment.

Regrettably, computers are often seen as a business solution rather than a business tool and in this light, cause more confusion, cost and loss of goodwill than any form of benefit. Often computers are asked to do tasks which could be done more efficiently another way, or conversely, are simply under-utilized. Installed and used correctly, they can be a very useful management tool.

An essential ingredient of good decision making is information that is accurate and timely and is in a format that allows you to manage and control the day to day operations of your business simply. Information is important but does not eliminate the element of risk, however, with timely and accurate information, decision making is less subject to unknown risks.

REPORTING OBJECTIVES

In the past this has invariably centred around financial reporting but business information today should extend well beyond this limited scope. Business today is customers, profit and cash flow and you need to know detailed information about each of these three elements. Customer information is the basis of all marketing strategies. Who your customers are, why they are and their likes and dislikes are all essential components of management information as is the statistical efficiency of your business.

In respect to financial records, a large number of marina operators regrettably tend to see the administrative side of their business as less important and spend the majority of their time in the physical running of their establishment. Too often financial statements are prepared, often some months after year end, and the only motivation behind their preparation is to meet statutory Revenue Authority requirements. Management must accept that the compilation of the information for this purpose could just as easily have been done during the course of the year and the interim results used as a management tool. Like any other business, management, to be effective, must get detailed timely financial statements and be able to compare these with budgeted forecasts.

Reporting by profit centres will give a greater understanding and awareness of the business. Computerization makes this form of reporting simple and allows management to easily identify those areas that are under-performing and why and those areas into which cash profits are being absorbed beyond predetermined levels. From this detailed information, it is a simple matter to put corrective measures in place during the course of the financial year. Too often, marina operators are not interested in trying to understand the benefits of management reporting and rely solely on their bank balance as the broad gauge of success during the course of a year. Such a system can be so misleading.

Strategic planning, in all facets, is essential. Before the commencement of each financial year, a detailed operating and capital budget should be prepared on a month by month basis and broken down to give reporting by the various income producing sectors. The plan binds management to giving effect to the planned targets and also allows directors and owners to give acceptance to the return projected on the capital invested. Such projections also allow management to determine their debt servicing requirements both in respect to interest and principal and to ensure that depreciation funds are used for their correct purpose. The implementation of new projects should only be allowed after budgeted net incomes have been determined and funding sources secured both in respect to capital costs and working capital requirements. With modern computerization, this information is not hard or expensive to secure.

PRIMARY SYSTEMS AND CONTROLS

The installation of a computer will not solve your problems if you have bad primary systems. Any form of reporting will only

be as good as the input available. Accordingly, simple but
effective primary systems, disciplines and controls must be in
place before any enhancement to computerization or expansion of
your present computer system can be considered. The adage:
"Rubbish in - rubbish out" has never been truer. A computer is
simply a tool which assists to do a job. All financial
information has to be gathered at some stage, accordingly,
collating the prime data regularly is not an onerous additional
task and is the means by which the information, when accumulated
during the course of the year rather than at year end, can be
used for regular management reporting. This hands-on approach
will provide management with:

1. Control of material, goods and labour sales.
2. Control of spending and cash collections.
3. Performance standards and facility utilization reporting.
4. Improved efficiencies.
5. Better profits.
6. Reduced working capital requirements.
7. Stronger cash flow.
8. Information that will allow corrective action during the
    year and allow management to measure its effectiveness.

To do this and get the best results, you will need a software
package, preferably designed for the industry.

BASIC SOFTWARE CONSIDERATIONS

Determining Requirements
Computers will not provide any easy answers to management
unless the software and primary systems have been designed to
give this. The basic concept of computers and their programs is
that they can quickly add, multiply, divide, subtract and handle
vast volumes of information in a very efficient manner. The key
to any installation is the software. This amounts, in simple
terms, to the program which will capture the data input and
present it in a meaningful format.

The following are the basic steps that should be taken when
considering computerization:

1. Determine your business reporting needs and what
    enhancements to the present system you want it to provide.

2. Research the features of the commercially available industry
    systems and the technology used in writing the software.

3. Match your existing and future business needs with the
    systems available in the marketplace.

4. Short list the systems that seem to best match your needs.

5. Get from the companies on your short list a customer contact

list and spend some time with users to determine their level of satisfaction or otherwise.

6. Reduce your short list and ask the companies in question to demonstrate their systems in respect to your business.

7. Request written proposals from each vendor. These should include: undertakings regarding performance, the ability of the system to expand to cater for your future needs, the cost structure and the basis of payment.

8. The final selection should be linked to a performance guarantee and the undertakings, as dealt with elsewhere in this paper.

During the course of your review, you will no doubt be led to question your own style of management and whether the particular features are necessary or just nice. Try to focus on the least complex system that will meet your present needs and have an eye to future expansion. In certain of the above steps, you will probably need some additional support in helping to analyse the various features and benefits. People that can assist include your Chartered Accountant, who will not only have a deeper breadth of knowledge but will also have to be satisfied that the system provides the information that will be necessary for the statutory end of year accounting requirements. Most larger accounting firms have a separate EDP Section whose primary job is not only to provide bureau services but also to help clients taking this very important step. Generally speaking, they will be impartial, have a good perspective of what is available in the marketplace and be able to competently guide you with technical expertise. Computerization, if installed correctly, will become the heart of your business and therefore there can be no room for concessions in achieving what you want. As you compare system by system, you will develop an extensive questionnaire which will provide a valuable comparative analysis of the features and benefits of the systems available and will ensure that you are comparing "apples with apples". Do not be afraid to seek expert advice - their knowledge will unquestionably be to your benefit.

Software Considerations
Specific features to be considered include:

.   supplier's attitude towards support and upgrades

.   is it suitable for multi-user, networking, multi-location?

.   are the features practical or unnecessary?

.   can you buy the system in a module form that allows extensions to interface into the base program?

.   try and find some unsatisfied users and determine why - are their problems attributable to them or the suppliers?

- ensure the system is user-friendly, has help screens and a detailed easy to follow manual with plenty of examples - no special qualifications should be necessary to run the day to day operation of a computer system

- does the system have built in protection against deletion of information?

- how compatible are the source codes with what you are using at present?

- is the system an established system or are you the "guinea pig" for some programmer's imagination?

- does the system provide you with flexibility in the reporting format?

- what is the limitation of custom software modifications on upgrades, support and warranty?

- is there provision for various security levels?

- are price structures locked into the system?

- can you reopen previous periods?

- can you review on the screen and/or printout?

- are satisfactory hard copy audit trails produced?

- will the supplier agree to a contract of supply on the basis that at least a substantial portion of the purchase price is paid once the system is installed and up and running?

- what are the suppliers' attitudes towards training, support, service and maintenance?

- what are users' experiences in respect to these attitudes?

- is there lockout from the operating system whilst the printer is being used?

Software Modules

Software programs normally come in various modules which can be used to expand the base system and in many instances the various modules interface, thus providing a direct exchange of information within the system. Check on the availability, where appropriate, of the following in the various program suites:

Accounts Payable   Normally only used in larger facilities, however, if inventory management is being used, this module is highly desirable. Is there facility for a "hold" classification? This ensures that items in dispute whilst being acknowledged as a liability, are held in abeyance but are consistently monitored whilst being safeguarded against accidental payment.

Inventory Management   Inventory can absorb a large amount of working capital. Stock levels must be closely monitored to ensure that the appropriate stockturn is achieved and minimum order levels maintained and balanced.

Job Costs   Too often profitability is eroded by poor job costing procedures which effectively lead to incomplete chargeouts. Time sheets, which also link to the payroll, can easily control the labour content and this area linked to Inventory, ensures that no goods are requisitioned from stock without the appropriate job reference being detailed. To complete this module, the ability for management to review actual costs with quotes, both during work and at the end of the job, quickly identify the causes associated with cost overruns and under-quoting.

Funds Committed to Working Capital   Management should, as a matter of procedure, regularly (daily or weekly) receive a report detailing changes in balance sheet working capital items. The need to maintain a constant vidual on cash flow cannot be over stressed. Absorption of funds in balance sheet items, other than term deposits or reduced borrowings, have a direct cost when equated with the value of what money will earn.

Accounts Receivable   This module normally includes: billing, invoicing, sales analysis and ageing of debtors' accounts and credit control. The detailed information must clearly address management requirements in terms of working capital reporting, as dealt with above, and give a precise and accurate measurement of sales and margins being achieved against the various cost centres. In this area it is important that all sales go through the one system whether they be cash or credit. The system should also provide management controls regarding the issuing of credit notes, writing off balances, etc.

Customer Information   Management, to be informed, must have an easy way of obtaining full details of all their customers including tenants, berth licensees - both permanent and temporary, and those using dry storage. This customer data base will become an extremely valuable tool when establishing marketing strategies.

Wait List Management   From this, management should be able to glean not only the customer information, as detailed above, but also marketing information which should be solicited when someone joins the wait list. The benefit of knowing the reasons why people elect to use the facility cannot be underestimated.

Payroll and Staff Records   A simple way of monitoring all staff information, together with providing the information necessary for statutory returns, employee tax certifications, deductions, non taxable allowances and leave entitlements.

Travelift, Hardstand and Berth Availability   This module provides an efficient manner of controlling and maximizing the use of the travelift and hardstand facilities as well as marina berths available for lease both by way of short or permanent tenancies. The ability to maximize income from these sectors is essential. All have high fixed costs and under-utilization will take a heavy toll in respect to profitability.

General Ledger   This is the heart of the financial reporting system and should be suitably flexible to give the level of detailed reporting by the month and year-to-date, both in terms of actual and against budget and the resulting variances, as dealt with elsewhere in this paper.

Asset Register   Although it can be an onerous task to establish, from a management point of view and revenue reporting, the benefits cannot be underestimated. The ability to use both straightline and Tax Department depreciation rates should be available.

Productivity and Efficiency Reports   Both of these are key management indicators. These reports should extend not only to labour areas but also into the utilization of all facilities.

Maintenance Scheduling   Maintenance programmes for both short term and long term requirements should be planned to ensure the effective use of maintenance staff and the achieving of maintenance objectives. Where appropriate, long term maintenance schedules can also be drawn up covering maintenance requirements that fall due every, say, 2 - 5 years. Scheduling in this manner ensures the correct deployment of staff to maintain the standards required and achieve management goals.

Profit and Cash Flow Reconciliations - Source and Disposition of Funds   Actuals for both capital and revenue should be compared on a regular basis with projections. Working capital absorption is easy to achieve and hard to rectify.

Spread Sheets   A simple system to allow what-if calculations. These are normally used extensively during the preparation of annual budgets, etc.

Marina Security   Provision to monitor who moves on and off marinas with the ability to link back into your debtors' and credit control areas. This area can be extended to include monitoring of security systems on boats in the marina.

Report Writing and Word Processing   These provide the flexibility for editing and the integration of other information such as spread sheets, performance reporting, preparation of agreements, etc.

Mailing Lists   Often a byproduct of the accounts receivable module but in large facilities, the ability to extend beyond current and past customers and incorporate potential users of the marina for regular news bulletins and customer communication. The uses of this facility, on a selective basis, should not be understated.

Hardware
Having decided on the appropriate software modules, the hardware or machinery to run the system will normally be a relatively simple choice. Stand alone equipment is now relatively cheap. Be careful to ensure that the equipment you are purchasing is using current industry standards, is backed with support and a suitable warranty, provides sufficient storage for your current and estimated future needs and can be upgraded and expanded. The electronics industry is developing at an ever increasing rate. Do not be left behind at the start!

IMPLEMENTATION

Prior to the commitment of the capital cost of purchase, for either a new installation or an upgrade, the following must be considered and planned by management thus ensuring minimum disruption and maximizing the proposed benefits.

Timing
A most important element particularly when implementing an accounting system. If it is proposed to make the changeover at the end of an accounting period, then an adequate start must be made with an appropriate lead time to ensure that systems are allowed to run side-by-side prior to the commencement date. Experience has shown that this area is often grossly underestimated.

Management of Implementation
There must be a team approach involving finance, administrative and computer people. There is a need normally for one project co-ordinator or leader who has direct access to top management.

Education/Training Personnel
New staff may be required and "old" staff may need educating and training. A considerable portion of the budget should be reserved for these elements. If the budget only provides money for software and hardware, the system will not work effectively.

Accounting Policies
These must be decided initially as they can effect the format of the general ledger. There are plenty of guides for general accounting policies, however, particular policies will have to be determined for particular departments.

Audit and Quality Control Requirements
These must be designed as an integral part of the system and professional advice such as from the audit office or your chartered accountant should be obtained early in the process. Clear policies should also be determined from the outset in respect to system backups and security of data files.

Performance Evaluation
These issues must be thought out at the initial stage and the
way they will effect design. Some of the issues are:

(a) Measurement criteria, qualitative and quantitative.
(b) Responsibility centre, profit centre, investment centre.
(c) Transfer pricing.
(d) Allocation of overheads.
(e) Cost of funds.
(f) Valuation of assets.

Co-ordination
The crucial factor to the whole process. This co-ordination is
with the components of the system and with the components of the
whole process associated with the installation or its upgrading.

Loading Information
A substantial amount of base material has got to be initialised
into the system. If it is not capable of being transferred from
an existing system, management will have to give consideration
as to whether this should be done by their own staff or become
part of the contract with the suppliers. Again, this is an area
which is often grossly underestimated in terms of time and
serious consideration should be given to getting the supplier to
build this feature into his basic cost quotation.

Watch and Consider

1. Work with your Chartered Accountant - the system should
    provide his required information as well as your management
    information.

2. The ability to expand the system both in size and to other
    modules that interface.

3. Ensure before commitment that your prime system is adequate
    and will work with the chosen product.

4. Get from each vendor their commitment, in writing, to their
    specifications and continued customer support.

THE DEATH CYCLE

Many organisations have gone through the turmoil and expense of
upgrading or installing computerized information systems only to
find that new systems are immediately subject to severe
criticism and demands for replacement. As every parent knows,
it often seems that, no sooner has the long awaited skateboard
been placed under the Christmas Tree than, it is usurped in its
role as the one and only thing the recipient ever wanted by
something larger and far more expensive. Similarly with
computer systems, as management experience the benefits of

improved facilities, they begin to perceive ways in which further refinements will enable them to perform their functions even better. Clearly, this can be viewed positively as part of the process by which management of an organisation becomes more sophisticated and effective in their roles. Unfortunately, however, it is not usually the principal source of dissatisfaction about new computer systems. The following are the principal causes of negative sentiment:

<u>Think I Will Delegate This One</u>   Computer system problems are first and foremost reflections of management ineffectiveness. Despite all the lessons that organisations have learnt, computers still retain much of their aura of mystery, glamour and danger. As a result, it is still a rule rather than exception that major system development projects are permitted to be conducted under conditions and environments which no senior manager would contemplate for other forms of major investment. Senior management must not duck this important issue and must be responsible for the project. If the ultimate responsibility for the project is vaguely defined or even worse, assigned to a committee, then the resulting system will be indifferent, to say the least. Management has to take the ultimate responsibility for ensuring that the project is always orientated properly, that the project is tightly co-ordinated and that people are held responsible for any failure to perform the assigned role.

<u>We Never Dreamed It Would Cost This Much</u> Cost justification is more often done poorly than done well. The intangible nature of the benefits, which computer systems will bring the organisation, are used as an excuse for not performing a proper cost justification and calculation on return of investment, yet everybody knows and accepts that there are intangible benefits from all forms of investment. There is nothing magic about computers and there is no reason why a formal detailed thorough cost benefit analysis cannot be performed on any computer project. Management must ensure this.

<u>If Only They Had Told Us</u> There is a slightly bitter little catch phrase which has been doing the rounds. It goes as follows: "This is exactly what I asked for but not what I want". Although computer professionals have to accept a substantial share of the blame in most computer fiascoes, there is usually an element - often, a large element - of failure to properly define what was required and the blame for this can be laid squarely at the feet of the users (and management). Few of us would be silly enough to invite a builder into our home to perform extensive renovations without requiring a fully detailed plan at the outset. We know that, unless the builder is given a precise specification, the result will not be satisfactory. Do not be fooled that new data bases are so flexible that any form of request or report can be defined at any stage and delivered almost at no cost. Changes will almost invariably result in

cost escalation. The only solution is to get the design as nearly correct as possible at the outset. To do this requires resources and hard work. System design is a combination of mind stretching, lateral thinking and endless wading through minute detail. The key to success is the motivation of users in telling the system developers what is needed. Management must facilitate this.

<u>They Keep Blowing The Deadline</u> Having considered common shortcomings in the contributions of management it has to be clearly stated that, in the absence of proper planning, delivery dates will drift and cost overruns are inevitable. It is a team effort. Truly successful systems only come from a properly controlled, planned and co-ordinated team effort. Management and users have vital roles to play in the project and the team is the organisation as a whole. If the level of teamwork is not achieved, the organisation is likely to find that it has endured the provocations of implementing or replacing major systems without achieving the full potential benefit. Experience clearly indicates that if a system is botched as to design, implementation or cost, effectiveness will be received negatively and the people will immediately begin to look forward to a day when it is replaced.

CONCLUSION

This topic is one which, particularly to non users who are often not administratively minded, one they would prefer to avoid and not become involved in. Management, without the appropriate information, will not maximize profit and cash flow. These are the core elements to a business' success and will lead you, the operator, to a superior lifestyle. If you wish to maintain the edge on your competitor and maximize returns, you must have this information readily available in a detailed, timely and cost effective manner. Seek advice if in doubt - a bad choice can be expensive to rectify.

Computerised Integrated Systems for Management and Control of Marinas

A.F. Rendell, P.A. May

MIMICS Ltd., 1 Gate Farm Road, Shotley Gate, Ipswich, Suffolk IP9 1QH U.K.

ABSTRACT

This paper outlines the evolution of the marina over the last twenty years to the current scenario, where the wide range of functions involved in the running of the modern marina necessitates the use of computers. This automation often involves the use of many separate systems, even separate computers, and other electronic equipment, all working independently. The purpose of this paper is to outline how all these functions can be brought together, increasing efficiency, providing a better service to the customers, allowing a reduction in staff and improving management information and control.

THE MODERN MARINA CONCEPT

Today more people own boats than ever before. Boats are becoming ever more comfortable, faster and reliable and are therefore used extensively and travel further in a given time. The result is that more boats need places to berth and to visit. Thus many marinas now exist (with more still needed) and every marina receives an ever increasing number of visiting boats.

Today's requirements are for secure and comprehensive facilities with all the creature comforts to be found at home. Most boat owners are happy to pay for really good modern facilities. The equipment carried on board most modern vessels is becoming more sophisticated and vessels are becoming more dependent upon marinas for the ability to plug into electricity, and obtain service for electronic and mechanical systems. Environmental and pollution laws add to requirements for effluent pump-out facilities and the disposal of rubbish.

As with many other leisure industries, UK marinas have been compelled to greatly improve their standard and range of facilities to match those available on the Continent.

STAFFING IN TODAY'S MARINAS

While the numbers of leisure boats and marinas have greatly increased, the number of marina staff has not increased in the same proportion. This is partly due to the greatly increased cost of employing both skilled and unskilled staff to-day. The result is that it is quite normal to find a modern UK marina, berthing up to 1000 boats, being run by no more than one or two staff after 1800 hours during summer weekdays, and three or four staff at the week-ends - the busiest times for any marina. Many marinas are not able to employ reception staff in the evenings or at weekends, yet these are the only times that many resident boat owners can get to a marina and the times when most visiting boats arrive and leave.

Thus, in the possible absence of administration office staff, accounts and reception staff, a few dockmasters are expected not only to perform their normal duties of allocating berths, helping berthing, issuing power cables and fuel, maintaining security patrols and provision of the myriad of other services, but also to answer possibly hundreds of visiting residents' queries on accounts, berthing problems and changes, booking of boat hoist/other marina plant and innumerable other customer matters. The increasing number of visiting foreign craft adds further problems to the duty dockmasters' formidable number of tasks.

NEED FOR IMPROVED SERVICE

As the number of visitors, and their demands for an ever increasing range of services, continues to grow, current marina staffing levels will almost certainly result in a very inferior service, that would be considered unacceptable in an hotel. Since dockmasters are unable to access customers' accounts, berthing contracts, plant schedules, short/long term berth availability, etc. frustrated residents (and would-be residents) are left to raise their problems over the telephone during the following week. Discussion of berthing, financial and other personal matters over the telephone during the customer's office hours can be an extremely inconvenient and expensive exercise for the customer. Marina management should avoid such unsatisfactory methods of doing business. Delays and confusion in providing berths and other services for visiting boats discourage them from returning and word soon spreads to other potential visitors.

LOSS OF REVENUE

Because marina staff who are under pressure are most unlikely to keep accurate or reliable records of details of boats, their owners, berth and plant usage and availability, etc., inevitably a considerable loss of revenue results. (It is rumoured that many yachtsmen know where it is possible to get free berthing and power in various south coast marinas. It is obvious that tighter controls are required in these circumstances).

Delays in providing services such as fuel or LPG persuade visiting yachtsmen to abandon such requests until they return to their home port, thus losing the marina valuable earnings.

Many yachtsmen, both resident and visiting, frequently need quotations for work required on their craft or changes in berthing periods, etc. The time and information to compile such figures and modify schedules is simply not available to dockmasters on duty. Thus customers take the work elsewhere or postpone it. (Precise printed quotations are available to any passing motorist in most modern garages).

Even where services are provided satisfactorily, dockmasters will almost certainly not raise immediate invoices, which means that the marina will have to wait far too long for payment.

TYPICAL MARINA STAFF DUTIES

The comprehensive facilities in the modern marina result in a long list of tasks which must be carried out efficiently and swiftly by the marina staff, e.g. consider the following:

Allocation of berths
Assistance in berthing
Monitoring all boat movements
Calculation of berthing fees
Preparation of invoices
Reception of existing and future customers
Word Processing and selective mailshots
Preparation of management information and statistics
Production of berthing contracts
Production of quotations and automatic invoicing for yard work
Metering of electricity, water, telephones, television on berths
Metering of fuel
Control of access to car parks, pontoon access, ablution blocks etc.
Security for boats, houses, marina buildings etc
Customer Accounts, Purchase Ledger and General Ledger
Travel Hoist scheduling and billing
Issuing of electricity cables, access control cards and keys
Use of access control cards for credit facilities in marina
Pumping out vessel grey-water tanks
Checking mooring lines
Sales of LPG
Towing/rescuing vessels with failed engines
Disposal of refuse

All the above functions are interrelated in some way. For instance, the various metering systems need to know which boats are in which berths; the word processing function needs to have access to customer and boat information, and the accounts system needs to have details of all invoices issued to boats and customers. These details should be held in a database, accessible simultaneously by every marina department.

INTEGRATED COMPUTERISATION

Assuming that marina staffing has to remain at present levels (many marinas are currently reducing staff still further !), the only possible solution to the serious conflict between workload and staffing, must be to computerise all aspects of a marina's operation and for all marina staff to have access to authorised programs at all times.

Computerisation would enable marina staff to access all necessary boat / berth / customer records instantly. They would then be able to make instant berth allocations / decisions / plant bookings and raise and print immediate and accurate invoices for visitors' berthing and other marina services.

For example, the dockmaster who sees a boat leaving the marina will ask "Has that customer paid all dues ?". If there has been a change of watch, he may not have first-hand knowledge. The computer would answer that question instantly, whereas the commonly-used manual record system would necessitate tedious checking of invoices and payments etc., by which time the vessel would be beyond recall.

A computer screen can provide visual representation of the marina, showing which berths are occupied/available, and indicate which berths are occupied by visitors or residents. Expected departure or arrival dates for any boat or group of boats, could be displayed and a summary of any day's boat movements listed or printed as required.

Automatic end-of-month invoicing of residents would be carried out from system contract files, from service charges keyed in by marina staff and automatic power monitoring inputs. Instant quotations could be raised and printed for all berthing and marina services. Annual contracts would be automatically updated, printed and mail-merged before each season. Marketing statistics would be instantly available from the system database. Computerisation would provide management with instant access to details of revenue sources and enable the charges in those areas not showing profit to be adjusted by means of a few keystrokes.

All the operational database and customer invoicing information inserted by the marina operations staff would be linked to any existing or future accounting system. Thus, even if the marina's accounting section or parent company are sited in another country, every aspect of the marinas' operation can be accessed instantly.

In addition to performing the above duties, the ideal computer system must be "modular" so that as extra facilities are added to the marina, the appropriate feature can be integrated into the computer system.

As an example of true integration, an electrical power metering system should transmit details of power consumption automatically into a single account for the customer, along with berthing fees, boatyard charges etc.   If any problem / alarm situation arises, warnings would be shown on any combination of VDU screens, allowing the staff to provide the appropriate automatic responses (e.g. boat / house owner warned of fire by telephone).

INFORMATION

Any properly designed computer system would have a built in "query" facility, so that managers could extract almost any statistic, characteristic or list from the system database. Such information as "How many visiting boats over 12 metres LOA arrived after 1800 hours in June and July?" or "How many resident boats left the marina more than 20 times this year?" or "List those visiting boats that used shore power last month" can be displayed in a few seconds and printed out if required - the range of facts available is almost limitless!

With Management Software, flexibility is very important, and it must be designed so that it adapts initially to the requirements of different marinas, and thereafter, to the ever-changing day-to-day requirements of any business. It must also be able to cope with changes in management policy without the need to make changes to the computer program.

Software should be capable of running on a marina's existing hardware where appropriate. This is vital where a smaller marina is concerned. As the operation expands more sophisticated hardware can be purchased or leased to suit each owner's particular needs.

In short, all information about the marina, its customers and their boats would be available at the touch of a few keys thus avoiding the primitive, slow and unreliable use of hand written records and ensuring maximum use of the marina's assets. There should be the ability for simultaneous access to information by the manager, accountant, boatyard and reception staff, dockmaster etc., password-protected where necessary.

Such a system must not only pay for itself in a short time, greatly improve customer satisfaction and hence marina/customer relations, but significantly improve the overall profitability of any marina.

THEORY OR PRACTICE ?

Many attempts have been made to produce the ideal system for a marina manager. Virtually all have failed to satisfy what has to be a very specialised requirement. Modification of existing accounting software programs has been the usual technique used by software companies because of the prohibitive cost of writing a program specifically for management of marinas. This is rather like trying to toast bread with a typewriter! Many a marina has been landed with a bastardised inflexible package which provided very little help and was inevitably less than " user friendly ". Regrettably, as in many pioneering situations, these poor quality attempts to computerise marina operations have given the whole concept of computerised marina management a bad name. Quite understandably many a manager is now cynical about claims made in advertisements for computer systems.

A SUCCESSFUL MARINA COMPUTER MANAGEMENT SYSTEM

As in many situations where a specialised task exists the solution has to be to use a specialist tool produced specifically for that task.   However, in order to produce any tool its future use must be fully understood by its maker - in this case the writer of the software program. Thus the software writer must work extremely closely with an experienced marina manager and himself visit a variety of marinas and understand the problems facing marina management today.

CONCLUSION

There can be little doubt that with the modern yachtsman increasingly demanding efficient and value - for - money service, he will expect the standard of service that only a marina employing a properly configured, reliable, and well - proven computer management system can provide.

Modern Marina Developments and Maritime Museums

J.J.F. Elwin

Oxford Centre for Tourism and Leisure Studies, Oxford Polytechnic, Oxford OX3 OBP U.K.

INTRODUCTION

Over the last two decades, in Britain, the marina has developed from the simple boat storage area, by way of the marina village, to todays modern comprehensive and varied development of which boat storage is only one facet. The modern marina is frequently part of a integrated redevelopment scheme taking place in a disused commercial dock. Shops, offices and residential accommodation, in the form of flats and houses, are provided alongside the traditional marina enterprises of boat storage, sales and repair. To this list of activities the maritime museum is invariably added. This may take the form either of a display of model ships together with various artefacts rescued from long scrapped or wrecked ones, or of live exhibits which may range from the ubiquitous lifeboats and lightships to paddle steamers and warships. In contrast in America this multi-faceted waterfront development is well established as a technique for revitalizing run down parts of a city.

This paper will explore the relationship between these two contrasting uses, that of yacht storage and museum with a view to providing a set of guide-lines that proprietors of both marinas and museums can consider before embarking on a joint venture. The paper is prepared primarily from the viewpoint of the marina operator but it is important that both sides understand the objectives of the other. The term marina is taken to include the overall development of a waterspace area of which the traditional pontoon berthing area is only one element. The term maritime museum includes the traditional display of artefacts in a building but for the purposes of this paper concentrates on the display of historic ships.

BACKGROUND

There are currently more ships under preservation and restoration than ever before in Britain. Only thirty years ago the Victory and Cutty Sark were the only two major ships that could be considered as having been deliberately saved and displayed for the benefit of a major maritime nation. Today there are some 100 ship preservation societies and maritime museums. Excluded from this list are many yachts and barges be they J class yachts or Thames sailing barges that have remained in private hands and while no longer used for their original purpose are still seaworthy and active . In addition there are a plethora of boats from fishing boat size downwards that have a regional or local significance and may form part of a local maritime collection.

It is the major ships that are the prime concern of this paper. That is to say those that exceed the dimensions that can normally be accommodated within a traditional marina, because it is these that will require special consideration if they are to be included in a modern multi-faceted marina and waterside development.

The number of such vessels to be preserved is also likely to rise. This is partly as wrecks are found and raised - such as the *H.M. Submarine Holland 1*, or when ships come to the end of their useful lives, a recent example, being the frigate *H.M.S. Plymouth*. Enthusiasts are always aware of old ships that might be preserved for instance, apparently, there is a victorian cross-channel packet steamer, reputedly lying in Turkey. Some third world navies have recently disposed of escort ships that were in service with the Royal Navy in the Second World War. If these ships are to be preserved and displayed to the public they will need a venue. It should be noted that, in Britain, there is not a national policy on ship preservation and each ship depends on the enthusiasm of individuals in the first instance. In contrast, in America certain ships can be designated National or State memorials and as such attract public funds.

The preservation of these ships is invariably the result of accident and the enthusiasm of a dedicated group of enthusiasts. The final resting place of the ship frequently has little direct connection with the ship itself. Exceptions to this being *H.M.S. Victory* at Portsmouth, *S.S. Great Britain* at Bristol and *R.R.S Discovery* at Dundee. In the case of smaller ships with a local relevance there is often a more direct correlation in the case of lifeboats, lights ships, tugs and fishing vessels.

The restoration and preservation of an historic ship will cost hundreds of thousands of pounds. Even then once completed the on-going maintenance costs will continue to absorb large sums of money, particularly if they are floating exhibits that will have to be periodically dry docked. The labour costs alone of employing people to collect entry fees and clean the ship so that the public can be shown round will normally exceed ø30,000. This means that most ships are owned and managed by charitable trusts and even if they could do so are unlikely to be considered as profitable ventures.

MARITIME MUSEUMS IN MARINAS.

The question must then be asked why should any marina consider the presence of a historic ship in its midst.

There are several answers to this question.

Firstly, as has already been indicated the marina may be only one element in a larger redevelopment and amenity enhancement scheme. The overall plans and ownership of the waterside area may be in the hands of other bodies and, as such, the marina operator has little choice in the matter, other than possibly to have some influence on the siting and day to day operations and management of the historic ship. The situation at Bristol Docks is one such example. Here the marina areas occupy only a small part of the total area of the Floating Harbour. The area as a whole is being redeveloped by the City Council and Port Company and the presence of the historic ships is seen as adding to the ambience of the area. While *S.S. Great Britain* is the principal attraction there are many others, some of which are used as restaurants, others are associated with the industrial museum, others are in the hands of private owners. Similarly at Swansea the overall ownership of the Maritime Quarter is by the City Council who have planned a comprehensive mixed redevelopment of the site to include a marina, offices, workshops, and housing. The Maritime Museum together with its floating exhibits is seen as a way of increasing the variety and adding to the visitor experience. The same is also true of the proposals to redevelop the Salford Docks in Manchester where Prince Charles's former command *H.M.S. Bronnington* will be an attraction.

Secondly, even under private sector ownership, a mixed redevelopment may be sought either to satisfy planning conditions or to utilize existing buildings and docks. Under these circumstances, as at St. Katherine's Dock, the resulting scheme is a mixed development of shops, offices and residential units. In these cases the presence of the public is needed to make certain elements of the scheme attractive and viable. The presence of historic ships can therefore act as a magnet and bring people in to an area. More recent examples of this are Brighton Marina, Port Solent and Ocean Village. These are all mixed developments which require the presence of people to make them successful.

Thirdly, in the initial stages of a marina or waterside development the presence of interesting or historic ships will create an interest and an atmosphere that in turn will lead to the marina element itself becoming fuller. The publicity surrounding the historic ship might also make it easier to sell or let other elements of the scheme and it might be possible to use the presence of such ships as part of the marina's own advertising campaign. In these circumstances, the presence of such ships might not always been seen as a permanent feature and the space may be more profitably used for other more commercial purposes in the long run.

Finally, a historic ship might be included in a marina purely for commercial reasons. It may be that it is the most economic way of using that particular area of the marina.

FACTORS TO BE TAKEN INTO ACCOUNT WHEN CONSIDERING THE INCLUSION OF HISTORIC SHIPS IN A WATERSIDE DEVELOPMENT

INCOME

Some of the smaller boats such as rowing boats and small fishing boats may take up only a single berth space and if they are in private ownership can be charged a commercial rate or given a discount if it is felt that their presence is of mutual benefit to the marina. In the case of most major ships of over a 150ft and with a beam of 20ft or more then in crude terms this size of ship is taking up space that could be occupied by ten or so yachts which at todays berthing rates would generate a minimum income in excess of ø10,000. It is highly unlikely that any historic ship could afford such an outlay on berthing charges. Therefore, it is important that the presence of such ships is not seen purely from the commercial viewpoint. In fact, it is unlikely that any form of fixed rental (other than a minimal or token amount) would be welcome by the owners of a historic ship which relies purely on visitor entrance fees and donations for its income. A better approach is to take a percentage of the visitor fee income. This also has the advantage that when setting their admission charges the owners of the ship can take into account this proportion and set the entrance fee accordingly. Other than for the most important historic ships, that have an international or national reputation, will the entrance fee be in excess of £2.50.

OWNERSHIP

Most historic ships are owned by charitable trusts. This in itself should be of no concern to the marina operator. However the financial management and well-being of the trust is of prime importance to the marina operator. It is important that it is run in a professional and business-like manner with ability to raise funds and effectively promote and market the ship. Although the technical expertise and enthusiasm available to carry out maintenance and restoration are obviously important assets. With the running costs alone capable of reaching tens or hundreds of thousands of pounds each year, before restoration costs are considered, the ability to raise funds from sources beyond pure entrance fees is often critical. As such most trusts rely on grants and donations to remain solvent.

From the marina view-point it is important that the ship is an asset and not an eyesore or liability. If it is going to deteriorate because of insufficient maintenance it will quickly become a rust stained hulk or worse. The last thing any marina operator wishes to be saddled with is such a ship which is no longer even seaworthy enough to be removed.

In many ways, possibly, the best option is to utilize a ship from a well established trust, such as the Maritime Trust or the Warship Preservation Trust, that may be looking for berths for its ships. This provides an excellent and interesting exhibit together with the expertise and resources of such an organization. In several instances local trusts, which are sometimes backed by local authorities, have chartered ships to help provide an attraction as part of a waterfront development. *The Gannet* at Chatham Dockyard is an example.

TEMPORARY OR PERMANENT

Whether the exhibit is going to be of a temporary or permanent nature could well determine its location in the marina.

Short Term. If the exhibit is going to be of a short term nature particularly, if the facility has recently been opened and the prime reason for the ship being there is to help create publicity and interest, then it has to be capable of easily being removed. This may coincide with Boat Shows, Regattas or the beginning of the sailing season. Similarly, if the ship is going to be active and make frequent excursions and visits then it will require a clear channel to the sea an example being the *Shieldhall* at Ocean Village.

Intermediate Term. If it is going to be in place for an intermediate term, which may range from a few months to a few years, then it is important that both parties clearly understand this. It is not necessary to stipulate the exact period of time in advance, although any agreement will have clauses that review the situation at a regular period. It may be that the exhibit is there while a permanent home is being found or until the marina grows sufficiently to demand the space occupied by it for more commercial purposes. Alternatively it may be felt that the purpose of the marina is better served by having a series of interesting and historic ships visiting it. Each one can then be used to regenerate interest. In most cases this will require the ships to be capable of moving under their own power as any short term visit requiring tugs is likely to be uneconomic. Brighton Marina has adopted this approach. The berth originally used by *H.M.S. Cavalier* is now used by visiting ships such as the paddle steamer Waverley which often visits south coast marinas and harbours out of her main cruising season. Ideally what is required is a set of such ships that can be moved round a series of harbours on a regular basis. To some extent *H.M.S.Plymouth* has recently been doing this having been berthed in Plymouth, Glasgow and shortly to be moved to Merseyside.

Long Term. If a ship is going to be displayed on a permanent basis then to some extent the marina can be built around her and other considerations will take precedence. Although, provision must be made for periodic repairs within the dock area.

TYPE OF SHIP

The romantic notion of a tall ship with masts and spars is often seen as being the most suitable to provide a central attraction. The masts and spars can often be seen above the buildings and catch the eye. However there are relatively few such ships around and the maintenance of the rigging is expensive. Warships, while attractive to small boys and their fathers, are extremely costly to maintain. They are complex and have no large areas such as promenade decks, saloons or holds. Wooden hulls survive better in sea-water while iron hulls are better in fresh water. Ultimately, ships have not been designed for display in a museum after the ends of their useful lives. Possibly the best answer is a replica.

SIZE

While this paper is primarily concerned with ships that exceed the dimensions of those normally associated with a marina, this to some extent depends on what was originally envisaged at the design stage. With the size of marinas and yachts growing it is not unusual for a marina to have several berths capable of taking vessels of 50ft in length and over 100 tons displacement, the size of a large luxury motor yacht. It also equates to the size of tugs, lightships and fishing boats. These together with smaller yachts, Edwardian steamers and rowing boats can often be readily accommodated within the normal confines of a marina. The main consideration then is one of public access and viewing. If direct public access is not required because they are privately owned and their presence adds more to the ambience of a marina, then they need to be suitably berthed so that they are best displayed. For larger ships the following factors need to be taken into account.

Firstly, there is the basic problem of getting the ship into and if necessary out of the marina. This is particularly the case if it is going to be involved in regular sailings, then it needs to be located so that it provides the minimum disruption to the normal smooth running of the marina. This may be especially relevant if the marina has restricted tidal access controlled by a lock-gate. If the ship can only just fit into the lock then this could take time that could hindered the use by other boats on that tide. Most large ships will from time to time have to be dry docked or slipped so that a full inspection of the hull can be carried out and normal maintenance undertaken. As this will only happen every five years or so then to some extent the marina can be built round her. But it might necessitate the temporary removal of some pontoons and other fixtures and the dredging of a channel. It may also mean that such departures are restricted to spring tides. Furthermore if the ship is no longer able to provide her own power then the use of a tug might be needed further adding to the space required. In Swansea Marina, for instance, some of the ships have to be poled out when being moved to a dry dock.

Secondly, there is a need to find a sufficiently large berth. A warship of destroyer or frigate proportions such as *H.M.S. Cavalier* or *H.M.S. Plymouth* are over 300ft in length, have a displacement in excess of 2000 tons and draw 14ft or so. In fact *H.M.S. Cavalier* used to regularly take the bottom in Brighton. Fortunately her propellers had been removed. A ship of this size will place a considerable load on any mooring structure especially under certain conditions of wind or tide. Her alignment may then need to be taken into account.

Thirdly, the size of the ship should be in keeping and sympathy with her surroundings. It will not improve the attractiveness of an area if the exhibit dwarfs everything else. *H.M.S. Belfast* in the Pool of London is the largest ship of any kind preserved in Britain. Being located in the centre of London with its tower blocks and nearby Tower Bridge means that she does not swamp her surroundings.

Overall, large ships are best located in docks that are being redeveloped. As these were designed from the onset to handle large ships. In fact, all ships should be placed on moorings that are commensurate with their size.

PUBLIC ACCESS AND SECURITY

Security is of paramount importance in a marina. Ten years ago in many marinas the public could often wander about the pontoons unchallenged. Today most of the boats are protected by security fences and electronic access systems. This means that the location of a ship, needs careful planning, so that the public can gain direct access to her. If a special jetty or wharf frontage is being provided then that can often be isolated from the rest of the marina security system. If the ship is purely there for display purposes and to create interest in the marina, such as the J class yacht Valhalla at Ocean Village, then it is purely a case of allocating her a suitable berth so that she can be seen by the public.

In a multi-purpose modern marina, especially if it is part of a dockland redevelopment scheme, then not only is public access required to the shops and offices but often it is also a planning condition as well. This means the historic ships can be located adjacent to the areas of public access. The concern then becomes one of the ship owner who wishes to ensure that the ship can only be boarded by bone-fide visitors. There is, though, the additional problem of vandalism and theft from the ship when she is not open to the public. Sometimes this can in part be covered by any security arrangements that are already in place.

MAINTENANCE AND REPAIRS

Most historic ships (when they are first saved for preservation) are invariably in a poor state of repair. This means that a comprehensive restoration programme needs to be embarked upon. While this is of interest to the general public who will make frequent visits to the ship to see progress on her, it can lead to problems in a marina.

The process of repair and restoration may involve the use of heavy equipment such as cranes and compressors. In addition items such as steel plates may need to be delivered and welded. This requires a different level of access to that which would normally be expected if the ship was purely open to view. The land area needs to be able to accommodate these pieces of equipment without disruption to the normal operation of the marina or inconvenience and danger to the general public. If the marina is part of general redevelopment scheme including shops and offices then a lot of crashing and banging is not going to endear the ship to the board of directors of a local firm, which has offices there, especially if they are going to subsequently seek sponsorship from them. This last point may control their activities. Similarly if a steam ship wishes to raise steam either to test the boilers or as a normal part of its operation the clouds of sulphurous smoke and hot ashes are not pleasant. The bunkering of the ships will probably take place elsewhere, although, the steam puffer VIC 32 used to coal at Crinan Basin. More normal aspects of maintenance such as repainting can often be carried out in a marina with a minimum of interference to the normal operations of the marina or to the public.

The dock that *S.S.Great Britain* is in at Bristol is an excellent example of a self contained site which is part of a larger whole. Here restoration can proceed without detrimental effect to her surroundings.

ANCILLARY SERVICES

While the numbers visiting a ship may range from a few thousand to over a hundred thousand the needs of the general public will have to be catered for. These can be summarized as follows:

Car Parking. Inevitably most visitors will arrive by car. At the design stage, when considering the parking requirements of the various elements of the marina development, the presence of a historic ship could increase the spaces required. This may only be marginal because if most visitors come at a weekend, office car parking might be available at this time.

Toilets. In most cases those that are on the ship itself will not be available to the public for general use and will need to be linked to the local sewage system to avoid discharge directly into the water basin. Depending on the juxtaposition of the ship in the marina to these facilities, the existing public conveniences may be adequate.

Restaurants. Where there are large numbers of visitors to ship and they are spending some time there, this may be seen by the owners as an opportunity to increase income. Although it is probably better if such facilities there can serve several elements of the marina.

Souvenir Shops. The opportunity for the public to buy souvenirs and other memorabilia associated with the ship is an important aspect of raising additional revenue. Again this is not always possible on board the ship itself. In the Portsmouth naval dockyard *H.M.S.Victory, the Warrior* and *the Mary Rose*, all have additional sales outlets which in some cases also utilize some of the historic buildings on the site.

HEALTH AND SAFETY

Various health and safety regulations will have to be complied with when the public board the ship. These are primarily the concern of the ship owner but fire regulations, in particular, that require additional exits and the presence of fire hydrants could affect the sitting and berth of the ship.

THE SHIP OWNERS VIEW POINT

Many of the previously mentioned facets are as important to the ship owner as they are to the operators of the museum. As well as restoring and preserving the ship the owners other prime consideration is one of income. The first source of this is the public through admission fees, sales and donations. The location and accessibility of the marina or maritime quarter can therefore be of importance to them especially if they are trying to attract the casual visitor as well as the enthusiast. Furthermore once in the marina area the owner is concerned to attract the visitor onto the ship and pay an admission charge. Therefore if the ship is berthed in such a way that she can easily be seen without boarding her a degree of income will be lost. The location of *S.S. Great Britain* and *H.M.S. Cavalier* in Bristol and Brighton are good examples of where the ships can act as a focal point can be seen from a distance but closer inspection will demand payment. In contrast when the Maritime Trusts collection was in St. Katherine's Dock income from admission fees suffered because it could be easily seen by the public without payment.

If the ship is located in a dock area with which it has some direct association and relevance this is also beneficial and it can be used to make it an additional promotional point. *The Discovery* was able to make great use of the slogan "Home to Dundee" when she returned there from London which has helped keep visitor numbers around the 100,000 mark. A local connection also means it can be part of a maritime museum or quarter theme and it is easier to involve local people in support and fund raising activities. The move of *H.M.S. Cavalier* to Hartlepool where she has been associated with the World War 2 destroyer *H.M.S. Kelly* which was loosely associated with the north east, demonstrates this.

CONCLUSION

The presence of historic and interesting ships together with a maritime museum in a waterside development, of which a marina is an integral element, is beneficial to both parties. It is best if such ships are in good condition and of local or even national significance. The presence of a major ship is not always essential as has been demonstrated at Albert Dock in Liverpool. Here the museum has several smaller ships of local and regional interest which are augmented by periodic visits of larger vessels. They provides an opportunity to generate interest in the scheme and a focal point. However the problems of continuing repair work, size and public access need to be taken into consideration at the early design stage. As more docks are redeveloped as comprehensive schemes the opportunities for berthing large ships will increase. The public seem to have remained interested in Britain's maritime heritage but the cost of preservation will always be an obstacle to be overcome.

REFERENCES

1. Simper, R. Britain's Maritime Heritage. David & Charles, Newton Abbot, Devon. 1982.
2. Williams, J and Associates. Tourism and the Inner City. English Tourist Board, London. 1980
3. Swansea City Council. Swansea Maritime Quarter. Swansea City Council. 1988
4. Smyth, M. and Ayton, B. Visiting the National Maritime Museum. H.M.S.O. London. 1985

Books are to be returned on or before
the last date below